Recombinant Proteins From Plants

Series Editor
John M. Walker
School of Life Sciences
University of Hertfordshire
Hatfield, Hertfordshire, AL10 9AB, UK

For other titles published in this series, go to
www.springer.com/series/7651

METHODS IN MOLECULAR BIOLOGY™

Recombinant Proteins From Plants

Methods and Protocols

Edited by

Loïc Faye and Véronique Gomord

CNRS, Université de Rouen, Mont Saint Aignan, France

Humana Press

Editors
Loïc Faye
CNRS, Université de Rouen
Mont Saint Aignan
France

Véronique Gomord
CNRS, Université de Rouen
Mont Saint Aignan
France

ISBN: 978-1-61737-864-5 e-ISBN: 978-1-59745-407-0
ISSN: 1064-3745 e-ISSN: 1940-6029
DOI: 10.1007/978-1-59745-407-0

Printed on acid-free paper

springer.com

Dedication

In memory of Jean-Philippe Salier (1948--2008). In addition to being a brilliant scientist, he was also my best friend

L. Faye

Preface

Altogether, the biochemical, technical and economic limitations on existing prokaryotic and eukaryotic expression systems and the growing clinical demand for complex therapeutic proteins have created substantial interest in developing new expression systems for the production of therapeutic proteins. To that end, plants have emerged in the past decade as a suitable alternative to the current production systems, and today their potential for production of high quality, much safer and biologically active complex recombinant pharmaceutical proteins is largely documented.

The chapters in this volume, contributed by leaders in the field, sum up the state-of-the-art methods for using a variety of different plants as expression hosts for pharmaceutical proteins. Several production platforms are presented, ranging from seed- and leaf-based production in stable transgenic plant lines, to plant cell bioreactors, to viral or *Agrobacterium*-mediated transient expression systems. Currently, antibodies and their derived fragments represent the largest and most important group of biotechnological products in clinical trials. This explains why the potential of most production platforms is illustrated here principally for antibodies or antibody fragments with acknowledged potential for immunotherapy in humans. In addition, a comparison of different plant expression systems is presented using aprotinin, a commercial pharmaceutical protein, as a test system.

Although multiple books and monographs have been recently published on molecular pharming, there is a noticeable dearth of bench step-by-step protocols that can be used quickly and easily by beginners entering this new field. This volume aims to fill the void by presenting detailed protocols for using the main plant expression systems, for rapidly detecting and quantifying recombinant proteins in a crude plant protein extract. Several chapters feature methods to improve the yield and stability of recombinant proteins using targeting to different subcellular organelles, expression of protease inhibitors or fusion to carrier sequences.

Most biopharmaceutical products have the potential to be immunogenic in at least a small population of human subjects. In this respect, the immunogenicity of plant N-glycans in mammals is a major concern. Protocols, extra notes and problem-solving tips are presented to define whether, how and where a pharmaceutical protein expressed in plants is glycosylated. However, until a number of plant-derived therapeutic glycoproteins have completed their clinical development and registration process, the risk related to parenteral administration of these products in humans remains purely theoretical. Principles and methods of biosafety and risk assessment of plant-derived therapeutic proteins for humans and for the environment are detailed in the last two chapters, which are dedicated to the safe development of plant-made pharmaceuticals.

We thank all the authors who have made this book possible, and the Humana staff and Professor John Walker for their assistance during the final stages of editing, proofreading and collating.

L. Faye, V. Gomord

Foreword

Laws are like sausages. It's better not to see them being made.
Ottovon Bismarck

Good Reasons to Dare Plant-Factories

On April 1, 2015, there will be solemn reminders of the 200th anniversary of the birth of Otto von Bismarck. Kids will commit silly April Fool pranks. And thousands of patients will be treated with recombinant drugs not made in Chinese hamster ovary (CHO) cells. Several of these non-CHO drugs will come from this book.

It is a reassuring thought if you plan to grow old enough to need serious health care, because when you look at it, biotech is stuck. Twenty-five years after recombinant insulin, the whole sector is still at the introduction stage. It has launched a quartet of product classes that, in 2007, owns less than 11% of the combined markets of seed and pharmaceuticals.[1] Most of biotech's output is so expensive that only the richest can enjoy it. In the meantime, fax machines, Starbucks coffee and cell phones took over the Earth; avian flu promises to do it even faster, while we can produce only 350 million doses of flu vaccines per year (WHO says we need 2 billion to contain a pandemic).

Plant-factories are part of the solution, and this book illustrates their rich versatility. It also shows that the plant's toolbox replicates the strengths and weaknesses of CHO cells: getting better and better at expression, which is by far the dominant topic of the following chapters. This raises two questions.

First, where are the non-expression, plant-specific or plant-useful tools? The authors do address product quality (control of post-translational modifications), touch gains in downstream processing (one chapter) and evoke speed (several rapid transient systems). But there is much more out there. What about somatic embryogenesis for rapid scale-up, intrinsic advantages of generally recognized as safe (GRAS) material, bioconfinement genetic tools? What about demonstrations of the celebrated cost and flexibility advantages plants will bring?

Second, what are the health, nutrition and industrial needs best served by plant-factories? We know there is tremendous pressure from investors to emulate the current trends of pharma, but one can envision plant-factories enabling new classes of drugs, affordable treatments, edible vaccines, ultra-large capacity, improved food and feed, industrial products, etc.

[1] CHO-made monoclonal antibodies, microbial system's insulin-like proteins, and, in four plants only, insect-resitant and heribicide-tolerent crops.

Over the past 15 years, plant-factories delivered a rich harvest of promises. They were insufficiently funded – a napkin calculation suggests that much less than $1 billion were invested in plant-factories, all included. This is not even 60% of the cost of taking one biodrug to market.

More than lacking cash, plant-factories might have suffered from being overly technology minded. The first three generations of plant-factory companies have been elegant one-trick ponies ran by people sincerely enamoured with their plant host (I know what I am talking about, I co-founded one of these!). Many have found the hard way that one good technology is not enough.

Bismarck, according to Wikipedia, 'engineered the unification of Germany'. If he was around, I would invite him to our next conference to help unify diverse plant tools around specific products. Maybe the era of haute-couture enabling has begun – certainly, the idea of a best-in-the-word, one-size-fits-all, cheaper-than-yours, plant-based solution has been proven unrealistic. I suggest that the most successful plant-factories of the emerging generation will be product-focused; its makers will assemble a specific, multi-technology, tailor-made production plant platform to enable that product. And yes, they will have the financial resources to optimize their processes, like everybody else does.

Of course, one cannot discuss plant-factories and avoid the fear factor. Genes, nanotechnologies, night stalkers, strangers, aliens, income tax auditors… well, it is not that bad. People do not want to know how laws and sausages are made, said Bismarck. Try this: poll people during dinner and ask if they wish to be injected with a substance produced in the reproductive organs of Chinese rats. I bet you will get a rather emotive reaction. Try the same survey within a population of arthritics – it is Embrel they want. Their desire, their immediate need to stop the joint pain would overcome disgust for the ovaries of Chinese hamsters, and will alleviate their fear of rice seeds, tobacco leaves, algae or carrot cells. Today, it is easy for cunningly sold fear to trump the safety plant-factories are promoting. This emerging industry is developing good products in broad daylight, and it showed it controls its production processes. This will work.

When young Otto turned 11 years old, his eventual compatriot Friedrich Wohler was discovering synthetic organic chemistry. The modern dye industry and small-molecule drugs were around the corner, both enabled by the capacity to control the manufacture of identical, minuscule compounds.

On April 1, 2015, Bismarck's 200th anniversary, I like to think there will be five, maybe ten plant-made biopharmaceuticals on the market and – why not! – one of them a billion dollar drug. We know some of these – they are in phase III, we read press releases about them. Other products, possibly more innovative and owning more to the specific advantages of plants, will be in clinical trials; plant-factories will also begin to contribute to bioremediation, industrial products, energy and protein-based nutrition.

On the same day, it will probably snow in Québec City, where I was born (global warming is not so quick), and it might rain in Barcelona where I write tonight, although the sun shall not be far behind the clouds. I have no idea what the weather will be, wherever I am on that day. But I certainly intend to be around, along with my

7.2 billion most intimate earthian friends. You, who read this, and me both know there is no way CHO cells in bioreactors and BT corn can churn out enough proteins for everyone. So, let us use the recipes in this book, and grow products in plant-factories.

François Arcand
Barcelona, April 14, 2007

Mr. Arcand is the CEO of ERA Biotech, developing a disruptive protein-production technology originating from the seed of corn, and being deployed in most eukaryotic cell production systems. He organized the 2003 and 2005 Conferences on Plant-Made Pharmaceuticals, is the co-founder of Medicago and was its first CEO. The opinions formulated in this Foreword are strictly personal.

Contents

Contributors

ELSA ARCALIS • *Institute for Molecular Biology, RWTH Aachen, Aachen, Germany*
FRANCOIS ARCAND • *ERA Biotech, Parc Científic de Barcelona, Barcelona, Spain*
JORGE AYALA • *Arkansas Biosciences Institute, Arkansas State University, Jonesboro, AR*
MARTA AYALA • *Center for Genetic Engineering and Biotechnology, Havana, Cuba*
AMINE M. BADRI • *Département de phytologie, Université Laval, Québec, Canada*
MURIEL BARDOR • *Faculté des Sciences, Université de Rouen, Mont Saint Aignan, France*
NICOLE BECHTOLD • *Medicago inc., Québec, Qc, Canada*
JULIE BELLES-ISLES • *Medicago inc., Québec, Qc, Canada*
MERIEM BENCHBANE • *Meriem Département de phytologie, Université Laval, Québec, Canada*
JOHAN BOTTERMAN • *Bayer BioScience NV, Gent, Belgium*
GLEYSIN CABRERA • *Department of Carbohydrate Chemistry, Center for Genetic Engineering and Biotechnology, Havana, Cuba*
PAUL D. CHAMBERLAIN • *bioLOGICA Consulting, France*
KARINE COENENE • *Département de phytologie, CRH/INAF, Université Laval, Québec, Canada*
UDO CONRAD • *Leibniz Institute of Plant Genetics and Crop Plant Research, Molecular Genetics/Phytoantibodies, Gatersleben, Germany*
CARIOLE L. CRAMER • *Arkansas Biosciences Institute, Arkansas State University, Jonesboro, AR, U.S.A.*
JOSÉ CREMATA • *Center for Genetic Engineering and Biotechnology, Havana, Cuba.*
MARC-ANDRÉ D'AOUST • *Medicago inc., Québec, Qc, Canada*
HENRY DANIELL • *Department of Molecular Biology and Microbiology, University of Central Florida, Biomolecular Science, Orlando, FL, U.S.A.*
KIRSTEN DE WILDE • *Department Plant Systems Biology, Flanders Institute for Biotechnology (VIB) and Department of Molecular Genetics, Ghent University, Gent, Belgium*
ANN DEPICKER • *Department Plant Systems Biology, VIB, Ghent University, Gent, Belgium.*
YI DING • *Department of Molecular Biology and Microbiology, University of Central Florida, Biomolecular Science, FL, U.S.A.*
MAUREEN C. DOLAN • *Arkansas Biosciences Institute, Arkansas State University, Jonesboro, AR, U.S.A.*
MANUEL DUBALD • *Bayer BioScience SA, Lyon, France*
GIL ENRÍQUEZ • *Center for Genetic Engineering and Biotechnology, Havana, Cuba.*
JOSE M. ESCRIBANO • *Departamento de Biotecnología, Madrid, Spain*
LOÏC FAYE • *CNRS, Université de Rouen, Mont Saint Aignan, France*

ANNE-CATHERINE FITCHETTE • *Faculté des Sciences, Université de Rouen, Mont Saint Aignan cedex, France*

ALEJANDRO FUENTES • *Center for Genetic Engineering and Biotechnology, Havana, Cuba*

MANFRED GHARTZ • *Dept. Cell Biology and Plant Physiology, University of Regensburg, Regensburg, Germany*

JORGE GAVILONDO • *Center for Genetic Engineering and Biotechnology, Havana, Cuba*

CÉCILE GIRARD • *Département de phytologie, CRH/INAF, Université Laval, Québec, Canada*

VÉRONIQUE GOMORD • *CNRS, Université de Rouen, Mont Saint Aignan, France*

CHARLES GOULET • *Département de phytologie, CRH/INAF, Université Laval, Québec, Canada*

GHISLAINE GRENIER DE MARCH • *Institut Polytechnique LaSalle Beauvais BEAUVAIS Cedex, France*

ROY JEFFERIS • *Immunity & Infection, University of Birmingham, B15 2TT UK*

PIERRE-OLIVIER LAVOIE • *Medicago inc., 1020 Route de l'Église, Québec, Qc, Canada, G1V 3V9*

PATRICE LEROUGE • *Faculté des Sciences, Université de Rouen, Mont Saint Aignan, France*

DAVID LIENARD • *Faculté des Sciences, Université de Rouen, Mont Saint Aignan, France*

LI LIU • *Department of Biological Chemistry, John Innes Centre, Colney Lane, Norwich, UK*

JIANYUN LIU • *Arkansas Biosciences Institute, Arkansas State University, Jonesboro, AR; Department of Plant Pathology/Physiology, Virginia Tech, Blacksburg, VA*

IMMA LLOP • *Consorci CSIC-IRTA, Jordi Girona Barcelona. Spain*

GEORGE P. LOMONOSSOFF • *Department of Biological chemistry, John Innes Centre, Colney Lane, Norwich, UK*

BRIAN MARSHALL • *MolecularFarming.com, Ballyhaskey, Newtowncunningham, Lifford, Donegal Ireland*

MICHÈLE MARTEL • *Medicago inc., 1020 Route de l'Église, Québec, QC, Canada*

GIULIANA MEDRANO • *Arkansas Biosciences Institute, Arkansas State University, Jonesboro, AR, U.S.A.*

DOMINIQUE MICHAUD • *Département de phytologie, CRH/INAF, Université Laval, Québec, Canada*

RÉMY MICHEL • *Institut Polytechnique LaSalle Beauvais BEAUVAIS Cedex, France*

FABIEN NOGUE • *Station de Génétique et Amélioration des plantes, INRA, Route de St Cyr, Versailles, France*

LINCIDIO PÉREZ • *Center for Genetic Engineering and Biotechnology, Havana, Cuba*

DANIEL M. PEREZ-FILGLUEIRA • *Consejo Nacional de Investigaciones Científicas y Técnicas (CONICET) - Buenos Aires, Argentina*

JÖRG PETERS • *Bayer Healthcare AG, Wuppertal, Germany*

CAROLE PLASSON • *Faculté des Sciences, Université de Rouen, Mont Saint Aignan, France*

MERARDO PUJOL • *Center for Genetic Engineering and Biotechnology, Havana, Cuba*

THOMAS RADEMACHER • *Institute for Molecular Biology, RWTH Aachen, Worringerweg, Aachen, Germany*

MICHAEL J. REIDY • *Arkansas Biosciences Institute, Arkansas State University, Jonesboro, AR; Department of Plant Pathology/Physiology, Virginia Tech, Blacksburg, VA, U.S.A.*

DANIEL RIVARD • *Département de phytologie, Université Laval, Québec, Canada*

MEILYN RODRÍGUEZ • *Center for Genetic Engineering and Biotechnology, C. P. 10600, Havana, Cuba.*

FRANK SAINSBURY • *Department of Biological Chemistry, John Innes Centre, Colney Lane, Norwich, UK*

CLAUDE SAINT-JORE • *CNRS, Université de Rouen, Mont Saint Aignan, France*

WERNER SCHRÖDER • *Bayer Healthcare AG, Wuppertal, Germany*

N. DOLENDRO SINGH • *N. Department of Molecular Biology and Microbiology, University of Central Florida, Biomolecular Science, Orlando, FL, U.S.A.*

CHRISTOPHE SOURROUILLE • *Faculté des Sciences, Université de Rouen, Mont Saint Aignan, France*

PENELOPE A. C. SPARROW • *John Innes Centre, Norwich Research Park, Norwich, UK*

JOHANNES STADLMANN • *Glycobiology Division, Department of Chemistry, University of Natural Resources and Applied Life Sciences, Vienna, Austria*

EVA STOGER • *Institute for Molecular Biology, RWTH Aachen, Worringerweg, Aachen, Germany*

MARGARITA TORRENT • *Consorci CSIC-IRTA, Jordi Girona, Barcelona, Spain*

RICHARD M. TWYMAN • *Department of Biology, University of York, Heslington, York, UK*

LOUIS-PHILIPPE VAILLANCOURT • *Département de phytologie, CRH/INAF, Université Laval, Québec, Canada*

BART VAN DROOGENBROECK • *Department Plant Systems Biology, Flanders Institute for Biotechnology (VIB) and Department of Molecular Genetics, Ghent University, Gent, Belgium*

GUY VANCANNEYT • *Bayer BioScience NV, Gent, Belgium*

LOUIS-P. VÉZINA • *Medicago inc., Québec, QC, Canada*

Chapter 1

From Neanderthal to Nanobiotech: From Plant Potions to Pharming with Plant Factories

Christophe Sourrouille, Brian Marshall, David Liénard, and Loïc Faye

Summary

Plants were the main source for human drugs until the beginning of the nineteenth century when plant-derived pharmaceuticals were partly supplanted by drugs produced by the industrial methods of chemical synthesis. During the last decades of the twentieth century, genetic engineering has offered an alternative to chemical synthesis, using bacteria, yeasts and animal cells as factories for the production of therapeutic proteins. After a temporary decrease in interest, plants are rapidly moving back into human pharmacopoeia, with the recent development of plant-based recombinant protein production systems offering a safe and extremely cost-effective alternative to microbial and mammalian cell cultures. In this short review, we will illustrate that current improvements in plant expression systems are making them suitable as alternative factories for the production of either simple or highly complex therapeutic proteins.

Key words: Glycosylation, Molecular farming, Plant-made pharmaceutical, Recombinant protein, Transgenic plant, Therapeutic protein.

1.1. Introduction

From 60,000 BC to the nineteenth century, plants were the main source for human drugs. For instance, when sick and obliged to stay in his cave, the Neanderthal man already used centaury to fight his fever. The first known text on medicinal plants, the Pen Tsao, was written more than 4500 years ago under the direction of emperor Shen-Nung in China, and describes 365 medicinal plants, including opium, ephedra and hemp. More recently, around 1500 BC, the Ebers papyrus describes 700 remedies made from plants, including mandrake, castor bean and hemp, illustrating that plants had a major place in Egyptian medicine. In the Middle ages, places such as Salagon abbaye became famous for

Loïc Faye and Véronique Gomord (eds.), *Methods in Molecular Biology, Recombinant Proteins From Plants, vol. 483*

DOI: 10.1007/978-1-59745-407-0_1

their specialization in the culture of medicinal plants and universities were created in Montpellier or Salerne to improve plant therapeutics, extraction and characterization.

There was a great turn in medicament history, starting at the beginning of the nineteenth century until the early 1970s, when pharmacy turned to be dominated by scientific chemistry with both the development of more and more sophisticated processes for extraction, purification and the synthesis of active pharmaceutical compounds. The twentieth century became a triumph for drugs produced at an industrial level by chemical synthesis. This evolution probably started with the production of aspirin, a synthetic analogue of salicylic acid previously extracted from willow bark. In parallel, more and more sophisticated extraction and purification procedures were developed resulting, for example, with the extraction of insulin from pig pancreas in 1922.

As a complement of synthesis and extraction chemistry, modern biology enters the world of pharmaceutical industry with the development of genetic engineering in the early 1970s, allowing biosynthesis of complex molecules too difficult to extract and purify from living material and inaccessible to synthesis chemistry. In the following decades, genetic engineering has offered an alternative to chemical synthesis and extraction procedures with the production of therapeutic molecules in transgenic bacteria, yeast and animal cells. More recently, molecular farming has forced plants into the business thinking of the major players in recombinant protein production systems. Indeed, plants offer several advantages over other expression systems for therapeutic protein production. For instance, plant expression systems can produce, for a lower cost, large amounts of biopharmaceuticals free of human infective viruses and prions and, unlike microbial fermentation, plants are capable of carrying out post-translational modifications (PTMs) often required for functionality of therapeutic proteins.

1.2. Some Good Reasons to Stake on Plant-Factories Production

The pharmaceutical industry needs cheap and efficient expression systems for therapeutic protein production. The ideal heterologous system having to fulfil different functions allowing to obtain recombinant proteins with reproducible quality and for a low cost together with a recognized capacity to carry out co- and post-translational modifications often required for a biopharmaceutical protein to be biologically active *(1)*.

Currently, no heterologous expression system of production satisfies all of these requirements. For instance, complex therapeutic proteins produced in prokaryotes are not always properly folded or processed to provide the desired degree of biological activity. Consequently, microbial expression systems are generally used for expression of relatively simple therapeutic proteins that do not require folding or extensive post-translational processing to be biologically active such as insulin, interferon or human growth hormone *(2)*. Because of the limitations of prokaryotes for production of complex therapeutic proteins, the pharmaceutical industry had focused efforts towards optimization of two main eukaryotic expression systems, yeasts and mammalian cell cultures. These production systems, however, suffer from many disadvantages such as inappropriate PTMs for yeast, or high operating costs, difficulties in scaling up to large volumes and potential contamination by virus or prion for cultured mammalian cells.

Altogether, the biochemical, technical and economic limitations on existing prokaryotic and eukaryotic expression systems, the growing clinical demand for complex therapeutic proteins and the lack of bioreactor capacity have created substantial interest in developing new expression systems for large-scale production of therapeutic proteins. To that end, plants have emerged in the past decade as a suitable alternative to the current production systems of therapeutic proteins and today their capacity in low-cost production of high quality, much safer and biologically active mammalian proteins is largely documented (for recent reviews, *see* **refs.** *1–4)*.

For instance, the use of transgenic plants could be a solution to the need for a rapid increase in production capacity of therapeutic antibodies *(5, 6)*. Indeed, even with relatively low expression levels for therapeutic proteins, the production capacity of recombinant antibodies in transgenic plants is almost unlimited, as it depends only on the surface dedicated to the plant culture. A plant "bioreactor" will allow the production of recombinant proteins up to 20 kg per hectare, regardless of the plant material considered: tobacco, corn, soybean or alfalfa *(7, 8)*.

Another major advantages of transgenic plants over other production systems available for large-scale and low-cost production, such as *Escherichia coli* or yeasts, is their ability to perform most PTMs required for therapeutic protein's bioactivity and pharmacokinetics *(2, 9)*. This is illustrated by their capacity to produce many therapeutic proteins requiring proteolytic cleavage(s), oligomerization and glycosylation for their bioactivity, pharmacokinetics, stability and solubility, with some of these proteins already in clinical trials (**Table 1.1**). The production of immunoglobulins in plant cells is a good illustration of plant capacity to produce complex proteins. Indeed, transgenic plant cells are able to correctly synthesize, mature and assemble, *via* disulphide bridges, the light and heavy polypeptide chains constitutive of an antibody.

Table 1.1
Status of molecular farming at the end of 2006.

	Plant -made pharmaceuticals – Pproduct status with regard to clinical trials – December. 2006			
Company/Institution	**Plant used**	**Product**	**Indication**	**Clinical Sstage**
Meristem Therapeutics	Maize	Lactoferrin	Gastrointestinal disorders	Phase 1
	Maize	Gastric lipase	Cystic fibrosis	Phase 2 (×2 "unformulated"and "formulated")
	Meristem Therapeutic's Gastric Lipase has passed phase 2 trial for the unformulated dosage. The formulated dosage trial is ongoing.			
Planet Biotechnology	Tobacco	slgA	Prevention of tooth decay	Phase 2 but already granted an EU licence as a medical device
	Tobacco	ICAM1	Receptor for common cold	Phase 1 ready
Prodigene	Maize	Lt-B vaccine	Traveller's diarrhoea	Phase 1 complete – not ongoing
	Maize	TGE vaccine	Piglet gastroenteritis	Phase 1 complete – not ongoing
	Maize	Avidin	Diagnostic use	Available in Sigma catalogue in 2006
	Maize	Trypsin	Wound care/insulin manufacture	Available in Sigma catalogue in 2006
	Maize	Aprotinin	Non-clinical use	No longer in Sigma catalogue
	Maize	GUS	Non-clinical use	No longer in Sigma catalogue
LSBC (in Chap. 11 bankruptcy)	Tobacco	Vaccine	Non-Hodgkin's lymphoma	Phase 1 successful in 2002
	Tobacco	Aprotinin	Non-clinical use	Was in Sigma catalogue in 2005
	Tobacco	Vaccine	Feline parvovirus	"very advanced" – Status confidential
	LSBC's Feline Parvovirus vaccine trials were being partnered by Schering-Plough Animal Health, and are believed to have continued during the Chap. 11 proceedings.			

Arizona State University	Potato	Vaccine	*E. Coli*	Phase 1
	Potato	Vaccine	Hepatitis "B"	Phase 1
	Potato	Vaccine	Norwark virus	Phase 1
	Tobacco	Vaccine	Norwark virus	Phase 1/2
	Undisclosed	Oral vaccine	Undisclosed	Phase 1 planned start mid-2007
Ventria Biosciences	Rice	Lactoferrin	Undisclosed (see comments)	Undisclosed
	Rice	Lactoferrin	Cell culture media	Available from company
	Rice	Lysozyme	Undisclosed (see comments)	Undisclosed
	Rice	Lysozyme	For research purposes	Available from company
	Ventria – A 135 children study conducted by investigators in the US and Peru found that by adding Ventria's Lactoferrin and Lysozyme proteins to the standard treatment of diarrhea, oral rehydration solution, both the length and the severity of diarrhea decreased.			
	Conducted following World Health Organization protocols, the trial found that those consuming oral rehydration solution with both proteins added were sick for 3.67 days on average, as compared to 5.21 days for children receiving oral rehydration solution without the proteins.			
Biolex	*Lemna*		Fibrinolytic "Clot-buster"	Phase 1 ready
	Lemna	Alpha interferon	Hepatitis "B" & "C" and cancer	Phase 1 successful 2 trials – "general" + "formulated dose"
	Biolex's Alpha-Interferon is of equal or better quality than that produced by CHO cell *(66)*			
Cobento AS (formerly Cobento Biotech AS)	Aribidopsis	Human intrinsic factor	Vitamin B12 deficiency	Approved Coban product launched. Successful 37 patient clinical trial cGMP production certified
	Aribidopsis	Transcobalamin	Diagnostic/research	Available from company

(continued)

Table 1.1 (continued)

Company/Institution	Plant -made pharmaceuticals – Pproduct status with regard to clinical trials – December. 2006			
	Plant used	Product	Indication	Clinical Sstage
Dow AgroSciences	Plant cell non-nicotine tobacco	Vaccine	Newcastle disease in poultry	USDA approved in January 06
	Plant cell -tobacco	Vaccines	"diseases of horses, dogs and birds"	Undisclosed
	Dow has not stated exact disease or animal targets but is believed to be developing several vaccines and to be at clinical trial stage in at least one (see Table legend for comments)			
Protalix	Plant cell	Glucocerebrosidase	Gaucher's disease	Phase 3
Guardian Biosciences	Canola	Edible vaccine	Coccidiosis in poultry	CFIA phase 2
Nexgen Biotech	Oriental Melon	Thyroid stimulating hormone receptor	Diagnosis of Graves' disease	Available from company
	Oriental Melon	Viral antigens	Rapid detection of Hantaan and Puumala	Available from company
CIGB (Cuba)	Tobacco	Recombinant monoclonal (Mab)	Purification reagent in Hep."B" vaccine	Approved mid-2006 in Cuba
D.Yusibov and others	Spinach	Vaccine	Rabies	Phase 1 successful in 2002
Farmacule	Tobacco	Virtonectin	Research use	Available from company mid-2007 distribution under negotiation
NeoRx/Monsanto	Maize	Antibody	Cancer	Phase 2 (failed)
Applied Biotech. Institute & partner	Maize	Undisclosed	Undisclosed	Animal trials underway
SemBioSys	Safflower	Insulin	Diabetes	Phase 1 trial planned for early 2008
	SemBioSys has announced commercial levels of production of Insulin [http://www.sembiosys.com/news2.aspx?id = 5295&secId = 7]			

Medicago	Tobacco	Vaccine	H5N1 influenza	Advanced pre-clinical
Chlorogen & partner	Tobacco	a TGF-beta protein	Ovarian cancer	Advanced animal trials

PMIPs – some advanced plant made industrial protein products

Company/Institution	Plant used	Product/purpose	Development stage
Aresa (Denmark)	Thale cress	Landmine detection	Successful field trials. Launch 2007
University of Wyoming	Alfalfa	Spider silk	"Spinning fibre" – Ready to licence
Linnaeus (Canada)	Canola/rape	Non-polluting engine oil and 2-stroke engine oil	Tested and proven at Lab scale – Ready to licence

PMIPs – Some advanced phytoremediation plants

Company/Institution	Plant used	Phytoremediation target/pollutant	Development stage
University of California – Berkley	Mustard	Selenium	Successful field trials
Hungarian Academy Sciences.	Poplar	Cadmium zinc chromium copper	Awaiting field trial permission
	The Hungarian Academy of Sciences had trial permission granted, only to have it withdrawn again. They still await trial permission		
AIST, Japan	Tobacco	BPA, PCP and other chlorophenols	Early field trials
Peking University College of Life	Tobacco and algae	Heavy metals especially Mercury	Awaiting field trial permission
University of York (England)	Tobacco	TNT and RDX explosive	Early trial

(continued)

Table 1.1 (continued)

Company/Institution	PMIPs – some advanced plant made industrial enzymes		
	Plant used	Product/purpose	Development stage
Applied Biotechnology Institute	Maize	Peroxidase – paper bleaching	Product verification trials completed – available to licence
	Maize	Laccase – paper bleaching	Product verification trials completed – available to licence
	Maize	Cellulase – ethanol production	Product verification trials completed – available to licence
	Maize	Embryo specific promoters	3 different types – available to licence
	Maize	Constitutive promoter	Available to licence
	Maize	Scoreble/electable marker	Available to licence

Dow AgroSciences LLC, on January 27th 2006 received the world's first regulatory approval for a plant-made vaccine from the United States Department of Agriculture (USDA) Center for Veterinary Biologics. (Licence No.642) http://www.aphis.usda.gov/vs/cvb/notices/2006/10.pdfof April 10th ™ Concert is a trademark of Dow AgroSciences LLC

It was for a vaccine against Newcastle disease, a virulent and contagious disease of fowl, although the company has stated that because of several competing vaccines already in the market it does not intent to commercialize the vaccine, but rather has used it to prove technology and method.

The Concert™ Plant-Cell-Produced System which Dow used to produce the vaccine utilizes non-nictotine tobacco plant cells grown in steel tank reactors. The Newcastle disease virus vaccine contained the immunological protein of the virus, derived from plant sources, not the live virus, or virus products. In the USDA approved trials chickens' immune systems quickly recognized the protein, producing an appropriate and easily measurable immune response. An added advantage was that the vaccine did not need cold storage

Dow now has several animal vaccines in development using Concert™ which it intends to bring to market, possibly by 2010

Since the first production of a functional antibody in tobacco *(5)*, many antibodies, or antibody fragments, have been produced for therapeutic or diagnostic purposes in various plant expression systems. Antibodies produced in plants are correctly assembled, proteolytically matured and glycosylated. Indeed, antibodies produced in plants bear both high-mannose and biantennary complex-type N-glycans. The high-mannose-type N-glycans have the same structure in plant and mammalian glycoproteins. But complex-type N-glycans are structurally different in plants and mammals *(10, 11)*.

Despite these differences in the N-glycan structures, antibodies produced in plants have similar antigen binding capacity as their homologs produced in mammalian cells. Furthermore, an antibody half-life in the bloodstream, as well as its ability to be recognized by Fc receptors, which are both determined by heavy chains N-glycosylation, are not strongly affected when a plant N-glycan is present instead of a mammalian N-glycan *(7, 12)*.

1.3. Stategies to Increase Yields

Current limitations of plant expression systems are the low yields observed for some therapeutic proteins. To achieve higher yields, different stages of therapeutic protein expression in plants can be optimized from foreign gene sequence optimization to subcellular targeting of foreign proteins (for a detailed review, *see* **ref.** *13)*.

1.3.1. Codon Optimization

In order to improve the rate and fidelity of translation in a plant expression system, it can be important to adapt the coding sequence of the gene of interest to the codon bias of the host plant. Little information related to the strategy is yet available for many plants, but a 5–100 times increase in protein expression has already been observed after codon optimization *(14–17)*. As an example, expression in tobacco and tomato of a bacterial insecticide gene (*cryIA*) either partially (3% nucleotide difference) or fully modified (21% nucleotide difference) were compared to non-codon-optimized gene transformed plants. Plants transformed with either the partially or fully codon-optimized gene, respectively, expressed 10 and 100 times more insecticidal protein than plants transformed with the wild-type gene *(14)*.

1.3.2. Suppressor of Post-Transcriptional Gene Silencing

RNA silencing, also named post-transcriptional gene silencing (PTGS) in plants, plays a key antiviral defence role in many eukaryotic organisms by influencing virus replication in cells *(18)*. Viruses produce proteins capable of suppressing host cell RNA silencing *(19)*. For instance, each plant virus seems to produce its own suppressor of silencing and the characterization of a large number of suppressors such as HC-Pro, 2b, p25 is currently in progress. Today, the best characterized suppressor is the p19 protein, encoded by *Tomato Bushy Stunt Virus* (TBSV) *(20)*. RNA silencing can be initiated not only by the presence of virus RNA but also by the presence of exogenous genes. As a result, in transformed plants, PTGS is occasionally targeted against transcripts of the transgene, so that corresponding gene products accumulate at a low level *(21)*. This phenomenon can be avoided by expressing simultaneously the protein of interest and a suppressor of silencing. This provides a new tool for molecular farming in plants to obtain high-level expression of some transgenes, as recently illustrated when p19 co-expression with a broad range of recombinant proteins increases their yield up to 50-fold *(21, 22)*.

1.3.3. Targeted Expression of Recombinant Proteins

Targeted expression of plant-made pharmaceuticals (PMPs) into specific organs and subcellular compartments represent a plant-specific strategy to increase yields and simplify the first steps of purification. In this way, different plant organs (leaves, seeds and root) and plant cell compartments (endoplasmic reticulum, (ER); chloroplast, vacuole and oil body) have been efficiently used to express many therapeutic proteins *(23, 24)*.

Generally, recombinant proteins are targeted into plant organs, which allows high biomass yield. For example, in plants with large foliage volume, such as tobacco, alfalfa and some other legume plants, expression is performed in leaves, whereas, in potato, maize, rapeseed, safflower, soybean, wheat or rice, the production and accumulation of recombinant proteins occur in tubers or in seeds *(25, 26)*. Both systems have their own advantages and drawbacks. Table 1.2 provides an overview of current production systems being explored by the major companies involved in PMP production. Using aprotinin as a model, Vancanneyt et al. *(27)* have recently compared the established expression systems for product quality and quantity. As well as organ-specific storage of PMPs, many subcellular compartments are available for accumulation of large amounts of recombinant therapeutic.

1.3.3.1. Targeted Expression of Therapeutic Proteins into the Secretory Pathway

Most recombinant proteins produced so far in plants have been secreted into the intercellular space or apoplast *(28, 29)*. This targeting is only dependent on the presence of an N-terminal signal peptide cleaved during the co-translational insertion of the nascent protein in the ER *(4)*. It has been shown in many plant expression systems and for many PMPs that plant and human signal peptides are recognized with a same efficiency.

Interestingly, recombinant proteins targeted to the secretory pathway can be secreted by suspension-cultured plant cells in their culture medium where they accumulate. This technology,

Table 1.2
Major crops and companies involved in plant-made pharmaceuticals

	Production systems											Expression		
Company	**Cell culture**	**Nicotiana**	**Alfalfa**	***Lemna***	**Peas**	**Moss**	**Corn**	**Safflower**	**Barley**	**Arabidopsis**	**Rice**	**Nuclear**	**Viral**	**Chloroplast**
Dow AgroSciences	■											■		
Protalix Biotherapeutics	■											■		
Planet Biotech		■										■		
Bayer		■											■	■
Chlorogen		■												■
Medicago, Inc.		■	■									■		
Biolex				■								■		
Novoplant					■							■		
Greenovation						■						■		
Meristem Therapeutics	■						■					■		
SemBioSys								■				■		
ORF Genetics									■			■		
Cobento										■		■		
Ventia									■		■	■		

Adapted from Vancanneyt et al. (27)

which avoids cropping, brings a great simplification to the purification process, as recently illustrated for house dust mite allergen production in BY-2 tobacco cells *(30)*.

While soluble protein secretion in the extracellular compartment is a default pathway only depending on the presence of a signal peptide, targeting to other compartments of the secretory pathway, such as ER or vacuoles, needs additional signals. Many examples illustrate that the H/KDEL-mediated protein retention in the ER could strongly increase the stability and consequently the yield of recombinant proteins as compared with secretion *(31, 32)*. Also of note, as developed below, retention into the ER also prevents addition of immunogenic complex N-glycans on plant-made glycoproteins.

Plant seeds store large amounts of proteins in membrane-bound organelles called protein bodies (PBs). A first class of PBs, also described as protein storage vacuole (PSV), is an intracellular organelle where proteins are transported *via* the Golgi apparatus in cereal endosperm cells and also in many other different types of plant cells, including leaf and root cells. Prolamins, a family of storage proteins of maize and rice seeds, remain in the ER, where they aggregate in a second class of ER-derived PBs. Because of their low proteolytic activity, these two types of protein storage organelles are attractive compartments for recombinant protein accumulation *(33)*. For example, human serum albumin has been expressed and delivered into the PSVs of wheat endosperm where it shows a good stability *(34)*. In the objective of a better exploitation of these subcellular compartments for storage of therapeutic proteins, a better knowledge of protein targeting to these organelles will help further investigation into their advantages and limitations for PMPs storage *(35)*. The capacity of proteins to remain and aggregate in ER-derived PBs has recently been adapted to production of recombinant proteins in non-seed plant tissues. This strategy relies on a fusion of the protein of interest with g zein domains responsible for ER-derived PB formation in maize seeds *(36)*. Preliminary results indicate that very high expression levels are obtained using these "artificial" PBs as PMP storage compartments. However, it is not known if major PTMs required for PMPs biological activity could occur on proteins aggregating co-translationally in these ER-derived PBs.

Oilseeds accumulate lipids to supply the energy required for seedling development in organelles arising from the ER: the oilbodies. Seed oilbodies are limited by a protein-rich phospholipids monolayer. Oleosins, the major proteins at the periphery of oilbody membrane, are anchored by their hydrophobic domain exposing their N- and C-terminal ends to the cytoplasm. Targeting of PMPs as oleosin fusions to oilbodies enables both high levels of expression and cost-effective recovery *(37)*. The recombinant protein fused with oleosin is separated with oilbodies from other seed components by liquid–liquid phase separation. This mild process reduces the number of chromatography steps required to obtain a purified PMP and thereby significantly reduce their purification cost. Recently, human insulin-expressed *Arabidopsis* seed oilbodies was recovered as an active molecule at commercially relevant levels *(38)*.

1.3.3.2. Production of Therapeutic Proteins in Chloroplasts

PMP(s) expression in the chloroplasts allows accumulation of very large amounts of recombinant proteins in plant leaves *(39–41)*. As an example, transgenic tobacco chloroplasts produce 300-fold higher amounts of human somatotropin that their nuclear transgenic counterparts *(42)*. Resulting from high expression levels and low proteolytic activity in this organelle, a foreign protein expressed in chloroplasts could represent up to 46% of total leaf proteins *(40)*.

With a limited protein maturation capacity, the chloroplast looks particularly well adapted for production of simple molecules *(43)*, but quite surprisingly, tobacco chloroplasts are also capable of properly fold some complex proteins *(44, 45)*. However, expression in the chloroplasts cannot be considered as a panacea for PMPs expression *in planta*, as a number of clinically useful proteins necessitate extensive post-translational processing. For instance, oligosaccharides attached to polypeptide chains by N- or O-glycosylation, in particular, have a strong impact on the activity of several therapeutic proteins and unfortunately, although they could import glycoproteins *(46)*, chloroplasts do not have the capacity to glycosylate proteins.

1.3.4. Engineering Plant Expression Systems for Reduced Protease Activity

In plants, as in any other heterologous expression system, recombinant protein yield not only depends on an efficient expression rate of the transgene but also on the stability of the resulting protein during the whole expression/recovery process *(47)*.

Proteases found in the different compartments of plant cells may dramatically alter the stability of foreign proteins either in vivo or in vitro during their recovery from plant tissues *(47, 48)*. Vacuolar proteases active in mildly acidic conditions, in particular, were readily identified as potentially damaging for the integrity of recombinant proteins expressed in vegetative organs of transgenic plants. As described above, targeting strategies based on the fusion of appropriate targeting signals to the therapeutic proteins have been used to avoid unwanted proteolysis in vivo by directing PMPs accumulation in compartments such as ER *(49, 50)* or chloroplasts where proteolytic activity is low *(51)*.

Transgenic plant lines with reduced protease activity levels in vivo could also help to maximize protein yields by slowing cellular hydrolytic processes. In particular, recent evidence in the literature suggests that hindering endogenous protease activities *in planta*, through expression of recombinant protease inhibitors, could help enhance protein levels in vegetative organs without compromising growth and development of the host plant. Illustrating this point, the rice cysteine proteinase inhibitor, oryzacystatin I, for instance, was shown to increase total soluble protein levels by 40% in leaves of transgenic tobacco lines expressing this inhibitor in the cytosolic compartment *(52)*. More recently, transgenic lines of potato expressing either tomato cathepsin D inhibitor or bovine aprotinin, both active against trypsin and chymotrypsin, show a decrease in Rubisco hydrolysis by 30–40% relative to control plants *(53)*. Based on current knowledge and progress to come on plant cell proteolytic processes, the design of transgenic plant lines deficient in specific protease activities in the secretory pathway could provide plant expression systems optimized for the production of complex proteins in “mild” cellular environments.

1.4. Post-Translational Modifications

The vast majority of therapeutic proteins undergo several PTMs, which are the final steps in which genetic information from a gene directs the formation of a functional gene product. One of the major advantages of plant expression systems is their capacity to perform most PTMs required for pharmaceutical protein bioactivity and pharmacokinetics (*see* **ref.** *2* for a recent review). For instance, several proteins synthesized as preproforms in animal cells are shown to be correctly matured into their biologically active forms when produced in a plant expression system (*see* **ref.** *30* for illustration).

However, there are also some difficulties with plant-specific PTMs as illustrated, for example, by the inability of transgenic plants to correctly mature human collagen *(54)* or to reproduce a human-type N- and O-glycosylation on plant-made antibodies *(55)*. Indeed, when a mammalian glycoprotein is produced in a plant expression system it is N-glycosylated on the same Asn residues as it would be in mammals, but its complex-type N-glycans are structurally different. In plant N-glycans, the proximal N-acetylglucosamine of the core is substituted by an α1,3 fucose in place of an α1,6 fucose in mammals, and the β-mannose of the core is substituted by a bisecting β1,2 xylose in plants, in place of a β1,4 *N*-acetylglucosamine in mammals. In addition, β1,3 galactose and fucose α1,4-linked to the terminal N-acetylglucosamine of plant N-glycans form Lewis a oligosaccharide structures instead of β1,4 galactose combined with sialic acids in mammals *(2)*.

Together with Lewis a, bisecting β1,2 xylose and core α1,3 fucose residues are constitutive of three glycoepitopes described on complex plant N-glycans. In fact, plant complex N-glycans are immunogenic in most laboratory mammals and elicit glycan-specific IgE and IgG antibodies in humans *(9, 56, 57, 58, 59, 60)*. As a result, as observed for any other eukaryotic system currently used for therapeutic protein production such as yeasts, insect cells or mammalian-cultured cells, because of their structural differences with human N-glycans, glycans N-linked to PMPs would be immunogenic in humans when delivered parenterally.

In order to fully exploit the potential of plants for the production of recombinant therapeutic glycoproteins, it is necessary to control the maturation of plant-specific N-glycans and thus prevent the addition of immunogenic glycoepitopes onto PMPs. One of the most drastic approaches is to prevent N-glycosylation, through the mutation of Asn or Ser/Thr residues constitutive of N-glycosylation sites. Generally, this strategy neither influence IgG folding and assembly in the plant ER, nor the antigen-binding capacity of an antibody *(61)*. However, many pharmaceuticals, including antibodies used for Fc-dependent functions, require

N-glycosylation for in vivo activity and longevity. This is why most efforts in glycoengineering of plant expression systems were focused on the production of glycosylated therapeutic proteins bearing non-immunogenic N-glycans. In plants as in other eukaryotic cells, proteins that reside in the lumen of the plant ER contain high-mannose-type N-glycans and, in contrast with complex-type N-glycans, high-mannose-type N-glycans have the same structure in plants and in mammals. We have recently shown that antibodies expressed in tobacco plants with a KDEL ER retention signal fused at the C-terminal ends of their heavy and light chains contain exclusively non-immunogenic high-mannose type N-glycans *(62, 63)*. Another strategy to get non-immunogenic N-glycans on PMPs is based on the inhibition of plant-specific Golgi glycosyltransferases. Knocking out α1,3 fucosyltransferase and β1,2 xylosyltransferase genes, to eliminate the plant-specific glycoepitopes was successful in several plant expression systems using either insertional mutation in *Arabidopsis* mutants *(64)* or targeted gene inactivation in the moss *Physcomitrella patens (65)*. RNA interference was also used for a knockout of α1,3 fucosyltransferase and α1,2 xylosyltransferase in *Lemna minor* and *Medicago sativa (66, 67)*. This strategy has allowed the production of plantibodies harbouring non-immunogenic N-glycans in several plant expression systems *(66–69)*.

In addition to approaches involving glycosyltransferase inactivation, another attractive strategy to humanize plant N-glycans is to express mammalian glycosyltransferases in plants, which would complete and/or compete with the endogenous machinery of N-glycan maturation in the plant Golgi apparatus. As part of these complementation strategies, it has been shown that the human β1, 4 galactosyltransferase, expressed in plant cells, transfers galactose residues onto the terminal N-acetylglucosamine residues of plant N-glycans *(11, 70, 71)*. These results are very promising and several laboratories are currently working to increase the performance of heterologous glycosyltransferases through better control of their targeting in the Golgi cisternae. Encouragingly, the analysis of several plant glycosyltransferases is currently providing a panel of specific signals sufficient for a targeted expression of heterologous glycosyltransferases within the different Golgi subcompartments of a plant cell *(55, 72, 73)*.

The presence of sialic acid residues at the termini of N-glycan antennae is very important for the clearance of many mammalian plasma proteins of pharmaceutical interest. Indeed, the absence of these residues on circulating proteins results in their rapid elimination from the blood, by interactions with galactose-specific receptors on the surface of hepatic cells. Sialic acids are not detectable in plant glycoproteins *(74, 75)*. The production of sialylated PMPs is not yet feasible in plants as previously shown in insect cells *(76)*. However, most of the complex sialylation pathway located both in the Golgi lumen and in the cytosol of mammalian cells was already rebuilt in plants *(77–79)*.

Engineering N-glycosylation in plants could improve the efficiency of PMPs not only by reducing structural differences between plant and mammalian N-glycans but also by producing glycovariants of therapeutic proteins showing a higher biological activity than those expressed in cultured mammalian cells *(55)*.As an example, removal of the α1,3 linked fucose from the N-glycans of a plant-made antibody has the same effect on its ADCC activity as the removal of the α 1,6 fucose on the same antibody produced in Chinese hamster ovary (CHO) cells *(68, 69, 80)*.

1.5. Emerging Plant Expression Hosts

The most frequently used plant expression systems for therapeutic protein production are "seed crops" such as maize, rice and safflower and "leaf biomass" plants like tobacco, *Arabidopis thaliana* and alfalfa. However, some emerging plant expression systems, like duckweeds, algae, mosses or higher plant cell suspension cultures, are offering new opportunities for molecular farming and are currently being investigated by companies involved in PMP production (**Table 1.2**). These expression systems would have a double benefit of being (i) much more consistent with public demand for high containment of genetically modified plants and (ii) more compliant with regulatory issues for the production of therapeutic proteins, since they are grown in a controlled environment.

1.5.1. Moss

Mosses are higher multicellular eukaryotes and therefore perform extensive post-translational processing of proteins including disulfide bridge formation and glycosylation. Transgenic *P. patens* are generated *via* the polyethyleneglycol-mediated transfection of protoplasts. Generation of stable transgenic *P. patens* takes about 8 weeks after transformation *(81–83)* and cultivation of this moss in glass bioreactors is well established. Several therapeutic proteins have been already produced in this expression system *(69, 84)*

P. patens is unique among all multicellular plants analyzed to date in exhibiting a very effective homologous recombination process in its nuclear DNA. This allows targeted knockouts and knockin of genes, a highly attractive tool for production of strains designed for PMP production *(85)*.

For instance, while N-glycosylation is very similar in *P. patens* and in higher plants *(86)*, this moss is currently one of the most advanced plant expression systems for glycoengineering due to the ease with which knockout and knockin of glycosylation enzyme genes can be performed by homologous recombination in this system *(65)*. Thus, *P. patens* has been engineered to produce a strain that does not add β1,2 xylose or α1,3 fucose, but produces PMPs bearing a core heptasaccharide identical to that of a human IgG *(69)*.

1.5.2. Algae

Algae are currently emerging as an alternative system for production of recombinant therapeutic proteins. Unicellular eukaryotic green algae such as *Chlamydomonas reinhardtii*, *Phaeodactylum tricornutum*, *Tetraselmis suecica* and *Odontella aurita* can produce a significant amount of recombinant proteins *(87)*. Freshwater algae *C. reinhardtii* is the best-studied type for recombinant protein production *via* chloroplast transformation *(88)*. *C. reinhardtii* contains a single large chloroplast that occupies ~40% of the cell volume and its transformation through homologous recombination was first described in 1988 *(89)*.

C. reinhardtii can be grown in a cost-effective manner at a large scale, in 500,000-l containers. Compared to land plants, it grows at a much faster rate, doubling its cell number every 4–8 h *(87)*. Purification of recombinant proteins should be simpler in algae than in terrestrial plants. The cellular population of algae is uniform in size and type, thus there is no gradient of recombinant protein distribution, a fact which simplifies purification and reduces the loss of biomass. *C. reinhardtii* has also the ability to secrete proteins in the culture medium, which could further reduce protein production costs *(90)*. A human mAb produced in transgenic algae was correctly assembled and has the same capacity to bind herpes virus proteins as its mammalian homolog *(45)*. Having said this, chloroplast-encoded proteins are not glycosylated and this mAb has shown no evidence for glycosylation required for the Fc-dependent functions. In addition, the codon bias in algae constitutes an additional difficulty for foreign protein expression in this system due to the need of an extensive optimization of the gene sequences.

1.5.3. Lemna

Lemna gibba and *Lemna minor*, commonly named duckweeds, are free floating plants which develop on water and are found all over the world. With their naturally simple growth conditions, duckweeds are well adapted for intensive culturing methods. Duckweed allows very high rates of biomass accumulation per unit of time – it can double in size every 24–48 h. Recombinant proteins produced in duckweeds after either *Agrobacterium tumefaciens*-mediated or by biolistic transformation can be extracted and purified or the plant containing the protein can be used directly, dry or fresh. As for other plant expression systems, secretion into the extracellular media is dependent on the presence of a signal peptide. *Lemna* recognizes plant and human signal sequences with the same efficiency *(91)*.

The capacity of this expression system for biosynthesis and maturation of complex therapeutic proteins was recently illustrated in Cox et al. *(66)*, with the production of a human monoclonal antibody in a glycoengineered *lemna*. This antibody exhibited a single major N-glycan species without any detectable plant-specific N-glycans and shows an higher antibody-dependent cell-mediated cytotoxicity and effector cell receptor binding activities than its homolog expressed in cultured CHO cells *(66)*.

1.5.4. Higher Plant Suspension-Cultured Cells

Higher plant cell cultures offer many advantages over field grown plants or even plants grown in greenhouses for PMP production. Among these advantages, plant cells are grown in highly controlled and sterile in vitro conditions. Some plant cells grow very fast, for instance, BY-2 tobacco cells number is doubling every 12 h in optimal growth conditions, thus rapidly providing an important biomass. Many therapeutic proteins have already been successfully expressed in suspension-cultured plant cells. The potential of plant cells for biopharmaceutical production was recently illustrated with the use of suspension-cultured tobacco cells to synthesize correctly matured and highly immunoreactive recombinant house dust mite allergens that could be used for allergy diagnostic and immunotherapy *(30)*. These results, together with the recent production of human glucocerebrosidase in suspension-cultured carrot cells *(92)* and the first regulatory approval for a plant cell-produced animal vaccine, exemplify the high potential of plant suspension-cultured cells as bioreactors for the production of therapeutic proteins under controlled and environmentally safe conditions. In addition, this production system allows for an efficient secretion of PMPs into an inorganic culture medium offering substantial cost advantages in downstream purification. This could counterbalance an increased production cost due to the use of fermentors for production instead of field or greenhouse production with whole plants.

Another advantage for downstream processing is that plant cells are uniform in size and types, which leads to a low PMP heterogeneity as compared to production in whole plants. For instance, it has been reported that glycosylation patterns of an antibody expressed in tobacco plants, differ from young to old leaves *(93)*. In contrast, glycan patterns are reproducible from batch to batch in BY-2 tobacco cell cultures and, interestingly, complementation of the culture medium could strongly reduce N-glycan heterogeneity (Faye et al. unpublished results).

1.6. Conclusion

Plants offer a safe and extremely cost-effective alternative to microbial or mammalian expression systems for the production of biopharmaceuticals. Current strategies to improve plant expression systems will rapidly result in increased yield and simplification of downstream processing of plant-made therapeutic proteins. These promising results, together with rapid progression in the control of post-translational maturations, will allow human-like maturations on PMPs and hence make plant expression systems suitable alternatives to animal cell factories.

Acknowledgement

Work on glycobiology at the University of Rouen was supported by the Centre National de la Recherche Scientifique (CNRS) and by the "Ministère de la Recherche". We thank present and former colleagues who contributed to the work described in this review and for their critical reading of the manuscript.

References

1. Walsh, G., and Jefferis, R. (2006) Post-translational modifications in the context of therapeutic proteins. *Nat. Biotechnol.* 24, 1241–1252.
2. Gomord, V., and Faye, L. (2004) Posttranslational modification of therapeutic proteins in plants. *Curr. Opin. Plant Biol.* 7, 171–181.
3. Twyman, R.M., Stoger, E., Schillberg, S., Christou, P., and Fischer, R. (2003) Molecular farming in plants: host systems and expression technology. *Trends Biotechnol.* 21, 570–578.
4. Gomord, V., Sourrouille, C., Fitchette, A.-C., Bardor, M., Pagnt, S., Lerouge, P., and Faye, L. (2004) Production and glycosylation of plant-made pharmaceuticals: the antibodies as a challenge. *Plant Biotechnol. J.* 2, 83–100.
5. Hiatt, A., Cafferkey, R., and Bowdish, K. (1989) Production of antibodies in transgenic plants. *Nature* 342, 76–78.
6. Ma, J.K., Hiatt, A., Hein, M., Vine, N.D., Wang, F., Stabila, P., van Dolleweerd, C., Mostov, K., and Lehner, T. (1995) Generation and assembly of secretory antibodies in plants. *Science* 268, 716–719.
7. Khoudi, H., Laberge, S., Ferullo, J.M., Bazin, R., Darveau, A., Castonguay, Y., Allard, G., Lemieux, R., and Vezina, L.P. (1999) Production of a diagnostic monoclonal antibody in perennial alfalfa plants. *Biotechnol. Bioeng.* 64, 135–143.
8. Austin, S., Bingham, E.T., Koegel, R.G., Mathews, D.E., Shahan, M.N., Straub, R.J., and Burgess, R.R. (1994) An overview of a feasibility study for the production of industrial enzymes in transgenic alfalfa. *Ann. N. Y. Acad. Sci.* 721, 234–244.
9. Gomord, V., Chamberlain, P., Jefferis, R., and Faye, L. (2005) Biopharmaceutical production in plants: problems, solutions and opportunities. *Trends Biotechnol.* 23, 559–565.
10. Cabanes-Macheteau, M., Fitchette-Laine, A.C., Loutelier-Bourhis, C., Lange, C., Vine, N.D., Ma, J.K., Lerouge, P., and Faye, L. (1999) N-Glycosylation of a mouse IgG expressed in transgenic tobacco plants. *Glycobiology* 9, 365–372.
11. Bakker, H., Bardor, M., Molthoff, J.W., Gomord, V., Elbers, I., Stevens, L.H., Jordi, W., Lommen, A., Faye, L., Lerouge, P., and Bosch, D. (2001) Galactose-extended glycans of antibodies produced by transgenic plants. *Proc. Natl. Acad. Sci. U. S. A.* 98, 2899–2904.
12. Bouquin, T., Thomsen, M., Nielsen, L.K., Green, T.H., Mundy, J., and Hanefeld Dziegiel, M. (2002) Human anti-rhesus D IgG1 antibody produced in transgenic plants. *Transgenic Res.* 11, 115–122.
13. Streatfield, S.J. (2007) Approaches to achieve high-level heterologous protein production in plants. *Plant Biotechnol. J.* 5, 2–15.

14. Perlak, F.J., Fuchs, R.L., Dean, D.A., McPherson, S.L., and Fischhoff, D.A. (1991) Modification of the coding sequence enhances plant expression of insect control protein genes. *Proc. Natl. Acad. Sci. U. S. A.* 88, 3324–3328.
15. Jensen, L.G., Olsen, O., Kops, O., Wolf, N., Thomsen, K.K., and von Wettstein, D. (1996) Transgenic barley expressing a protein-engineered, thermostable (1,3-1,4)-beta-glucanase during germination. *Proc. Natl. Acad. Sci. U. S. A.* 93, 3487–3491.
16. Batard, Y., Hehn, A., Nedelkina, S., Schalk, M., Pallett, K., Schaller, H., and Werck-Reichhart, D. (2000) Increasing expression of P450 and P450-reductase proteins from monocots in heterologous systems. *Arch. Biochem. Biophys.* 379, 161–169.
17. Hamada, A., Yamaguchi, K.-I., Ohnishi, N., Harada, M., Nikumaru, S., and Honda, H. (2005) High-level production of yeast (Schwanniomyces occidentalis) phytase in transgenic rice plants by a combination of signal sequence and codon modification of the phytase gene. *Plant Biotechnol. J.* 3, 43–55.
18. Hamilton, A.J., and Baulcombe, D.C. (1999). A species of small antisense RNA in posttranscriptional gene silencing in plants. *Science* 286, 950–952.
19. Voinnet, O. (2005) Induction and suppression of RNA silencing: insights from viral infections. *Nat. Rev. Genet.* 6, 206–220.
20. Scholthof, H.B., Scholthof, K.B., and Jackson, A.O. (1995) Identification of tomato bushy stunt virus host-specific symptom determinants by expression of individual genes from a potato virus X vector. *Plant Cell* 7, 1157–1172.
21. Voinnet, O., Rivas, S., Mestre, P., and Baulcombe, D. (2003) An enhanced transient expression system in plants based on suppression of gene silencing by the p19 protein of tomato bushy stunt virus. *Plant J.* 33, 949–956.
22. Voinnet, O., Pinto, Y.M., and Baulcombe, D.C. (1999) Suppression of gene silencing: a general strategy used by diverse DNA and RNA viruses of plants. *Proc. Natl. Acad. Sci. U. S. A.* 96, 14147–14152.
23. Goldstein, D.A., and Thomas, J.A. (2004) Biopharmaceuticals derived from genetically modified plants. *Q. J. Med.* 97, 705–716.
24. Hellwig, S., Drossard, J., Twyman, R.M., and Fischer, R. (2004) Plant cell cultures for the production of recombinant proteins. *Nat. Biotechnol.* 22, 1415–1422.
25. Masumura, T., Morita, S., Miki, Y., Kurita, A., Morita, S., Shirono, H., Koga, J., and Tanaka, K. (2006) Production of biologically active human interferon-alpha in transgenic rice. *Plant Biotechnol. J.* 23, 91–97.
26. Shirono, H., Morita, S., Miki, Y., Kurita, A., Morita, S., Koga, J., Tanaka, K., and Masumura, T. (2006) Highly efficient production of human interferon-alpha by transgenic cultured rice cells. *Plant Biotechnol. J.* 23, 283–289.
27. Vancanneyt, G., Dubald, M., Schröder, W., Peters, J., and Botterman, J. (2008) A case study for plant-made pharmaceuticals comparing different plant expression and production systems. This issue.
28. Goldstein, D.A., and Thomas, J.A. (2004) Biopharmaceuticals derived from genetically modified plants. *Q. J. Med.* 97, 705–716.
29. Streatfield, S.J. (2007) Approaches to achieve high-level heterologous protein production in plants. *Plant Biotechnol. J.* 5, 2–15.
30. Lienard, D., Tran Dinh, O., van Oort, E., Van Overtvelt, L., Bonneau, C., Wambre, E., Bardor, M., Cosette, P., Didier-Laurent, A., Dorlhac de Borne, F., Delon, R., van Ree, R., Moingeon, P., Faye, L., and Gomord, V. (2007) Suspension-cultured BY-2 tobacco cells produce and mature immunologically active house dust mite allergens, Chapter 12 in this book. *Plant Biotechnol. J.* 5, 93–108.
31. Conrad, U., and Fiedler, U. (1998) Compartment-specific accumulation of recombinant immunoglobulins in plant cells: an essential tool for antibody production and immunomodulation of physiological functions and pathogen activity. *Plant Mol. Biol.* 38, 101–109.
32. Wandelt, C.I., Khan, M.R., Craig, S., Schroeder, H.E., Spencer, D., and Higgins, T.J. (1992) Vicilin with carboxy-terminal KDEL is retained in the endoplasmic reticulum and accumulates to high levels in the leaves of transgenic plants. *Plant J.* 2, 181–192.
33. Vitale, A., and Pedrazzini, E. (2005) Recombinant pharmaceuticals from plants: the plant endomembrane system as bioreactor. *Mol. Interv.* 5, 216–225.
34. Arcalis, E., Marcel, S., Altmann, F., Kolarich, D., Drakakaki, G., Fischer, R., Christou, P., and Stoger, E. (2004) Unexpected deposition patterns of recombinant proteins in post-endoplasmic reticulum compartments of wheat endosperm. *Plant Physiol.* 136, 3457–3466.

35. Rademacher, T., Arcalis, E., and Stoger, E. (2008) Production and localization of recombinant pharmaceuticals in transgenic seeds. Chapter 5 in this book
36. Torrent, M., Llop, I., and Ludevid, M.D. (2008) Protein body induction: a new tool to produce and recover recombinant proteins in plants. This issue
37. van Rooijen, G.J., and Moloney, M.M. (1995) Structural requirements of oleosin domains for subcellular targeting to the oil body, Chapter 11 in this book. *Plant Physiol.* 109, 1353–1361.
38. Nykiforuk, C.L., Boothe, J.G., Murray, E.W., Keon, R.G., Goren, H.J., Markley, N.A., and Moloney, M.M. (2006) Transgenic expression and recovery of biologically active recombinant human insulin from Arabidopsis thaliana seeds. *Plant Biotechnol. J.* 4, 77–86.
39. Daniell, H., and Dhingra, A. (2002) Multigene engineering: dawn of an exciting new era in biotechnology. *Curr. Opin. Biotechnol.* 13, 136–141.
40. De Cosa, B., Moar, W., Lee, S.B., Miller, M., and Daniell, H. (2001) Overexpression of the Bt cry2Aa2 operon in chloroplasts leads to formation of insecticidal crystals. *Nat. Biotechnol.* 19, 71–74.
41. Singh, N.D., Ding, Y., and Daniell, H. (2008) Chloroplast-derived vaccine antigens and biopharmaceuticals: protocols for expression, purification or oral delivery and functional evaluation. This issue
42. Staub, J.M., Garcia, B., Graves, J., Hajdukiewicz, P.T., Hunter, P., Nehra, N., Paradkar, V., Schlittler, M., Carroll, J.A., Spatola, L., Ward, D., Ye, G., and Russell, D.A. (2000) High-yield production of a human therapeutic protein in tobacco chloroplasts. *Nat. Biotechnol.* 18, 333–338.
43. Jobling, S.A., Jarman, C., Teh, M.M., Holmberg, N., Blake, C., and Verhoeyen, M.E. (2003) Immunomodulation of enzyme function in plants by single-domain antibody fragments. *Nat. Biotechnol.* 21, 77–80.
44. Daniell, H., Streatfield, S.J., and Wycoff, K. (2001) Medical molecular farming: production of antibodies, biopharmaceuticals and edible vaccines in plants. *Trends Plant Sci.* 6, 219–226.
45. Mayfield, S.P., Franklin, S.E., and Lerner, R.A. (2003) Expression and assembly of a fully active antibody in algae. *Proc. Natl. Acad. Sci. U. S. A.* 100, 438–442.
46. Faye, L., and Daniell, H. (2006) Novel pathway for glycoprotein import into chloroplasts. *Plant Biotechnol. J.* 4, 275–279.
47. Michaud, D., Vrain, T., Gomord, V., and Faye, L. (1998) Stability of recombinant proteins in plants. *In Methods in biotechnology. Recombinant proteins from plants—production and isolation of clinically useful compounds, P. AJR, ed. (Totowa (NJ): Humana Press)*, pp. 177–188.
48. Michaud, D. (1998) Gel electrophoresis of proteolytic enzymes. *Anal. Chim. Acta* 372, 173–185.
49. Schouten, A., Roosien, J., van Engelen, F.A., de Jong, G.A., Borst-Vrenssen, A.W., Zilverentant, J.F., Bosch, D., Stiekema, W.J., Gommers, F.J., Schots, A., and Bakker, J. (1996) The C-terminal KDEL sequence increases the expression level of a single-chain antibody designed to be targeted to both the cytosol and the secretory pathway in transgenic tobacco. *Plant Mol. Biol.* 30, 781–793.
50. Pagny, S., Cabanes-Macheteau, M., Gillikin, J.W., Leborgne-Castel, N., Lerouge, P., Boston, R.S., Faye, L., and Gomord, V. (2000) Protein recycling from the Golgi apparatus to the endoplasmic reticulum in plants and its minor contribution to calreticulin retention. *Plant Cell* 12, 739–756.
51. Wong, E.Y., Hironaka, C.M., and Fischhoff, D.A. (1992) Arabidopsis thaliana small subunit leader and transit peptide enhance the expression of Bacillus thuringiensis proteins in transgenic plants. *Plant Mol. Biol.* 20, 81–93.
52. Benchabane, M., Rivard, D., Girard, C., and Michaud, D. (XXXX) Companion protease inhibitors to protect recombinant proteins in transgenic plant extracts. This issue
53. Rivard, D., Anguenot, R., Brunelle, F., Le, V.Q., Vézina, L.P., Trépanier, S., Michaud, D. (2006) An in-built proteinase inhibitor system for the protection of recombinant proteins recovered from transgenic plants. *Plant Biotechnol. J.* 4, 359–368.
54. Merle, C., Perret, S., Lacour, T., Jonval, V., Hudaverdian, S., Garrone, R., Ruggiero, F., and Theisen, M. (2002) Hydroxylated human homotrimeric collagen I in Agrobacterium tumefaciens-mediated transient expression and in transgenic tobacco plant. *FEBS Lett.* 515, 114–118.
55. Saint-Jore-Dupas, C., Faye, L., and Gomord, V. (2007) From planta to pharma with glycosylation in the toolbox. *Trends Biotechnol.* 25, 317–323.
56. Aalberse, R.C., Koshte, V., and Clemens, J.G. (1981) Immunoglobulin E antibodies that crossreact with vegetable foods, pollen, and Hymenoptera venom. *J. Allergy Clin. Immunol.* 68, 356–364.

57. Faye, L., and Chrispeels, M.J. (1985) Characterization of N-linked oligosaccharides by affinoblotting with concanavalin A-peroxidase and treatment of the blots with glycosidases. *Anal. Biochem.* 149, 218–224.

58. Aalberse, R.C., Koshte, V., and Clemens, J.G. (1981) Cross-reactions between vegetable foods, pollen and bee venom due to IgE antibodies to a ubiquitous carbohydrate determinant. *Int. Arch. Allergy Appl. Immunol.* 66, 259–260.

59. Bardor, M., Faveeuw, C., Fitchette, A.C., Gilbert, D., Galas, L., Trottein, F., Faye, L., and Lerouge, P. (2003) Immunoreactivity in mammals of two typical plant glyco-epitopes, core alpha(1,3)-fucose and core xylose. *Glycobiology* 13, 427–434.

60. Jin, C., Bencurova, M., Borth, N., Ferko, B., Jensen-Jarolim, E., Altmann, F., and Hantusch, B. (2006) Immunoglobulin G specifically binding plant N-glycans with high affinity could be generated in rabbits but not in mice. *Glycobiology* 16, 349–357.

61. Nuttall, J., Ma, J.K., and Frigerio, L. (2005) A functional antibody lacking N-linked glycans is efficiently folded, assembled and secreted by tobacco mesophyll protoplasts. *Plant Biotechnol. J.* 3, 497–504.

62. Sriraman, R., Bardor, M., Sack, M., Vaquero, C., Faye, L., Fischer, R., Finnern, R., and Lerouge, P. (2004) Recombinant anti-hCG antibodies retained in the endoplasmic reticulum of transformed plants lack core-xylose and core-alpha(1,3)-fucose residues. *Plant Biotechnol. J.* 2, 279–287.

63. Petruccelli, S., Otegui, M.S., Lareu, F., Tran Dinh, O., Fitchette, A.-C., Circosta, A., Rumbo, M., Bardor, M., Carcamo, R., Gomord, V., and Beachy, R.N. (2006) A KDEL-tagged monoclonal antibody is efficiently retained in the endoplasmic reticulum in leaves, but is both partially secreted and sorted to protein storage vacuoles in seeds. *Plant Biotechnol. J.* 4, 511–527.

64. Strasser, R., Altmann, F., Mach, L., Glossl, J., and Steinkellner, H. (2004) Generation of Arabidopsis thaliana plants with complex N-glycans lacking beta1,2-linked xylose and core alpha1,3-linked fucose. *FEBS Lett.* 561, 132–136.

65. Koprivova, A., Stemmer, C., Altmann, F., Hoffmann, A., Kopriva, S., Gorr, G., Reski, R., and Decker, E.L. (2004) Targeted knockouts of Physcomitrella lacking plant-specific immunogenic N-glycans. *Plant Biotechnol. J.* 2, 517–523.

66. Cox, K.M., Sterling, J.D., Regan, J.T., Gasdaska, J.R., Frantz, K.K., Peele, C.G., Black, A., Passmore, D., Moldovan-Loomis, C., Srinivasan, M., Cuison, S., Cardarelli, P.M., and Dickey, L.F. (2006) Glycan optimization of a human monoclonal antibody in the aquatic plant Lemna minor. *Nat. Biotechnol.* 24, 1591–1597.

67. Sourrouille, C. (2005) Inactivation de l'alpha (1,3)-fucosyltransférase et de la beta (1,2)-xylosyltransferase, en vue de la production de protéines recombinantes d'intérêt thérapeutique chez la luzerne *(Université de Rouen)*, pp. 164.

68. Schahs, M., Strasser, R., Stadlmann, J., Kunert, R., Rademacher, T., and Steinkellner, H. (2007) Production of a monoclonal antibody in plants with a humanized N-glycosylation pattern. *Plant Biotechnol. J.* 5, 657–663.

69. Schuster, M., Jost, W., Mudde, G.C., Wiederkum, S., Schwager, C., Janzek, E., Altmann, F., Stadlmann, J., Stemmer, C., and Gorr, G. (2007) In vivo glyco-engineered antibody with improved lytic potential produced by an innovative non-mammalian expression system. *Biotechnol. J.* 2, 700–708.

70. Misaki, R., Kimura, Y., Palacpac, N.Q., Yoshida, S., Fujiyama, K., and Seki, T. (2003) Plant cultured cells expressing human beta1,4-galactosyltransferase secrete glycoproteins with galactose-extended N-linked glycans. *Glycobiology* 13, 199–205.

71. Fujiyama, K., Palacpac, N.Q., Sakai, H., Kimura, Y., Shinmyo, A., Yoshida, T., and Seki, T. (2001) In vivo conversion of a glycan to human compatible type by transformed tobacco cells. *Biochem. Biophys. Res. Commun.* 289, 553–557.

72. Saint-Jore-Dupas, C., Nebenfuhr, A., Boulaflous, A., Follet-Gueye, M.L., Plasson, C., Hawes, C., Driouich, A., Faye, L., and Gomord, V. (2006) Plant N-glycan processing enzymes employ different targeting mechanisms for their spatial arrangement along the secretory pathway. *Plant Cell* 18, 3182–3200.

73. Bakker, H., Rouwendal, G.J., Karnoup, A.S., Florack, D.E., Stoopen, G.M., Helsper, J.P., van Ree, R., van Die, I., and Bosch, D. (2006) An antibody produced in tobacco expressing a hybrid beta-1,4-galactosyltransferase is essentially devoid of plant carbohydrate epitopes. *Proc. Natl. Acad. Sci. U. S. A.* 103, 7577–7582.

74. Seveno, M., Bardor, M., Paccalet, T., Gomord, V., Lerouge, P., and Faye, L. (2004) Glycoprotein sialylation in plants? *Nat. Biotechnol.* 22, 1351–1352; 1352–1353.

75. Zeleny, R., Kolarich, D., Strasser, R., and Altmann, F. (2006) Sialic acid concentrations in plants are in the range of inadvertent contamination. *Planta* 224, 222–227.
76. Jarvis, D.L. (2003) Developing baculovirus-insect cell expression systems for humanized recombinant glycoprotein production. *Virology* 310, 1–7.
77. Wee, E.G., Sherrier, D.J., Prime, T.A., and Dupree, P. (1998) Targeting of active sialyltransferase to the plant Golgi apparatus. *Plant Cell* 10, 1759–1768.
78. Misaki, R., Fujiyama, K., and Seki, T. (2006) ExpressionofhumanCMP-N-acetylneuraminic acid synthetase and CMP-sialic acid transporter in tobacco suspension-cultured cell. *Biochem. Biophys. Res. Commun.* 339, 1184–1189.
79. Paccalet, T., Bardor, M., Rihouey, C., Delmas, F., Chevalier, C., D'Aoust, M.-A., Faye, L., Vezina, L.P., Gomord, V., and Lerouge, P. (2007) Engineering of a sialic acid synthesis pathway in transgenic plants by expression of bacterial Neu5Ac-synthesizing enzymes. *Plant Biotechnol. J.* 5, 16–25.
80. Jefferis, R. (XXXX) Glycosylation of antibody therapeutics: optimisation for purpose. This issue
81. Schaefer, D., Zryd, J.P., Knight, C.D., and Cove, D.J. (1991) Stable transformation of the moss Physcomitrella patens. *Mol. Gen. Genet.* 226, 418–424.
82. Schaefer, D.G., and Zryd, J.P. (1997) Efficient gene targeting in the moss Physcomitrella patens. *Plant J.* 11, 1195–1206.
83. Lienard, D., and Nogué, F. (XXXX) *Physcomitrella patens*: a non-vascular plant for recombinant protein production. This issue.
84. Baur, A., Reski, R., and Gorr, G. (2005) Enhanced recovery of a secreted recombinant human growth factor using stabilizing additives and by co-expression of human serum albumin in the moss Physcomitrella patens. *Plant Biotechnol. J.* 3, 331–340.
85. Schaefer, D.G., Bisztray, G., and Zrÿd, J.-P. (1994) Genetic transformation of the moss Physcomitrella patens. *InPlant protoplasts and genetic engineering, V, Y.P.S. Bajaj, ed. (Berlin, Heidelberg, New York: Springer Verlag)*, pp. 349–364.
86. Vietor, R., Loutelier, B.C., Fitchette, A., Margerie, P., Gonneau, M., Faye, L., and Lerouge, P. (2003) Protein N-glycosylation is similar in the moss Physcomitrella patens and in higher plants. *Planta* 218, 269–275.
87. Franklin, S.E., and Mayfield, S.P. (2004) Prospects for molecular farming in the green alga Chlamydomonas. *Curr. Opin. Plant Biol.* 7, 159–165.
88. Leon-Banares, R., Gonzalez-Ballester, D., Galvan, A., and Fernandez, E. (2004) Transgenic microalgae as green cell-factories. *Trends Biotechnol.* 22, 45–52.
89. Boynton, J.E., Gillham, N.W., Harris, E.H., Hosler, J.P., Johnson, A.M., Jones, A.R., Randolph-Anderson, B.L., Robertson, D., Klein, T.M., Shark, K.B., et al. (1988) Chloroplast transformation in Chlamydomonas with high velocity microprojectiles. *Science* 240, 1534–1538.
90. Mayfield, S.P., and Franklin, S.E. (2005) Expression of human antibodies in eukaryotic micro-algae. *Vaccine* 23, 1828–1832.
91. Gasdaska, J.R., Spencer, D., and Dickey, L.F. (2003) Advantages of therapeutic protein production in the aquatic plant Lemna. *Bioprocessing J.* 3, 50–56.
92. Shaaltiel, Y., Bartfeld, D., Hashmueli, S., Baum, G., Brill-Almon, E., Galili, G., Dym, O., Boldin-Adamsky, S.A., Silman, I., Sussman, J.L., Futerman, A.H., and Aviezer, D. (2007) Production of glucocerebrosidase with terminal mannose glycans for enzyme replacement therapy of Gaucher's disease using a plant cell system. *Plant Biotechnol. J.* 5, 579–590.
93. Elbers, I.J., Stoopen, G.M., Bakker, H., Stevens, L.H., Bardor, M., Molthoff, J.W., Jordi, W.J., Bosch, D., and Lommen, A. (2001) Influence of growth conditions and developmental stage on N-glycan heterogeneity of transgenic immunoglobulin G and endogenous proteins in tobacco leaves. *Plant Physiol.* 126, 1314–1322.

Chapter 2

Cowpea Mosaic Virus-Based Systems for the Expression of Antigens and Antibodies in Plants

Frank Sainsbury, Li Liu, and George P. Lomonossoff

Summary

This chapter describes the use of Cowpea mosaic virus-based vectors for the production of foreign proteins such as antigens and antibodies in plants. The systems include vectors based on both full-length and deleted versions of RNA-2. In both cases, the modified RNA-2 is replicated by coinoculation with RNA-1. The constructs based on full-length RNA-2 retain the ability to spread systemically throughout an inoculated plant and the infection can be passaged. The vector based on a deleted version of RNA-2 can stably incorporate larger inserts but lacks the ability to move systemically. However, it has the added advantage of biocontainment. In both cases, vector constructs modified to contain a foreign gene of interest can be delivered by agroinfiltration to obtain transient expression of the foreign protein. If required, the same constructs can also be used for stable nuclear transformation. Both types of vector have proved effective for the production in plants of a diverse range of proteins including antigens and antibodies.

Key words: CPMV, Full-length viral vector, Deleted viral vector, Agroinfiltration, Suppressor of silencing, Virus passaging

2.1. Introduction

A number of RNA viruses have been developed over the last 20 years as vectors for the expression of foreign peptides and polypeptides *(1, 2)*, including antigens and antibodies *(3–5)*, in plants. The advantages of using viruses for expression include the fact that (i) viral genomes are small and infectious cDNA copies are, therefore, easy to manipulate, (ii) infection of plants with modified viruses is much simpler and quicker than the regeneration of stably transformed lines of plants, and (iii) a sequence inserted into a virus vector will be highly amplified during viral

Loïc Faye and Véronique Gomord (eds.), *Methods in Molecular Biology, Recombinant Proteins From Plants, vol. 483*

DOI: 10.1007/978-1-59745-407-0_2

replication. Unavoidably, there are also some potential disadvantages: there are likely to be size constraints on the sequences which can be inserted while retaining virus viability, the inserted sequence might be susceptible to "genetic drift" during virus replication, and there are biocontainment concerns over the use of vectors based on fully competent viruses. As a result, attention has recently turned toward the development of plant virus-based expression systems that can alleviate some of these disadvantages while retaining the advantageous features of viral vectors. These are often based on defective versions of viral RNAs which, though able to replicate, cannot spread in the environment *(6, 7)*.

Cowpea mosaic virus (CPMV) is the type member of the Genus *Comovirus* in the family *Comoviridae*. It infects a number of legume species and grows to particularly high titers in its natural host, cowpea (*Vigna unguiculata*). It also infects the experimental host, *Nicotiana benthamiana*. The genome of CPMV consists of two separately encapsidated positive-strand RNA molecules of 5889 (RNA-1) and 3481 (RNA-2) nucleotides. The RNAs each contain a single open reading frame (ORF) and are expressed through the synthesis and subsequent processing of precursor polyproteins (**Fig. 2.1a**). RNA-1 encodes proteins involved in the replication of viral RNAs and polyprotein processing. RNA-2 encodes the 48K movement protein and the two coat proteins, which are essential for cell to cell movement and systemic spread. The enzyme responsible for processing both the RNA-1- and the RNA-2-encoded polyproteins is the RNA-1-encoded 24K proteinase. Importantly, infectious cDNA clones of both viral RNAs are available, allowing manipulation of the viral genome and the generation of viral infection by agroinfiltration *(8)*. The fact that CPMV infects edible plants and grows to very high titers (yields of particles reaching ~1 g/Kg infected cowpea tissue) makes CPMV a very attractive system for the production of pharmaceutical proteins. It has already enjoyed considerable success as an "epitope presentation" system in which short antigenic sequences are expressed on the surface of assembled virus particles *(9, 10)*. More recent work has concentrated on the development of CPMV as a method of expression whole polypeptides in plants and this application is the subject of this chapter.

The development of CPMV-based systems for the expression of whole proteins has so far focused entirely on modifying the sequence of RNA-2. Replication functions are provided by coinoculating the RNA-2 constructs with unmodified RNA-1. Two types of system have been developed based on either full-length or deleted versions of RNA-2. In the first instance, the foreign protein to be expressed is added to the normal complement of RNA-2 proteins, resulting in an increase in size of RNA-2. Since all the normal functions are retained, the resulting modified

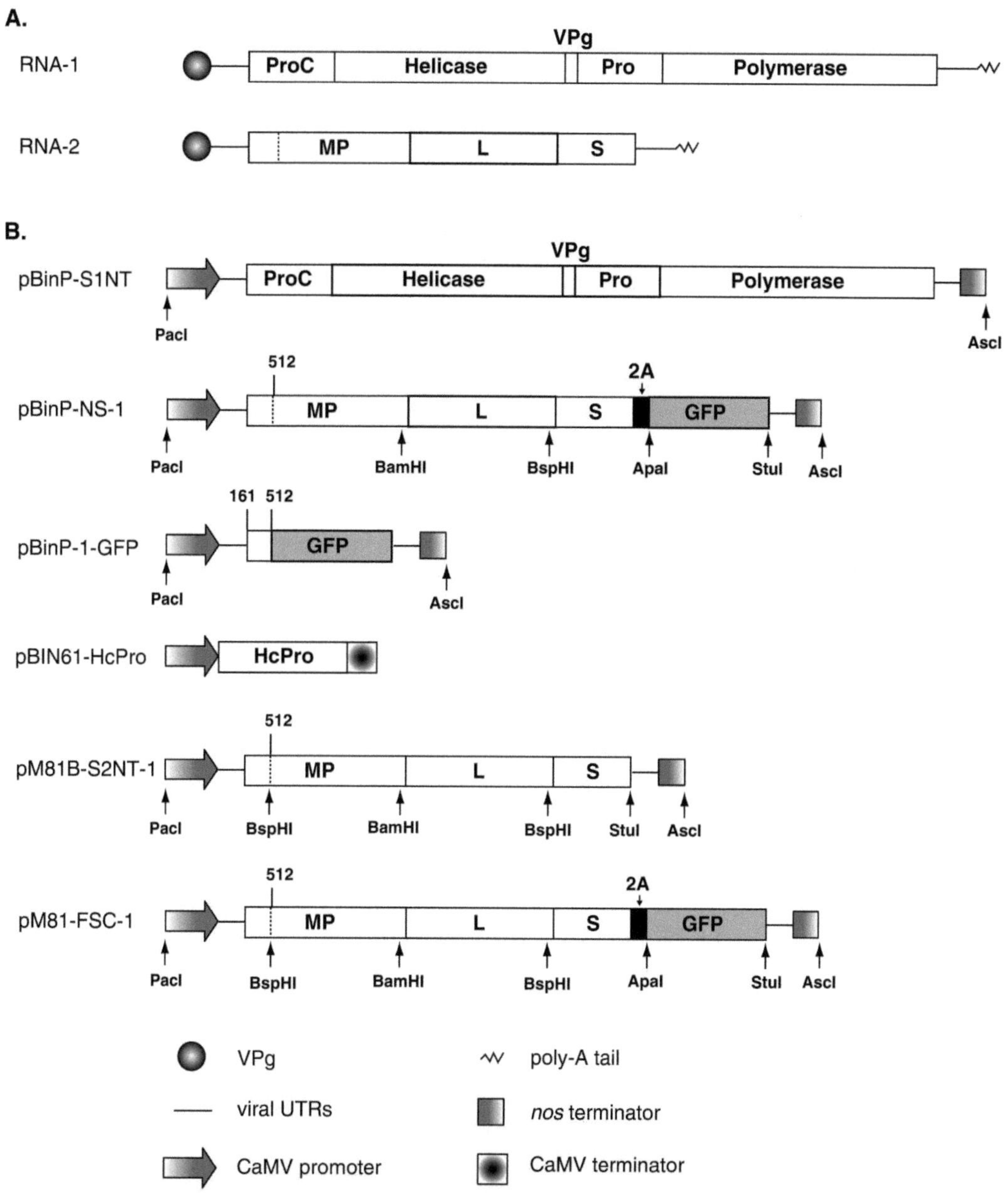

Figure 2.1. (**a**) Organization of CPMV genomic RNAs. The functions of the regions of the polyproteins are indicated. Abbreviations used are as follows: *ProC* proteinase co-factor, *Pro* proteinase, *MP* movement protein, *L* large coat protein, *S* small coat protein. (**b**) Structure of constructs used for the expression of foreign proteins. Relevant restriction sites are indicated.

virus can spread systemically within the inoculated plant and can be transmitted to other plants. In the second case, the foreign gene replaces most of the RNA-2 ORF, permitting larger inserts to be incorporated but abolishing the ability of the virus to spread both within and between plants.

2.1.1. Vectors Based on Full-Length CPMV RNA-2

In these systems, the sequence encoding the polypeptide is fused in-frame with the RNA-2-encoded polyprotein. This means that an appropriate site for the insertion of the foreign sequence had to be identified and a method developed for the release of the inserted protein from the viral polyprotein. While a number of approaches have been investigated, the most successful has involved the insertion of the foreign sequence at the C-terminus of the RNA-2 polyprotein (i.e. immediately after the S protein). Release of the foreign sequence is achieved through the action of 2A catalytic peptide sequence from foot-and-mouth-disease virus inserted between the S protein and the foreign sequence *(11)*. The utility of this approach was first demonstrated with GFP when the pUC9-based construct, pCP2/S-2A-GFP, was shown to able to systemically infect cowpea plants in the presence of RNA-1 and to express GFP at a level which represented ~1–2% of total soluble protein *(11)*. Plasmid pCP2/S-2A-GFP was designed so that the sequence of GFP could be removed by digestion with *Apa*I and *Stu*I and replaced by a sequence with *Apa*I- and *Stu*I (blunt)-compatible ends. The use of *Apa*I is particularly significant as its recognition sequence, GGGCCC, encodes the dipeptide G-P which is cleaved by the action of the 2A catalytic peptide. Thus, the inserted sequence (originally GFP) possesses only a single additional proline residue at its N-terminus.

To enable the pCP2/S-2A-GFP system to be used in conjunction with agroinfiltration, the modified RNA-2 sequence was inserted into a modified version of the binary vector pBINPLUS *(12)* in which the *Apa*I at position 9005 had been removed *(13)*. This gave rise to the vector pBinP-NS-1 in which the sequence of GFP can be replaced with any foreign sequence via the unique *Apa*I and *Stu*I restriction sites in a one-step procedure *(13)*. The pBinP-NS-1 derivative can then be used to agroinfiltrate leaves of *N. benthamiana* or cowpea in the presence of the RNA-1 construct pBinP-S1NT *(8)*. Initially, inoculated cowpea plants rarely, if ever, develop symptoms of infection, and it is generally necessary to passage the infections on to further plants to obtain sufficient material for characterization.

One of the advantages of using the foot-and-mouth-disease virus 2A sequence to achieve release of the foreign protein is that the processing is co- rather than posttranslational. This means that it is possible to achieve targeting of the expressed protein to the secretory pathway by the addition of an appropriate signal peptide. The expressed protein can also be retained in the endoplasmic reticulum if a retention signal is added to the C-terminus of the protein (Nicholson, Sainsbury, Porta, and Lomonossoff, unpublished data).

The full-length RNA-2-based vector pBinP-NS-1 has been successfully used to express assembled Hepatitis B virus core antigen (HBcAg) particles *(14)* and small immune proteins (SIPs;

15, 16). In the latter case, the plant-expressed protein was shown to be capable of passively immunizing pigs against challenge.

2.1.2. Deleted Versions of RNA-2

Two of the major limitations on the use of plant viral vectors based on full-length RNA molecules are concerns over biocontainment and size constraints on the inserted sequence. While the ability of infections derived from pBinP-NS-1 derivatives to spread both within a host plant and to other healthy plants is advantageous in terms of scaling-up the manufacture of the expressed protein, it simultaneously raises the possibility of unintentional environmental spread. The problem of the size constraints relates to the fact that longer RNA-2 molecules replicate more slowly than the corresponding wild-type molecule. This is reflected both in terms of low yield and the genetic stability of the inserted sequence. These problems have indeed been observed with full-length RNA-2 molecules modified to express both HBcAg and SIP molecules.

To address these concerns, a system based on a defective form of CPMV RNA-2 has been developed. The concept is based on the observation that the sequences necessary for replication of RNA-2 by the RNA-1-encoded replicase lie exclusively at the 5′ and 3′ ends of the RNA *(17)*. This allows most of the RNA-2 ORF to be deleted without affecting the ability of RNA-2 to be replicated. However, while the essential 3′ terminal sequence lies exclusively within the 3′ UTR, the 5′ region extends beyond the first in-frame AUG (position 161), which means that initiation of translation of the foreign gene should be driven by the second in-frame AUG at position 512.

The ability of a deleted version of RNA-2 carrying a foreign gene to be replicated by RNA-1 was tested by creating construct pBinP-1-GFP **(Fig. 2.1b)** in which expression of GFP was driven by the AUG at 512 *(7)*. It was found that RNA-1 alone was unable to replicate RNA derived from pBinP-1-GFP as the deleted version of RNA-2 lacked the region encoding the CPMV suppressor of silencing which resides at the C-terminus of the S protein *(18, 19)*. This problem could be relieved by simultaneously supplying the suppressor of silencing, HcPro, from *Potato virus Y* (pBIN61-HcPro in Fig. 2.1b; *20)*. The replication of pBinP-1-GFP-derived RNA in the presence of both RNA-1 and a suppressor allowed high levels of GFP to be expressed. Replication of pBinP-1-GFP by RNA-1 in the presence of a suppressor could be achieved in several different formats: by supplying all three components simultaneously to nontransgenic leaves by agroinfiltration, by transforming plants with some of the components and inducing expression by agroinfiltration with the others, or by transforming plants with all three components *(7)*.

To create a useful cloning vector for the expression of foreign proteins from a pBinP-1-GFP-based plasmid, the complete sequence of RNA-2 flanked by the Cauliflower mosaic virus

(CaMV) 35S promoter and nopaline synthase (*nos*) terminator from pBinP-S2NT *(8)* was inserted into mutagenesis plasmid pM81W *(21)* as an *Asc*I/*Pac*I fragment. The resulting plasmid, pM81W-S2NT, was subjected to a single round of mutagenesis which simultaneously introduced four changes (*see* method in **ref.** *21)* to give pM81B-S2NT-1. The mutagenesis removed two *Bsp*HI sites from the vector backbone and introduced a *Bsp*HI site (TCATGA) around AUG 512 and a *Stu*I site (AGGCCT) after UAA 3299, the termination codon for the RNA-2-encoded polyprotein. Subsequently, the *Bam*HI/*Asc*I fragment was excised from pBinP-NS-1 and ligated into similarly digested pM81B-S2NT-1, yielding pM81-FSC-1 **(Fig. 2.1b)**. This vector allows the whole of the RNA-2 ORF downstream of AUG 512 to be excised by digestion with *Bsp*HI and *Stu*I and replaced with any sequence with *Bsp*HI and *Stu*I (blunt)-compatible ends. The use of the *Bsp*HI site is important as it preserves the AUG at 512 and this initiator is used to drive translation of the inserted gene.

To express the foreign gene in plants, the pM81-FSC-1-derived plasmid is digested with *Asc*I and *Pac*I and the fragment containing the foreign sequences transferred to similarly digested pBINPLUS and the resulting plasmids are finally transformed into *Agrobacterium tumefaciens.* In its simplest format, an *Agrobacterium* suspension containing the deleted version of RNA-2 harboring the inserted gene can be coinfiltrated into leaves in the presence of suspensions of bacteria containing RNA-1 and HC-Pro. This approach has been used successfully to express HBcAg (Mechtcheriakova, personal communication) and to express two proteins simultaneously by coinfiltration with two separate RNA-2-based constructs. In this case, fully formed IgG molecules have been assembled in leaves by coinfiltrating with separate constructs containing the heavy and light chains systems *(22)*.

The use of deleted versions of RNA-2, lacking the regions encoding both the movement protein and the viral coat proteins, provides the high levels of gene amplification associated with viral vectors without the possibility of the modified virus spreading in the environment. Recent results show that even in the presence of wild-type RNA-2, the construct 1-GFP is not transencapsidated indicating that even accidental coinfection with wild-type CPMV would not lead to the spread of the foreign sequence (Sainsbury and Lomonossoff, unpublished data). Furthermore, this system should support replication of much larger inserts as the RNA-2-based replicon is intrinsically smaller than wild-type RNA-2.

The lack of ability of constructs based on pM81-FSC-1 to spread either within or between plants also has certain disadvantages. Infections are not able to spread systemically throughout a plant, with the site of expression of the foreign protein being limited to the infiltrated region. Similarly, it is not possible to bulk up material simply by passaging an infection to healthy plants.

However, these disadvantages can potentially be overcome by incorporating a construct based on the deleted version of RNA-2 as transgene and inducing expression by crossing with plants transgenic for RNA-1 and a suppressor *(7)*. Clearly, such an approach is more time-consuming than a purely transient system based on agroinfiltration but has the potential to generate lines of plants expressing high levels of a foreign protein throughout the plant.

2.2. Materials

2.2.1. Plants, Plasmids, and Bacterial Strains

1. *N. benthamiana* and cowpea plants (*Vigna unguiculata*) are grown in greenhouses with supplemental light to 16 h at a constant temperature of 24°C for *N. benthamiana* and 25°C for cowpea.
2. Plasmids described for cloning, subcloning, and *Agrobacterium* transformation are prepared using the QIAprep miniprep Kit (Qiagen, West Sussex, UK) for high purity.
3. *Escherichia coli* strain used is DH5α.
4. *Agrobacterium tumefaciens* strain LBA4404 *(23)* is used for all transformations.

2.2.2. Enzymes, Materials, and Commercial Kits

1. *Apa*I (New England Biolabs, Hertfordshire, UK)
2. *Bsp*HI (New England Biolabs)
3. *Stu*I (New England Biolabs)
4. *Pac*I (New England Biolabs)
5. *Asc*I (New England Biolabs)
6. Alkaline Phosphatase, Calf Intestinal (New England Biolabs)
7. T4 DNA ligase (Roche Diagnostics, East Sussex, UK)
8. QIAprep Miniprep Kit (Qiagen)
9. QIAquick Gel Extraction Kit (Qiagen)
10. RNeasy Plant Mini Kit (Qiagen)
11. Protoscript First Strand cDNA synthesis Kit (New England Biolabs)
12. Butter muslin
13. Carborundum (Sigma-Aldrich, Cambridgeshire, UK)

2.2.3. Buffers and Solutions

1. Luria-Bertani (LB) media: 10 g/L Bacto-tryptone, 10 g/L NaCl, and 5 g/L Yeast extract, pH 7.0.
2. MMA: 10 mM MES (2-[N-morpholino]ethanesulfonic acid; Sigma-Aldrich) pH 5.6, 10 mM $MgCl_2$, 100 μM Acetosyringone (Sigma-Aldrich).

3. Sodium Phosphate buffer, 0.1 M, pH 7.0.
4. Protein extraction buffer: 50 mM Tris-HCl pH 7.25, 150 mM NaCl, 2 mM EDTA, 0.1% (v/v) Triton X-100 (Sigma-Aldrich).

2.3. Methods

2.3.1. Restriction Site Cloning

Cloning and subcloning into both vector systems is straightforward with PCR-generated inserts amplified with appropriate restrictions sites at both ends (*see* **Note 1**). As mentioned in the Introduction, cloning into the full-length RNA-2 vector can be done in a single step with pBinP-NS-1 (*see* **Note 2**). For the deleted RNA-2 vector, the entire expression cassette including the CaMV promoter and *nos* terminator is released from pM81-FSC-1 following the insertion of the gene of interest and inserted into pBINPLUS *(12)*.

2.3.1.1. Cloning into Full-Length RNA-2

1. Digest both the insert and plasmid pBinP-NS-1 with *Apa*I and *Stu*I by mixing components including the appropriate buffer according to the enzyme manufacturer's recommendation in a 1.5 mL microcentrifuge tube (*see* **Note 3**).
2. Dephosphorylate vector with alkaline phosphatase following the manufacturer's instruction (*see* **Note 4**).
3. Resolve digests on a 1% agarose gel and purify vector and inserts using QIAguick gel extraction kit (*see* **Note 5**).
4. Combine in ligase buffer with T4 DNA ligase and incubate according to manufacturer's recommendations.
5. Transform competent *E. coli* and plate onto an LB agar plate with kanamycin (50 µg/mL) selection.
6. Colonies may be screened by PCR or restriction analysis. Positive clones are grown overnight and plasmids are extracted for sequencing, confirmation of insertion, and *Agrobacterium* transformation.

2.3.1.2. Cloning into the Deleted Vector

1. Digest both the insert and plasmid pM81-FSC-1 with *Bsp*HI and *Stu*I by mixing components including the appropriate buffer according to the enzyme manufacturer's recommendation in a 1.5 mL microcentrifuge tube and incubate at 37°C for 1 h.
2. Dephosphorylate vector with alkaline phosphatase following the manufacturer's instruction (*see* **Note 4**).
3. Resolve digests on a 1% agarose gel and purify vector and inserts using QIAguick gel extraction kit (*see* **Note 6**).

4. Combine in ligase buffer with T4 DNA ligase and incubate according to manufacturer's recommendations.
5. Transform competent *E. coli* and plate onto an LB agar plate with carbenicillin (100 μg/mL) selection.
6. Colonies may be screened by PCR or restriction analysis. Positive clones are grown overnight, and plasmids are extracted for sequencing, confirmation, and subcloning.
7. Upon confirmation of clones, digest plasmid with *Pac*I and *Asc*I by mixing components including the appropriate buffer according to the enzyme manufacturer's recommendation in a 1.5 mL microcentrifuge tube and incubate at 37°C for 1 h. Simultaneously digest pBINPLUS or pBINPLUS-derived plasmid, for example, pBinP-NS-1, in the same way.
8. Dephosphorylate vector with alkaline phosphatase following the manufacturer's instruction (*see* **Note 4**).
9. Resolve digests on a 1% agarose gel and purify vector and inserts using QIAguick gel extraction kit (*see* **Note** 7).
10. Combine in ligase buffer with T4 DNA ligase and incubate according to manufacturer's recommendations.
11. Transform competent *E. coli* and plate onto an LB agar plate with kanamycin (50 μg/mL) selection.
12. Colonies may be screened by PCR or restriction analysis. Positive clones are grown overnight and plasmids are extracted for sequencing, confirmation of insertion, and *Agrobacterium* transformation.

2.3.2. Agrobacteria Infiltration and Monitoring Expression

For CPMV-based expression of foreign proteins, agroinfiltration is an integral step. It is used to initiate viral infection of both cowpea and *N. benthamiana* plants in the case of full-length RNA-2 and for transient expression using the deleted RNA-2 system. In both case, RNA-1 must be provided and a suppressor of silencing is also required in the case of the deleted version of RNA-2.

2.3.2.1. Agrobacteria Infiltration (Agroinoculation)

1. Agrobacteria cultures are grown to stationary phase from glycerol stocks, fresh liquid, or solid media cultures.
2. Gently spin cells (~2,000 g) at 4–10°C for 20 min to pellet, and discard media.
3. Resuspend in MMA to an OD600 of 1.0 (*see* **Note 8**).
4. Incubate for at least 2 h at room temperature or overnight at 4°C if necessary.
5. Mix equivalent volumes of equal-density cultures, with or without a suppressor of silencing (*see* **Note 9**).
6. Prick underside of leaf (*see* **Note 10**) with a razor blade or small pipette tip and gently pressure infiltrate the wound

against a counter pressure (provided by a finger) with the Agrobacteria mixture using a 1 or 2 mL syringe (*see* **Note 11**).

2.3.2.2. Monitoring Expression

Protein production in infiltrated leaf tissue is generally at a very high level and in the case of GFP, is visible in vivo under UV illumination. Therefore, it is sensible to include inoculation with a GFP-expressing construct in each experiment. Using both the full-length and deleted RNA-2-based vectors, expression of GFP in *N. benthamiana* becomes visible after 2 days, reaches a peak at 5–6 days, and remains at or near this level for at least 12 days. For constructs based on full-length RNA-2, symptoms of systemic spread appear after 8–10 days and continue to spread throughout the plant for the life of the plant. A pattern of high-GFP expression levels detectable under UV illumination follows that of systemic symptoms. Because of the thick cuticle of cowpea leaves, considerable damage is sustained during agroinoculation rendering the fluorescence of GFP indistinguishable from that of autofluorescence caused by the damage. Therefore, expression levels of GFP can be monitored under UV illumination only when the viral particles produced are passaged. Fluorescent infection foci appear on leaves inoculated with virus particles after ~10 days and systemic symptoms develop after 2–3 weeks.

2.3.3. RT-PCR Confirmation of Genetic Stability

When using full-length RNA-2-based vectors, RT-PCR of RNA extracted from infected leaves provides a useful check on the genetic stability of the foreign sequence within the viral genome. This step will also check for any possible WT CPMV contamination.

1. Harvest up to 100 mg of leaf tissue showing systemic symptoms and grind to a powder under liquid nitrogen.
2. Extract total RNA using the RNeasy Plant Mini Kit.
3. cDNA is synthesized from 1 μg of RNA using the Protoscript First Strand cDNA synthesis Kit.
4. PCR is carried out in a 20 μL reaction using standard reaction conditions *(24)* with 4 μL of the cDNA synthesis reaction as a template.

2.3.4. Virus Preparations and Passaging (Cowpea)

Passaging of viral infections is an important means for scaling-up the production of foreign proteins in cowpea plants. Obviously, this is only relevant for constructs based on full-length CPMV RNA-2. For passaging, highly purified virus preparations are not required. The following protocol describes a procedure for extracting virions sufficient for passaging. A detailed protocol for obtaining preparations of CPMV of higher purity can be found in **ref.** *25.*

1. Harvest inoculated or systemically infected leaves from cowpea or *N. benthamiana* and grind to a powder under liquid nitrogen.
2. Add 2× volume of sodium phosphate buffer and vortex.

3. Filter through two layers of butter muslin and centrifuge at 3,500*g* for 25 min at 4°C.
4. Take supernatant and add ¼ volume of 1 M NaCl with 20% PEG 6000 (w/v, Sigma-Aldrich).
5. Incubate for 1 h at room temperature.
6. Centrifuge at 15,000*g* for 15 min at 4°C and discard supernatant.
7. Re-suspend pellet in 1/10 (10 mM) sodium phosphate buffer (0.1–1 mL/g of starting material, *see* **Note 12**)
8. Dust the first true leaves of young cowpea plants with carborundum (*see* **Note 13**).
9. Apply 100 μL of virus preparation and gently spread over the leaf with forefinger.
10. Symptoms should appear on inoculated leaves after 10–14 days and on upper trifoliate leaves 2–3 weeks of passaging.

2.3.5. Protein Extractions

Following extraction, recombinant proteins produced using the vectors described in this chapter are generally detectable by coomassie blue staining of SDS-PAGE separated samples. Other commonly used techniques for detection for antigens and antibodies expressed from CPMV include western blotting, enzyme-linked immunosorbant assays, or as in the case of HBcAg, electron microscopy *(14)*. Further detailed characterization of the recombinant proteins produced by CPMV-based vectors should be carried out using techniques specific to the protein. Below is a general protocol for the extraction of soluble proteins from tissue inoculated with CPMV vectors that is compatible with most protein analysis methods.

1. Harvest inoculated or systemically infected leaf tissue.
2. Grind tissue to a powder under liquid nitrogen.
3. Suspend in 1–2× volume (w/w) of EB.
4. Centrifuge at ~15,000*g* for 10 min at 4°C.
5. Collect supernatant and store at 4°C for up to 12 h or frozen for longer periods (*see* **Note 14**).

2.4. Notes

1. The *Bsp*HI site is also compatible with *Nco*I and *Pci*I. The choice of restriction enzymes permits different nucleotides to be incorporated immediately after the ATG start codon. The recognition sites for the enzymes are T/CATGA for *Bsp*HI, C/CATGG for *Nco*I, and A/CATGT for *Pci*I, permitting the incorporation of an A, G, or T after the ATG. The enzyme

*Fat*I (/CATG) allows any base to follow the start codon, but it is of limited use as this enzyme has a four base recognition site. To incorporate a C after the ATG, the insert may be digested with *Sph*I (GCATG/C) followed by removal of the 3′ overhang by "polishing." The blunt end fragment can then be ligated into *Bsp*HI/*Stu*I-cut pM81-FSC-1 after the 5′ overhang of the *Bsp*HI has been filled in.

2. Low copy number and the large size of pBINPLUS-based plasmids may increase the difficulty of directly cloning into them. Therefore, it is also possible to use pM81-FSC-1 for cloning into the full-length vector with subsequent subcloning into pBINPLUS as for the deleted version of RNA-2. To do this, follow 3.1.2 using *Apa*I and *Stu*I restriction sites.
3. Despite the manufacturer's recommendation to carry out ApaI digests at 25°C, we find that the *Apa*I/*Stu*I double digest is effective at 37°C.
4. Calf intestinal alkaline phosphatase from New England Biolabs may be used with any of the buffers supplied with the enzymes described in this chapter. Therefore, following heat inactivation of the restriction enzymes, phosphatase may be added to the digest reaction mixture and incubated for a further 60 min.
5. *Apa*I/*Stu*I-cut pBinP-NS-1 will yield a fragment ~15 kb long, corresponding to the vector, and a fragment of 0.7 kb corresponding to the GFP sequence.
6. *Bsp*HI/*Stu*I-cut pM81-FSC-1 will result in three fragments: the 4.5 kb vector and two fragments of 2.0 and 1.6 kb corresponding to the RNA-2 coding region (including the 2A and GFP sequences), which contains an internal *Bsp*HI site.
7. *Pac*I/*Asc*I-cut pM81-FSC-1 clones will result in a vector fragment of 2.9 kb. If the insert into the deleted RNA-2 is between 1.2 and 2.0 kb, it will be necessary to further digest the vector backbone. Provided there is no recognition sequence in the insert for *Bgl*I, this enzyme will reduce the backbone to three fragments of 1.4, 1.2, and 0.3 kb. *Pac*I/*Asc*I-cut pBINPLUS plasmids will result in a vector fragment of 11 kb. If pBin-NS-1 is used, another fragment of ~5.2 kb, corresponding to full-length RNA-2 plus GFP, will also be present.
8. Resuspension to OD600 of 1.0 usually requires 1–1.5 times volume of MMA to that of the starting culture. Although each culture tends to grow at a different rate, allowing cultures to grow to stationary phase generally ensures that all cultures have an equal density.
9. Including the suppressor of silencing will increase efficiency of viral infection initiated by the full-length vector and increase protein yield in infiltrated tissue. For the deleted

vector, the presence of a suppressor of silencing is essential. A number have been cloned into expression cassettes within pBIN61 (D. Baulcombe, pers. comm.). We have seen varying efficacy with different suppressors in transient assays using pBinP-1-GFP.

10. For cowpea plants, the first true leaves should be agroinoculated immediately before the first set of trifoliate leaves appear. This stage is normally reached about 10 days after sowing. For *N. benthamiana*, viral infection initiated by agroinoculation of autonomously replicating vectors is most efficient in young plants that have 4–6 true leaves and are ~8 cm across at their widest point. These plants are generally about 14 days old. Cotransformation with a suppressor of silencing broadens the window for successful initiation of systemic viral infection and allows transient transformation of infiltrated tissues in much older plants (potentially up to the floral transition) by both CPMV-based vector systems.
11. Videos of the agroinoculation technique can be seen at http://www.plantsci.cam.ac.uk/baulcombe/protocols.html.
12. Viral particles prepared by this method should be used within no more than 48 h and should be kept at 4°C if not used immediately.
13. Passaging of CPMV in cowpea is most efficient when the first trifoliate are just appearing and no more than 10 mm in size.
14. Optimal storage conditions of protein extracts will vary according to the recombinant protein of interest and the requirements of its future uses. For useful discussions on protein extraction and storage conditions and analysis *see* **ref.** *26*.

Acknowledgments

The authors thank D. Baulcombe for kindly providing the silencing suppressors. Part of the work described in this review was funded under the EU FP6 "PharmaPlanta" project. F.S. was supported by the John Innes Foundation.

References

1. Scholthof, H.B., Scholthof, K.B., and Jackson, A.O. (1996) Plant virus gene vectors for transient expression of foreign proteins in plants. *Annu. Rev. Phytopathol.* 34, 299–323.
2. Porta, C. and Lomonossoff, G.P. (2002) Viruses as vectors for the expression of foreign sequences in plants. *Biotechnol. Genet. Eng. Rev.* 19, 245–291.
3. Pogue, G.P., Lindbo, J.A., Garger, S.J., and Fitzmaurice, W.P. (2002) Making an ally from an enemy: Plant virology and the new agriculture. *Annu. Rev. Phytopathol.* 40, 45–74.

4. Cañizares, M.C., Nicholson, L., and Lomonossoff, G.P. (2005a) Use of viral vectors for vaccine production in plants. *Immunol. Cell. Biol.* 83, 263–270.

5. Lomonossoff, G.P. (2005) *Antigen Delivery Systems: Use of Recombinant Plant Viruses, in Mucosal Immunology*, 3rd ed. Mestecky J., Bienenstock, J., Lamm, M.E.., Mayer, L., McGhee, J.R., Strober, W., (eds.). Elsevier, Amsterdam, pp. 1061–1072.

6. Gleba, Y., Marillonnet, S., and Klimyuk, V. (2004) Engineering viral expression vectors for plants: The 'full virus' and the 'deconstructed virus' strategies. *Curr. Opin. Plant Biol.* 7, 182–188.

7. Canizares, M.C., Liu, L., Perrin, Y., Tsakiris, E., and Lomonossoff, G.P. (2006) A bipartite system for the constitutive and inducible expression of high levels of foreign proteins in plants. *Plant Biotechnol. J.* 4, 183–193.

8. Liu, L. and Lomonossoff, G.P. (2002) Agroinfection as a rapid method for propagating *Cowpea mosaic virus*-based constructs. *J. Virol. Methods* 105, 343–348.

9. Lomonossoff, G.P. and Hamilton, W.D.O. (1999) Cowpea mosaic virus-based vaccines, in Current Topics in Microbiology and Immunology 240 Hammond, J., McGarvey, P. and Yusibov, V., (eds.). Springer, Heidelberg, pp. 177–189.

10. Cañizares, M.C., Lomonossoff, G.P., and Nicholson, L. (2005b) Development of *Cowpea mosaic virus*-based vectors for the production of vaccines in plants. *Expert Review of Vaccines* 4, 687–697.

11. Gopinath, K., Wellink, J., Porta, C., Taylor, K.M., Lomonossoff, G.P., and van Kammen, A. (2000) Engineering cowpea mosaic virus RNA-2 into a vector to express heterologous proteins in plants. *Virology* 267, 159–173.

12. van Engelen, F.A., Moltoff, J.W., Conner, A.J., Nap, J.-P., Pereira, A., and Stiekema, W.J. (1995) pBINPLUS: An improved plant transformation vector based on pBIN19. *Transgen. Res.* 4, 288–290.

13. Liu, L., Cañizares, M.C., Monger, W., Perrin, Y., Tsakiris, E., Porta, C., Shariat, N., Nicholson, L., and Lomonossoff, G.P. (2005) Cowpea mosaic virus-based systems for the production of antigens and antibodies in plants. *Vaccine* 23, 1788–1792.

14. Mechtcheriakova, I.A., Eldarov, M.A., Nicholson, L., Shanks, M., Skryabin, K.G., and Lomonossoff, G.P. (2006) The use of viral vectors to produce hepatitis B virus core particles in plants. *J. Virol. Methods* 131, 10–15.

15. Monger, W., Alamillo, J.M., Sola, I., Perrin, Y., Bestagno, M., Burrone, O.R., Sabella, P., Plana-Duran, J., Enjuanes, L., Garcia, J.A., and Lomonossoff, G.P. (2006) An antibody derivative expressed from viral vectors passively immunizes pigs against transmissible gastroenteritis virus infection when supplied orally in crude plant extracts. *Plant Biotechnol. J.* 4, 623–631.

16. Alamillo, J.M., Monger, W., Sola, I., Garcia, B., Perrin, Y., Bestagno, M., Burrone, O.R., Sabella, P., Plana-Duran, J., Enjuanes, L., Lomonossoff, G.P., and Garcia, J.A. (2006) Use of virus vectors for the expression in plants of active full-length and single chain anti-coronavirus antibodies. *Biotechnology Journal* 1, 1103–1111.

17. Rohll, J.B., Holness, C.L., Lomonossoff, G.P., and Maule, A.J. (1993) 3′ terminal nucleotide sequences important for the accumulation of cowpea mosaic virus M-RNA. *Virology* 193, 672–679.

18. Liu, L., Grainger, J., Cañizares, M.C., Angell, S.M., and Lomonossoff, G.P. (2004) *Cowpea mosaic virus* RNA-1 acts as an amplicon whose effects can be counteracted by a RNA-2-encoded suppressor of silencing. *Virology* 323, 37–48.

19. Cañizares, M.C., Taylor, K.M., and Lomonossoff, G.P. (2004) Surface-exposed C-terminal amino acids of the small coat protein of *Cowpea mosaic virus* are required for suppression of silencing. *J. Gen. Virol.* 85, 3431–3435.

20. Brigneti, G., Voinnet, O., Li, W.-X., Ji, L.-H., Ding, S.-W., and Baulcombe, D.C. (1998) Viral pathogenicity determinants are suppressors of transgene silencing in *Nicotiana benthamiana*. *EMBO J.* 17, 6739–6746.

21. Liu, L. and Lomonossoff, G.P. (2006) A site-directed mutagenesis method utilising large double-stranded DNA templates for the simultaneous introduction of multiple changes and sequential multiple rounds of mutation: Application to the study of whole viral genomes. *J. Virol. Methods* 137, 63–71.

22. Sainsbury, F., Lavoie, P.-O., D'Aoust, M.-A., Vézina, L.-P., and Lomonossoff, G.P. (2008) Expression of multiple proteins using full-length and deleted versions of *Cowpea Mosaic Virus* RNA-2. *Plant Biotechnol. J.* 6, 82–92.

23. Hoekema, A., Hirsch, P.R., Hooykaas, P.J.J., and Schilperoort, R.A. (1983) A binary plant vector strategy based on separation of

vir- and T-region of the *Agrobacteria tumefaciens* Ti-plasmid. *Nature* 303, 179–180.
24. Sambrook, J., Fritsch, E.F., and Maniatis, T. (1989) *Molecular Cloning:* A Laboratory Mannual, 2nd ed. Cold Spring Harbor Laboratory, Cold Spring Harbor, NY.
25. Wellink, J. (1998) Comovirus isolation and RNA extraction, in *Plant Virology Protocols* Foster, G.D. and Taylor S.C., (eds.). Humana, Totowa, NJ.
26. Walker, J.M. (ed.) (2002) *The Protein Protocols Handbook.* Humana, Totowa, NJ.

Chapter 3

Transient Expression of Antibodies in Plants Using Syringe Agroinfiltration

Marc-André D'Aoust, Pierre-Olivier Lavoie, Julie Belles-Isles, Nicole Bechtold, Michèle Martel, and Louis-P. Vézina

Summary

The improvements in agroinfiltration methods for plant-based transient expression now allow the production of significant amounts of recombinant proteins in a matter of days. While vacuum-based agroinfiltration has been brought to large scale to meet the cost, speed and surge capacity requirements for vaccine and therapeutic production, the more accessible and affordable syringe agroinfiltration procedure still represents a fast and high-yielding approach to recombinant protein production at lab scale. The procedure exemplified here has proven its reproducibility and high-yield capacity for the production of proteins with varying levels of complexity, including monoclonal antibodies.

Key words: Agroinfiltration *Nicotiana benthamiana* Monoclonal antibody Transient expression.

3.1. Introduction

Transient protein expression, that is, the expression of proteins in a system where genomic integration of the transgene does not intervene, is arguably the fastest approach to the production of recombinant proteins. It has been used extensively to assist in vaccine and therapeutic development when small but significant quantities are required for efficacy studies and physicochemical characterization. The development of transient expression in plants has followed three major routes, each related to specific DNA or RNA delivery approaches. Direct delivery methods, include particle bombardment, protoplast transformation,

Loïc Faye and Véronique Gomord (eds.), *Methods in Molecular Biology, Recombinant Proteins From Plants, vol. 483*

DOI: 10.1007/978-1-59745-407-0_3

microinjection and others, and rely on mechanical or chemical disruption of the cell wall and/or cell membrane to introduce DNA of interest inside the cell. Viral vector systems are infectious viruses (usually RNA viruses) in which the viral genome is extended to include the sequence encoding the protein of interest, and the encoded protein is produced upon infection and replication of the virus in the plant cell. Finally, bacterial infection systems have been developed, in which the DNA of interest, cloned into a transfer plasmid, is actively transported into the nucleus of the infected plant cell by bacterial proteins in conjunction with host proteins.

For many years, however, the limits of direct transformation methods restricted the possibility of analysis to in situ observations of reporter gene expression or to immunological analysis of protein accumulation. The development of viral vector systems opened the way to the production of large-scale amounts of recombinant proteins, with major advances being concentrated on vaccine production *(1)*. However, the production of large vaccine proteins and multimeric proteins like antibodies by plant infection with recombinant virus vectors has encountered limited success although transient expression of fully assembled antibodies from infection with a plant virus vector has been successfully achieved using tobacco mosaic virus- or Potato virus X-based vectors *(2, 3)*.

Vacuum-based transient expression system (agroinfiltration) was developed 10 years ago and first published by Kapila et al. *(4)*. In this simple technology, vacuum was used to force the replacement of the apoplastic fluid of detached leaves by *Agrobacterium* cultures, and the infiltrated tissues were incubated to allow infection of the cells by surrounding bacteria. Initially developed using detached leaves, this technology has been used to produce recombinant antibodies with yet unattained productivity *(5–8)*. Developments and demonstrations of robustness and versatility of the method prompted the adoption and further adaptation of vacuum infiltration for the commercial production of pharmaceuticals. Today, with the rapidly evolving market demands, agroinfiltration represents a production method of choice when speed, surge capacity and cost are considered, as it is the case for vaccines.

Although vacuum agroinfiltration represents an efficient transient expression system when large amounts of proteins are needed, the quest for simpler means of expressing proteins at the milligram scale has led to the development of a syringe agroinfiltration method. This transient expression method, inspired by the agroinoculation method used to inoculate viruses to juvenile host plants *(9)*, is based on the infiltration of *Agrobacterium* cultures into a maximum number of leaves of a fully developed plant. In our laboratories, transient expression by syringe agroinfiltration of *Nicotiana benthamiana* plants has become the ideal method for screening product candidates and expression strategies, and developing analytical methods because it is exceptionally simple, rapid and productive.

The procedure presented here proposes a method for producing complex proteins in plants and is exemplified with the production of a heteromultimeric protein, the C5-1 murine antibody, by infiltrating a mixture of two independent *Agrobacterium* cultures for the expression of the light and heavy chain of the antibody. Using this method, a mean expression level of 2 mg of C5-1 per 100 g of fresh leaf weight could be obtained.

3.2. Materials

3.2.1. Seeds and Plant Growth Materials

1. *N. benthamiana* seeds (can be obtained from National tobacco germplasms) (*see* **Note 1**).
2. 11.5 cm plastic pots
3. Light peat moss-based substrate (e.g. AgroMix™, Fafard, Saint-Bonaventure, QC, Canada)
4. High nitrate water soluble fertilizer

3.2.2. Binary Plasmid and Agrobacterium Strains

1. In this example, the assembly of antibody light and heavy chains into expression cassettes (presented in **Subheading 3.3.2**) was performed using pCAMBIA 2300 as receptor binary plasmid (*see* **Notes 2, 3**).
2. *Agrobacterium tumefaciens* strain AGL1 (ATCC, Cat# BAA-101) (*see* **Note 4**).

3.2.3. Laboratory Materials

1. LB Broth MILLER (EMD Chemicals, Cat. No. 1.10285.5007, San Diego, CA)
2. Acetosyringone
3. Bacteria resuspension solution (10 mM $MgCl_2$, 10 mM 2-(4-morpholino) ethanesulfonic acid (MES), 100 µM acetosyringone)
4. 1 mL- or 3 mL-syringe without needle
5. Protein extraction buffer (50 mM Tris pH 7.4, 150 mM NaCl, 2 mM sodium metabisulfite, 1 mM phenylmethanesulfonyl fluride (PMSF), and 10 µM chymostatin)
6. Domestic blender
7. Miracloth™ (EMD Chemicals)
8. Anti-mouse IgG antibodies (peroxidase-conjugated)
9. HiTrap Protein G column
10. Antibody elution buffer (100 mM glycine pH 3.0)
11. Neutralization buffer (1 M Tris-HCl pH 7.5)

3.3. Methods

3.3.1. Plant Seeding and Growth

Six week-old-*N. benthamiana* plants were used for agroinfiltration.

1. Fill each pot with prewetted growing substrate, sow 3–5 seeds per pot and place the pots in a warm environment with plenty of sunlight (*see* **Note 5**).
2. Water the plants regularly and make sure to avoid drying of the substrate during the germination period (*see* **Note 6**).
3. Once cotyledons appear, nitrogen fertilization should be applied to maximize vegetative development (*see* **Note 7**).
4. Upon emergence of the second true leaves, thin extra plants leaving only one healthy plant per pot to ensure optimal development of the plants to be infiltrated.

3.3.2. Expression Cassette Assembly and Transformation of Agrobacterium

1. C5-1 antibody light chain and heavy chain cDNAs *(10)* were provided by Héma-Québec (Québec, Canada). To ensure high leaf expression of the antibody, regulatory elements should be carefully selected. We have observed that the combination of a strong viral promoter like the enhanced version of the CaMV 35S promoter *(11)* with the 5'UTR region of the tobacco etch virus (TEV) *(12)* used in conjunction with the nopaline synthase (NOS) terminator sequence *(13)* repeatedly promotes high expression levels of recombinant proteins in agroinfiltrated leaves of *N. benthamiana* (*see* **Note 8**). Antibody light and heavy chain coding regions are both assembled in expression cassettes and cloned into separate binary plasmids using standard molecular biology techniques as described by Sambrook and Russell *(14)*, and polymerase chain reaction (PCR)-based assembly techniques as presented by Darveau et al. *(15)*. The plasmids used are presented in **Fig. 3.1**. (*see* **Note 9**). *Agrobacterium tumefaciens* is transformed by electroporation (*see* **Note 10**).

3.3.3. Bacterial Culture and Suspension Preparation

1. Precultures are prepared 2 days before infiltration by inoculating 3 mL of LB medium containing 25 mg/L carbenicillin and 50 mg/L kanamycin with isolated colonies of *Agrobacterium* strains harbouring each of the light and the heavy chain expression plasmids.
2. Incubate precultures overnight at 28°C under constant agitation at 220 rpm and allow growth, preferentially to an OD (600 nm) >1.2.
3. Inoculate fresh LB medium cultures containing 25 mg/L carbenicillin, 50 mg/L kanamycin and 20 µM acetosyringone with the light and the heavy chain *Agrobacterium* precultures. For each plant to be infiltrated, a total of 100 mL

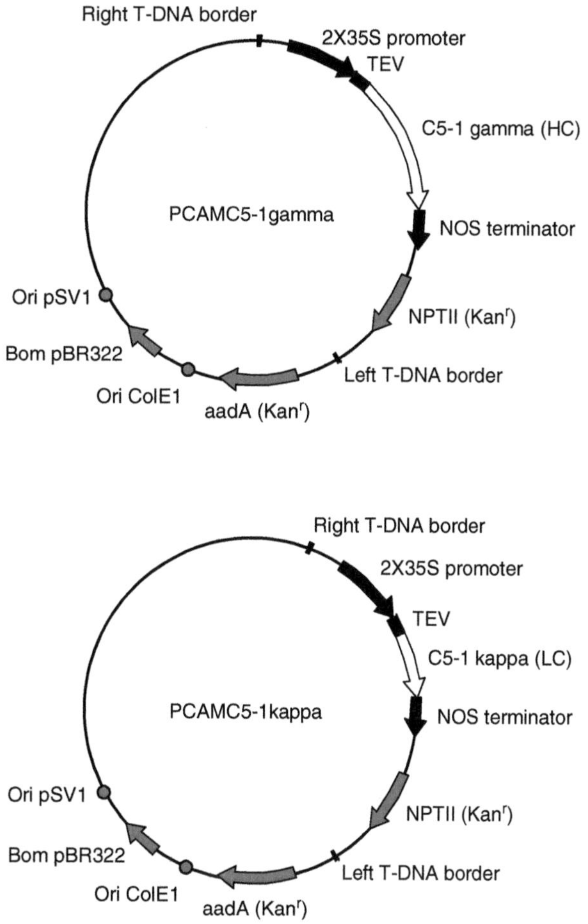

Fig. 3.1. Schematic representation of the binary plasmids used for C5-1 antibody expression.

of culture (50 mL of each strain) should be inoculated (*see* **Notes 11, 12**).

4. Incubate cultures at 28°C under constant agitation at 220 rpm to an OD (600 nm) of 0.8–1.2 (~18 h).
5. Centrifuge cultures (10,000*g*; 10 min) and discard supernatant (*see* **Note 13**).
6. Resuspend the bacterial pellets in 1 volume of bacteria resuspension solution, and incubate for 1 h at room temperature before use.
7. Mix the light and heavy chain bacterial suspensions in a 1:1 ratio.

3.3.4. Syringe Infiltration and Plant Incubation

1. Fill a 1 mL- or 3 mL-syringe (without needle) with bacterial suspension, and hold the leaf to be infiltrated between the index and the syringe, the syringe being on the abaxial side of the leaf. Gently push the piston to force the entry of the

bacterial suspension in the leaf and maintain an even pressure during the infiltration in the tissue. Wetting of the leaf surrounding the infiltration point is observed as the suspension enters the tissue in the apoplastic space. For each point of infiltration, a surface of ~7 cm^2 should be filled. Several points of infiltration may be necessary to completely inoculate each leaf (*see* **Notes 14, 15**).

2. Infiltrate a maximum number of leaves on each plant and remove all uninfiltrated leaves as well as apical and axillary buds to avoid growth of non-infiltrated leaves during the incubation period (*see* **Note 16**).
3. Incubate infiltrated plants in the greenhouse for 6 days, watering the plants as needed and continuing nitrogen fertilization (*see* **Note 17**).

3.3.5. Protein Extraction, Analysis and Antibody Purification

Described below are proposed steps to extract total soluble proteins from leaf biomass for analysis of antibody expression and purification purposes.

3.3.5.1. Protein Extraction and Western Blot Analysis

1. Harvest the infiltrated leaves and extract immediately, or place at –80°C until needed (*see* **Note 18**).
2. Place leaves in a domestic blender and add 3 volumes of ice cold protein extraction buffer. Blend at high speed for 30 s.
3. Filter the homogenate using Miracloth to remove insoluble debris.
4. Centrifuge the filtrate (20,000*g*; 20 min, 4°C) and collect the clarified supernatant.
5. Separate the proteins from the clarified supernatant on a sodium dodecyl sulfate polyacrylamide gel electrophoresis (SDS-PAGE) gel and evaluate antibody accumulation by immunodetection with an anti-mouse IgG antibody preparation (*see* **Fig. 3.2**) *(16)*.

3.3.5.2. Antibody Purification

1. Apply the clarified supernatant to a 1 mL HiTrap Protein G column (GE Healthcare, Piscataway, NJ) previously equilibrated with protein extraction buffer.
2. Wash with 5 column volumes of protein extraction buffer.
3. Elute the antibody with the antibody elution buffer and immediately neutralize each elution fraction with 1/10 volume of neutralization buffer.
4. Separate the proteins of elution fractions on a SDS-PAGE gel and assess antibody quantity and purity by Coomassie blue staining of the gel (*see* **Fig. 3.3**) *(17)*.

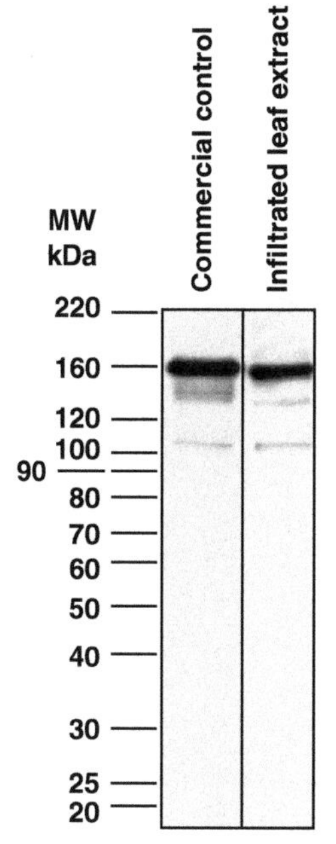

Fig. 3.2. Protein blot analysis of C5-1 accumulation in syringe-infiltrated *Nicotiana benthamiana* plants. Immunoblotting was performed with a peroxidase-conjugated goat-anti-mouse IgG (H+L), on a protein extract from agroinfiltrated leaf. As a control of molecular weight, 100 ng of commercial murine IgG1 (Sigma-Aldrich, St-Louis, MO, ; Cat. No. M9269) was immunoblotted in parallel.

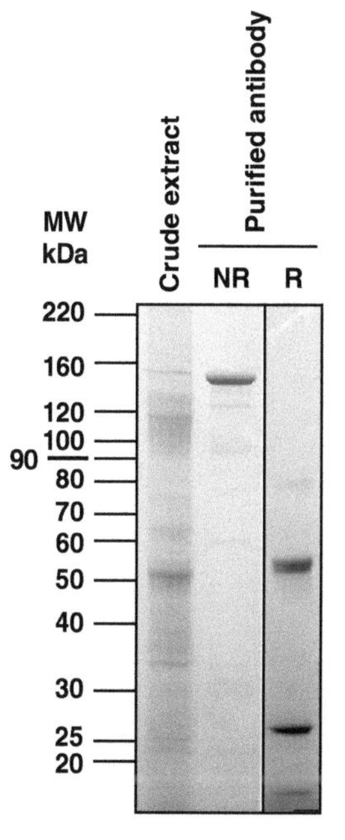

Fig. 3.3. Coomassie stained sodium dodecyl sulfate polyacrylamide gel electrophoresis (*SDS-PAGE*) gel of crude extracts and purified antibodies form syringe infiltrated leaves. The crude extract was separated under non-reducing (*NR*) conditions whereas the electrophoresis of purified product was performed in *NR* and reducing (*R*) conditions.

3.4. Notes

1. *Nicotiana tabacum* plants may also be used for syringe infiltration but usually result in lower expression levels. Although syringe infiltration is applicable to other plants including

plants from other genus, *N. benthamiana* remains among the easiest plants to infiltrate with a syringe, and has produced the highest levels of recombinant proteins to date.

2. The pCAMBIA series of plasmids has been developed and is distributed by CAMBIA (Canberra, Australia) to serve diverse applications in plant biotechnology. However, several plant transformation plasmids containing T-DNA borders compatible with commonly used *A. tumefaciens* strains (binary plasmids) have been developed for stable plant transformation and should be suitable for syringe agroinfiltration.
3. Agroinfiltration in *N. benthamiana* has been shown to induce post-transcriptional transgene silencing *(18, 19)*, resulting in suboptimal expression of the gene of interest. If available, the co-infiltration with an *Agrobacterium* stain harbouring a plasmid engineered for the expression of a suppressor of transgene silencing is suggested, which should improve the accumulation of the recombinant protein of interest.
4. Other *Agrobacterium* strains have been used successfully for agroinfiltration, including LBA4404, C58C1, EHA105 and AGL0.
5. Temperature is crucial at the germination stage, seedling pots should be incubated above 24°C for better results.
6. Incubators for seed trays (plastic domes) may be used to minimize water loss during germination.
7. During weeks 1 and 2, 50–100 ppm of nitrogen can be incorporated to the watering solution, followed by 150 ppm at week 3, 250 ppm at week 4, and 400 ppm at weeks 5 and 6.
8. Regulatory elements are key to the efficacy of the transient expression system. Although the enhanced 35S/TEV combination is proposed for convenience, availability and proven records of expression, other regulatory elements may prove better with certain proteins. Among the other elements, leaf-specific or light-inducible regulatory elements including those of Rubisco and plastocyanin promoters have been used successfully for agroinfiltration.
9. We use PCR cloning for most expression cassette assembly to ensure accurate assembly in the vicinity of the initial ATG, and to avoid insertion of sequences with unknown effect on expression.
10. Mattanovich et al. *(20)* proposes an efficient electroporation method for *Agrobacterium* spp.
11. Acetosyringone should be added immediately before use.
12. The volume of precultures to be inoculated in the fresh medium is dependant on the OD at the moment of inoculation. The following table gives an indication of the amount

of preculture to use for each 100 mL of fresh medium to obtain an OD between 0.8 and 1.2 after 18 h of incubation. These volumes should be used as indications only since growth kinetics are affected by strain, transgene, total volume of culture and size of the culture flask.

Preculture OD (600 nm)	Volume of preculture to inoculate for 100 mL of culture considering 18 h of growth
0.8	400 μL
1.2	300 μL
1.6 and over	200 μL

13. Once bacterial cultures are obtained, it is not necessary to work in sterile conditions.
14. A polycarbonate face shield (or bench shield) should be used for face and eye protection from projection of bacterial suspension during syringe infiltration. Gloves should also be worn during infiltration to avoid contamination.
15. Infiltration is a delicate manipulation and dexterity comes with experience. Practicing using water before wasting precious cultures is a good idea.
16. The wet infiltration zone will disappear shortly after infiltration, noninfiltrated leaves should be removed immediately after completing the inoculation of each plant.
17. Effluents from watering should be collected and devitalized to avoid contamination with *Agrobacterium*.
18. At the moment of harvest, the leaves are covered by necrotic lesions at the sites of infiltration caused by the pressure of the syringe on the leaf.

References

1. Yusibov, V., Rabindran, S., Commandeur, U., Twyman, R.M., and Fischer, R. (2006) The potential of plant virus vectors for vaccine production. *Drugs R&D* 7, 203–217.
2. Verch, T., Yusibov, V., and Koprowski, H. (1998) Expression and assembly of a full-length monoclonal antibody in plants using a plant virus vector. *J. Immunol. Methods* **220**, 69–75.
3. Alamillo, J.M., Monger, W., Sola, I., García, B., Perrin, Y., Bestagno, M., Burrone, O.R., Sabella, P., Plana-Durán, J., Enjuanes, L., Lomonossoff, G., and García, J.A. (2006) Use of virus vector for the expression in plants of active full length and single-chain anti-coronavirus antibodies. *Biotechnol. J.* **1**, 1103–1111.
4. Kapila, J., De Rycke, R., Van Montagu, M., and Angenon, G. (1997) An *Agrobacterium*-mediated transient gene expression system for intact leaves. *Plant Sci.* **122**, 101–108.
5. Vaquero, C., Sack, M., Chandler, J., Drossard, J., Schuster, F., Monecke, M., Schillberg, S., and Fischer, R. (1999) Transient expression of a tumor-specific single chain fragment and a chimeric antibody in tobacco leaves. *Proc. Natl. Acad. Sci. USA* **96**, 11128–11133.
6. Kathuria, S.R., Nath, R., Pal, R., Singh, O., Fischer, R., Lohiya, N.K., and Talwar, G.P.

(2002). Functional recombinant antibodies against human chorionic gonadotropin expressed in plants. *Curr. Sci.* 82, 1452–1456.

7. Rodriguez, M., Ramírez, N.I., Ayala, M., Freyre, F., Pérez, L., Triguero, A., Mateo, C., Selman-Houssien, G., Gavilondo, J.V., and Pujol, M. (2005) Transient expression in tobacco leaves of an aglycosylated recombinant antibody against the epidermal growth factor receptor. *Biotechnol. Bioeng.* **89**, 188–194.
8. Sriraman, R., Bardor, M., Sack, M., Vaquero, C., Faye, L., Fischer, R., Finnern, R., and Lerouge, P. (2004) Recombinant anti-hCG antibodies retained in the endoplasmic reticulum of transformed plants lack core-xylose and core-a(1–3)-fucose residues. *Plant Biotechnol. J.* **2**, 279–287.
9. Liu, L., and Lomonossoff, G.P. (2002) Agroinfection as a rapid method for propagating Cowpea mosaic virus-based constructs. *J. Virol. Methods* **105**, 343–348.
10. St Laurent, M., Marcil, A., Verrette, S., and Lemieux, R. (1993) Functional cooperation among human IgG-specific murine monoclonal antibodies for the detection of weak blood group antibodies in routine agglutination tests. *Vox. Sang.* 64, 99–105.
11. Fang, R.X., Nagy, F., Sivasubramaniam, S., and Chua, N.H. (1989) Multiple cis regulatory elements for maximal expression of the cauliflower mosaic virus 35S promoter in transgenic plants. *Plant Cell* **1**, 141–150.
12. Carrington, J.C., and Freed, D.D. (1990) Cap-independent enhancement of translation by plant potyvirus 5' non-translated region. *J. Virol.* **64**, 1590–1597.
13. Bevan, M., Barnes, W.B., and Chilton, M.D. (1983) Structure of the nopaline synthase gene region of T-DNA. *Nucleic Acids Res.* **11**, 369–385.
14. Sambrook, J., and Russell, D.W. (2001) *Molecular cloning: a laboratory manual.* Cold Spring Harbor Laboratory Press, Cold Spring Harbor, N.Y.
15. Darveau, A., Pelletier, A., and Perreault, J. (1995) PCR-mediated synthesis of chimeric molecules. *Methods Neurosc.* **26**, 77–85.
16. Bollag, D.M., Rozycki, M.D., and Edelstein, S.J. (1996) Protein methods (2nd edition). Wiley-Liss, New York, USA.
17. Harlow, E., and Lane, D. (1988) *Antibodies: a laboratory manual.* Cold Spring Harbor Laboratory, Cold Spring Harbor, USA.
18. Johansen, L.K., and Carrington, J.C. (2001) Silencing on the spot. Induction and suppression of RNA silencing in the *Agrobacterium*-mediated transient expression system. *Plant Physiol.* **126**, 930–938.
19. Voinnet, O., Rivas, S., Mestre, P., and Baulcombe, D. (2003) An enhanced transient expression system in plants based on suppression of gene silencing by the p19 protein of tomato bushy stunt virus. *Plant J.* **33**, 949–956.
20. Mattanovich, D., Rüker, F., da Câmara Machado, A., Laimer, M., Regner, F., Steinkellner, H., Himmler, G., and Katinger, H. (1989) Efficient transformation of Agrobacterium spp. By electroporation. *Nucleic Acids Res.* 17, 6747.

Chapter 4

Rapid System for Evaluating Bioproduction Capacity of Complex Pharmaceutical Proteins in Plants

Giuliana Medrano, Michael J. Reidy, Jianyun Liu, Jorge Ayala, Maureen C. Dolan, and Carole L. Cramer

Summary

Transgene product yield remains a key limitation in commercializing plant-derived pharmaceutical proteins. Although significant progress has been made in understanding the roles of promoters, enhancers, integration sites, codon usage, cryptic RNA sites, silencing, and product compartmentalization on product yield and quality, researchers still cannot reliably predict which proteins will be produced at high levels or what manipulations will guarantee enhanced productivity. We have optimized a simple transient expression system in *Nicotiana benthamiana* enabling rapid assessment of transgene potential for plant-based bioproduction. Briefly, intact *Nicotiana benthamiana* plants are vacuum-infiltrated with *Agrobacterium tumefaciens* cultures carrying the transgene of interest. After 48–96 h of further incubation, leaves are harvested for protein characterization. Using the immunomodulator interleukin-12 as a model pharmaceutical protein, we obtained bioactive recombinant protein at levels exceeding 5% of total soluble leaf protein. Appropriately assembled multimeric proteins have also been obtained following coinfiltration with *Agrobacterium tumefaciens* strains individually encoding each subunit. This system provides a rapid source of transgene product for assessing posttranslational modifications, purification strategies, and bioactivity as well as an effective system for optimizing construct elements. For vaccines, product purified from two to eight plants may support mouse vaccination trials providing efficacy and immune assessment data early in the development process.

Key words: Transgene product yield, *Agrobacterium tumefaciens, Nicotiania benthamiana*, Transient expression, Vacuum infiltration, Interleukin 12.

Loïc Faye and Véronique Gomord (eds.), *Methods in Molecular Biology, Recombinant Proteins From Plants, vol. 483*

DOI: 10.1007/978-1-59745-407-0_4

4.1. Introduction

Plant-based platforms for bioproduction of complex pharmaceutical proteins provide key advantages in scale, safety, and capitalization costs over current technologies involving mammalian cell-based production. Diverse production platforms ranging from seed-, leaf-, and tuber-based production in stable transgenic plant lines, to plant cell and root culture bioreactor systems, to viral or *Agrobacterium*-mediated transient expression systems have been developed and are being scaled-up and evaluated for commercialization *(1, 2)*. In all of these systems, the recoverable yield and quality of the transgene product is critical for commercialization and remains one of the key challenges in bringing plant-derived pharmaceuticals to market. Many factors influence the yield and quality of plant-synthesized proteins. For more than a decade, scientists have analyzed the roles of promoters, enhancers, introns, RNA instability elements and crytic-processing sites, gene silencing, codon usage, protease activity, subcellular targeting, tissue-specificity, storage, extraction methodologies, and other factors in mediating transgene product accumulation and recovery *(3–6)*. Although significant progress has been made in understanding many of these processes, we still cannot predict a priori which proteins or gene constructs will be produced at high levels in plants and which will be problematic. It is surprisingly common that a platform that has been optimized for expression and yield of one protein does not support equal productivity of a second closely related product in spite of identical promoters, vectors, and sites of accumulation. We have leveraged transient expression systems as a rapid assessment tool to quickly identify proteins supported for high level expression in plants and to compare variants (e.g., codon usage, signal peptides, N- vs C-terminal fusion partners or tags) for their ability to enhance yield of a particular product. Whether the ultimate plant-based bioproduction platform involves stable, transient, or tissue-specific expression, the ability to essentially "test drive" your transgene for plant production and bioactivity before significant investment of time and effort is expended, is highly advantageous.

Transient expression systems, vectored either by *Agrobacterium*, viral vectors, or a combination of these, offer a simple, rapid, and useful alternative method for analyzing a variety of different heterologous proteins whose respective precursor gene can be transcribed extrachromosomally without generation of stable transformed plants *(7–12)*. *Agrobacterium*-mediated transient expression platforms deliver protein product anywhere from 5 to 10 days after the transgene vector becomes available compared with 2 to 6 months for production of stable transgenic plants. Transient expression is often carried out in intact leaves of

Nicotiana benthamiana (5, 13). The higher disease susceptibility of this plant species supports high level of T-DNA-mediated transient expression following *Agrobacterium tumefaciens* infiltration (up to 5 mg of reporter protein per gram fresh weight; *(14)*. This transient expression method is generally used to verify transformation construct activity and to validate small amounts of recombinant protein *(5, 15)*. However, development of this agroinfiltration process for large-scale recombinant protein production is actively being explored by both academic and industry researchers *(1, 12)*.

In this chapter, we describe our optimized parameters for agrointroduction into intact *N. benthamiana* plants and efficient assessment of transgene product yields and quality. We routinely use this method to test the *in planta* production potential of a new transgene. Based on the presence of polyadenylated transgene mRNA, N-linked glycosylation of transgene product, and assembly of multimeric protein from transgenes delivered from distinct *Agrobacterium* strains, we are confident that the products of the transient system reflect synthesis within the plant cell. We have also exploited this system to compare *in planta* productivity of constructs that vary codon usage, signal peptides, targeting information, and orientation of fusion partners. For comparative analyses, standardization of promoters/vectors [e.g., we use the double-enhanced 35S promoter *(16)* with TEV enhancer *(17)* in a pBIB-Kan transformation vector *(18)*], infiltration parameters, and developmental stage of the plants is key to producing reliable data. Under these conditions, we have observed consistency of product yields (<10% variability) with the same construct but significant differences in yields among transgenes or transgene variants. Thus, we believe that products of transient expression effectively reflect that plant's intrinsic ability to effectively "read" the transgene and support transcript/protein stability. Because the transient system does not involve chromosomal integration, these differences do not reflect "position effects." However, Voinnet et al. *(5)* reported that codelivery of genes that reduce posttranscriptional gene silencing increase yields in *N. benthamiana* transient assays. Based on our experience, outcomes from transient expression qualitatively predict relative productivity in stably transformed plants or hairy roots (although yields are always greater per unit fresh weight or soluble protein in the transient system).

Herein, mouse interleukin-12 (mIL-12) serves as the model gene of interest for discussion of this transient-based protein expression method. This 70 kDa heterodimeric-glycosylated cytokine is an immunomodulator composed of an α-chain (p35 subunit) and a β-chain (p40 subunit) with promising therapeutic and vaccine applications due to its role in directing cell-mediated immunity *(19, 20)*. In an effort to promote stochiometric expression of these subunits, we have expressed a single chain mIL-12 (scmIL-12) form with a glycine–serine linker, which retains the full biological function of the native heterodimer *(21)*. Bioproduction of mIL-12 in this transient system has enabled us to produce sufficient quantities of purified IL-12 for extensive in vitro assays as well as for mouse vaccination trials. However, because mIL-12 is a heterodimer, it has served as a useful model to demonstrate the utility of the transient system for effectively producing and assembling IL-12 (and other multimeric proteins) based on coexpression of independent *A. tumefaciens* strains that encode for each subunit.

4.2. Materials

4.2.1. Plant Material

1. Source of plant material: 4–6-week-old *N. benthamiana* plants are grown in an environmental growth chamber (e.g., Conviron) under controlled conditions of 16 h light (12,000 lux)/8 h dark, 65% relative humidity, and a 25°C (day)/21°C (night) temperature cycle.
2. Germination procedure: 4½-inch pots containing a standard potting soil (e.g., 1 part Pro-mix BX /2 parts Pro-mix PGX; Premier Horticulture Ltd., Canada) fertilized with Osmocote 14-14-14 (Scotts, Canada) are seeded with 2–3 *N. benthamiana* seeds. Seeds are overlaid with vermiculite and are watered, and pots are covered with a humidity dome. At 2 weeks, if multiple seeds germinated, extra seedlings are transferred to new pots. At ~3 weeks, the domes are gradually vented over a 2–3-day period to avoid humidity shock to the young seedlings.

4.2.2. Bacterial Strain and Vector

1. Bacterial strain: *A. tumefaciens* strain LBA4404 *(22)* harboring the disarmed Ti plasmid pAL4404 bearing the *Vir* region and streptomycin selectable marker (*see* **Note 1**).
2. Plant expression vector: The "transgene of interest" needs to be inserted within the T-DNA region of a plant transformation vector. For the expression of scmIL-12, the coding sequence for this gene was constructed downstream of a strong, constitutively expressing, plant promoter *(16)*, and cloned into the multiple cloning site of the pBIB-Kan binary vector *(18)* with *Hind*III and *Sac*I restriction sites (**Fig. 4.1.**). This pBIB-Kan expression vector encoding the scmIL-12 gene was mobilized into *A. tumefaciens* strain LBA4404 using a modified freeze/thaw method *(23)* (*see* **Note 2**). It is important to construct the proper negative

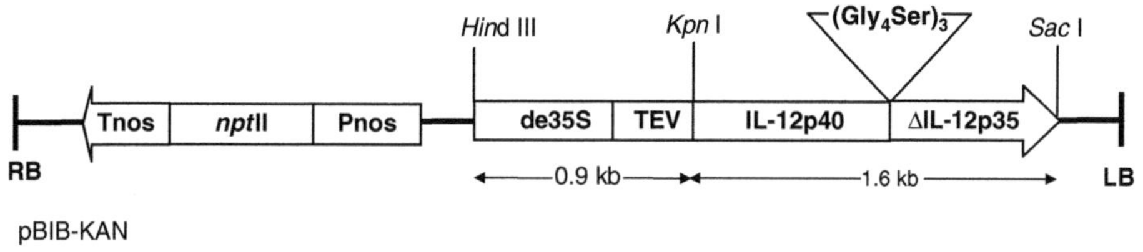

Fig. 4.1. The mouse IL-12 transient expression construct: Plant expression vector pBIB-Kan *(18)* is modified to encode the single chain form of mIL-12 (scmIL-12; *21)* such that the product contains the p40 (with its endogenous mouse signal peptide) and p35 subunits linked with a flexible $(Gly_4\text{-}Ser)_3$ spacer. This 1.6 kb scmIL-12 gene is under the control of the strong, constitutive plant promoter (dual-enhanced 35S Cauliflower Mosaic Virus promoter; *16)* and translational enhancer from the tobacco etch virus (TEV; *17)*. The neomycin phosphotransferase II (*nptII*) gene, under the control of the nopaline synthase (Pnos) promoter, supports kanamycin selection of stable transgenic plant lines but is not necessary for transient expression in *Nicotiana benthamiana*. *mIL-12* mouse interleukin-12, *scmIL-12* single chain mouse interleukin-12.

control strain for your transient assay (e.g., *A. tumefaciens* LBA4404 transformed with the pBIB-Kan "empty" vector).

4.2.3. Production of *A. tumefaciens* Inoculum

1. Yeast extract peptone (YEP) medium: 10 g/L of bacto-peptone, 10 g/L of yeast extract (Difco, Franklin Lakes, NJ), 5 g/L of NaCl (Sigma-Aldrich, St. Louis, MO), pH 7.0; autoclave.
2. Kanamycin and streptomycin sulfate stock solution (1,000×): Dissolve the antibiotics (Sigma-Aldrich) in distilled water to a final concentration of 100 and 30 mg/mL, respectively. Filter-sterilize using 0.2 mm-filter; do not autoclave. Aliquot and store at –20°C.
3. Benzylaminopurine stock solution (1,000×): Dissolve 25 mg of benzylaminopurine (Sigma-Aldrich) in a small amount of 0.5 M HCl, adding dropwise to the powder until solubolized. Dilute to 25 mL with distilled H_2O (final concentration of 1 mg/mL), filter-sterilize, aliquot, and store at –20°C.
4. Acetosyringone stock solution (1,000×): Dissolve acetosyringone (3,5-dimethoxy-4-hydroxy-acetophenone; Sigma-Aldrich) in chloroform or DMSO to a final concentration of 200 mM (39.2 mg/mL). Aliquot and store at –20°C.
5. Induction media (IM; 1×): 0.5× Murashige & Skoog basal medium salts, 1× of Murashige & Skoog vitamins *(24)*, 5% sucrose, 1× benzylaminopurine, pH 5.5; autoclave or filter sterilize.
6. Fixed angle centrifuge for 50-mL conical tubes (e.g., Eppendorf 5810R with fixed angle rotor with insert for 50-mL tubes).
7. Erlenmeyer flasks (50, 250, and 1,000 mL).
8. Polypropylene conical tubes (50 mL).

4.2.4. Vacuum Equipment

1. Vacuum pump (e.g., KNF Laboport vacuum pump model UN840.3, KNF Neuberger, Inc., Trenton, NJ).
2. Vacuum desiccator unit; 240-mm inner diameter (e.g., Bel-Art Products, Inc., Pequannock, NJ).
3. Glass beakers (400 mL).
4. Clear plastic wrap and rubber bands.

4.2.5. Protein Extraction, ELISA, and Western Blot

1. Tris-ascorbate (TA) extraction buffer: 100 mM Tris base, 100 mM ascorbic acid, 150 mM NaCl, 4 mM EDTA, 2.5% PVP-40, 0.1% Tween-20, pH 7.0.
2. 96-well high binding protein plates (e.g., Immunlon 4HBX plates, Thermo Electron Corporation, Milford, MA).
3. Mouse IL-12 p70 enzyme-linked immunoabsorbant assay (ELISA) kit (R&D Systems, Inc., Minneapolis, MN).
4. 1× PBS (phosphate-buffered saline): 137 mM NaCl, 2.7 mM KCl, 10.1 mM Na_2HPO_4, 1.76 mM KH_2PO_4, pH 7.4.
5. PBS-T: 1× PBS with 0.1% Tween-20.
6. Streptavidin HRP (R&D Systems, Inc.).
7. Microwell Peroxidase Substrate System – TMB (KPL, Gaithersburg, MD).
8. Bradford Protein Assay Reagent (Cytoskeleton, Inc., Denver, CO).
9. SDS protein loading buffer (3×): 2.4 mL 1 M Tris-HCl pH 6.8, 3 mL glycerol, 3 mL 20% SDS, 0.006 g bromophenol blue. Bring to 10 mL with distilled water. Store at 4°C.
10. 10% Tris-glycine precast gels (Invitrogen, Carlsbad, CA).
11. SDS-PAGE running buffer (1×): 3 g Tris base, 14.4 g glycine, 1 g SDS, pH 8.3.
12. Mini-format protein electrophoresis unit (e.g., Mighty Small II; 8 × 9 cm gel format; Amersham Biosciences, Piscataway, NJ).
13. Electrophoretic gel transfer unit (e.g., Mini Trans-Blot cell; Bio-Rad, Hercules, CA).
14. Electrotransfer buffer: 20 mM Tris-HCl, 150 mM glycine, 20% methanol.
15. Nitrocellulose membrane (Bio-Rad).
16. Antibodies (Western analyses): Primary antibody is goat antimouse IL-12 antibody (R&D Systems, Inc.); secondary antibody is alkaline phosphatase-conjugated rabbit antigoat IgG (Sigma-Aldrich).
17. Detection reagents (Western analysis): CDP-Star (Boehringer-Mannheim, Germany); Nitroblock Enhancer II (Tropix, Bedford, MA); CDP-Star Detection Buffer (0.1 M Tris-HCl, 0.1 M NaCl, pH 9.5).
18. X-ray film, X-ray cassette, darkroom, developer.

4.3. Methods

Experimental timeline of the complete process for transient infiltration is delineated in **Fig. 4.2.**

4.3.1. Production of *Agrobacterium* Inoculum

1. Day 1: Add 100 μg/mL kanamycin and 30 μg/mL streptomycin (final concentrations) to a flask containing 20 mL YEP media just before inoculation with *A. tumefaciens* carrying the appropriate plant expression construct (e.g., scmIL-12; pBIB-Kan negative control plasmid). Incubate at 28°C on an orbital shaker at 220 rpm for 24 h (**Fig. 4.3.**).
2. Day 2: Transfer the 20 mL culture to a new flask containing YEP media plus antibiotics to yield a final volume of 100 mL (*see* **Note 3**). Allow culture to incubate for an additional day or until culture is visibly turbid (*see* **Note 4**).
3. Day 3: Transfer the bacterial cultures to sterile 50-mL conical tubes and pellet the cultures by centrifugation at 5,000*g* for 10 min. For each 100 mL inoculum (spun in two 50-mL conical tubes), pellets are resuspended and seeded into 400 mL IM containing 200 μM acetosyringone (~4 volumes IM/1 volume culture) and incubated for 3 h at 28°C on an orbital shaker at 220 rpm (*see* **Note 5**). This induction step can be extended overnight if necessary.
4. Following this induction period, the optical density ($OD_{600\ nm}$) of the culture will be 0.1–2.0 (*see* **Note 6**). Transfer each IM culture harboring a given expression construct (e.g., scmIL-12 or pBIB-Kan empty vector) to an appropriately labeled 400-mL glass beaker in preparation for infiltration into plants; from this step forward, one does not need to maintain sterility.

4.3.2. Vacuum Infiltration

1. Cover the base of 4–6-week-old *N. benthamiana* plants (*see* **Note 7**) with clear plastic wrap and secure plastic wrap with a rubber band to retain the soil in the pot during the vacuum infiltration procedure (**Fig. 4.3.**).
2. Invert the plant into the beaker containing the IM (*see* **Subheading 4.3.1.4**) and place the beaker in the center of the

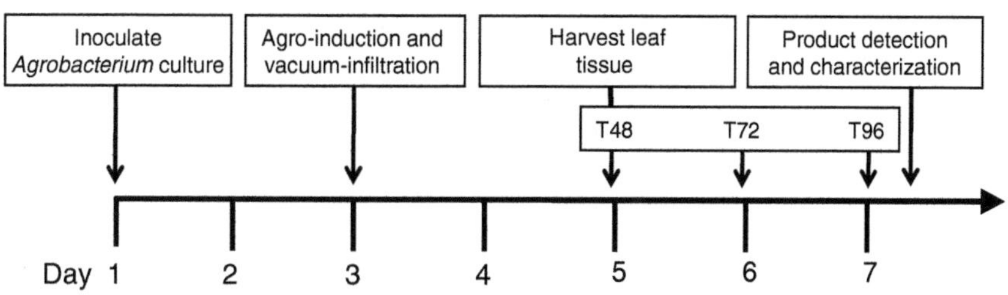

Fig. 4.2. Experimental timeline of the transient expression system for plant-based protein production.

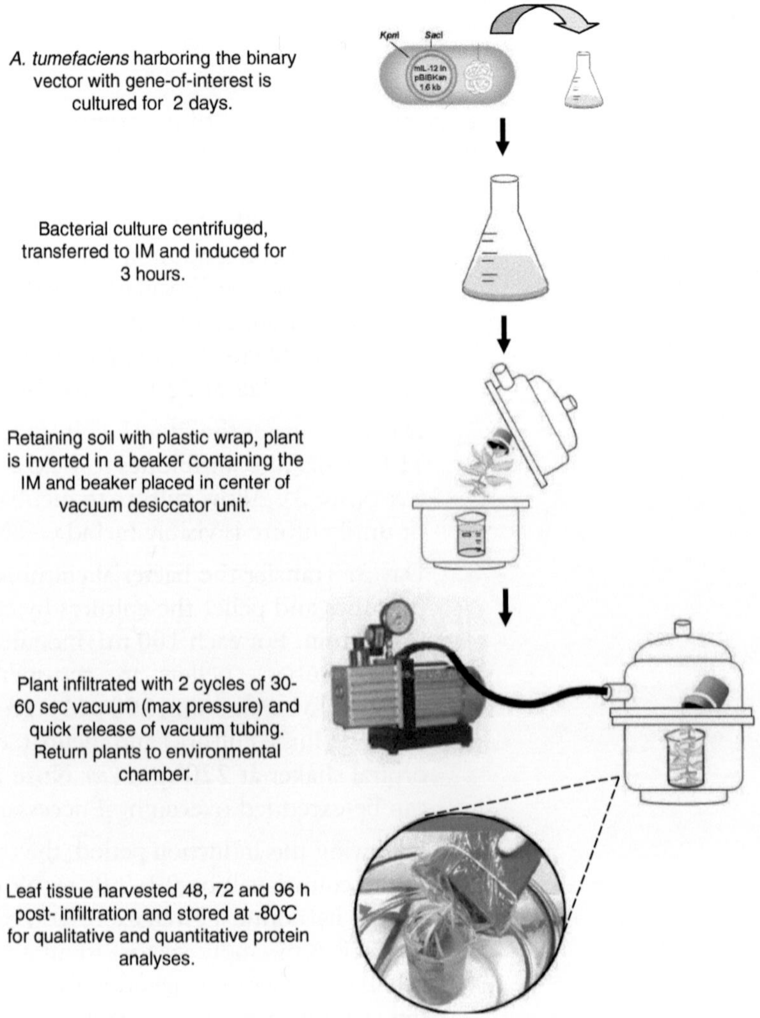

Fig. 4.3. Diagrammatic overview of the procedure for production of induced *Agrobacterium* cultures and vacuum infiltration into leaves of intact *Nicotiana benthamiana* plants.

vacuum desiccator unit. Check that all leaves of the plant are submerged in the IM solution.

3. Close and properly seal the vacuum desiccator unit before turning on the vacuum pump. Vacuum for 30–60 s at maximum pressure (~25 in. Hg), then quickly release the vacuum. Repeat the vacuum and release step. Darkening of the leaves is a good visual indicator that the induction media was successfully infiltrated. Remove the pot from the vacuum desiccator unit, place plant on a clean tray and label the pot with the designated construct (*see* **Note 8**).
4. Repeat **step 3.2.1** through **3.2.3** for additional plants using the same construct/culture or replace with a new beaker of

IM containing another construct/culture. IM cultures of a given construct can be reused for infiltration of multiple plants (*see* **Note 9**).

5. Return infiltrated plants to the environmental chamber at conditions established previously during plant growth.
6. If plant material is limited, an alternative syringe infiltration protocol can be used for agroinfiltration. Briefly, a 2-mL syringe (without needle) is pressed on the underside of a *N. benthamiana* leaf and IM culture is slowly injected. Multiple injections are required to "fill" a single leaf. If using a single plant for a time course or to test different expression constructs, properly label the base of each infiltrated leaf with the appropriate designation (*see* **Note 10**).

4.3.3. Harvesting Infiltrated Plant Tissue

1. The optimal postinfiltration time of harvest (i.e., yielding maximal product recovery) is transgene dependent. To establish these parameters, protein expression kinetics for each new transgene construct must be determined. Leaf tissue should be harvested 48, 72, and 96 h postinfiltration, placed in labeled, freezer-compatible bags and stored at −80°C for subsequent qualitative and quantitative protein analyses (*see* **Note 11**).
2. Tissue should be frozen in small aliquots to avoid subjecting the sample to undesired freeze–thaw cycles. In our experience, aliquots of 0.5–1 g fresh weight of leaf are sufficient for conducting initial verification and quantification of transgene expression in a crude protein extract. Typically, a single 5-week-old *N. benthamiana* plant yields ~15–20 g fresh weight of leaf material.
3. When using this system to compare different constructs or different infiltration times, it is important to collect "matched" samples reflecting leaf material corresponding to the same leaf age of the plant to minimize variability (*see* **Note 11**).

4.3.4. Containment and Disposal Procedures

As this procedure utilizes bacteria and plant tissues that contain recombinant DNA, it is important that proper containment and disposal procedures are followed to effectively decontaminate the equipment, workspace, and waste.

1. Disposal of bacteria: Decontaminate all residual *Agrobacterium* cultures, glass containers, and the vacuum desiccator unit with a 30% bleach treatment before general cleaning. Beakers containing the remaining IM culture must be either treated with bleach (>5% final concentration) or autoclaved to render cultures inviable before disposal.
2. Disposal of plant-related material: The soil and the remaining plant tissue should be rendered unviable before disposal.

3. Decontamination of work space: Bench paper and any plastic consumables that were in contact with bacterial cultures or infiltrated plants (e.g., 50-mL conical tubes, plastic wrap, tips, and gloves) should be autoclaved before disposal. All work surfaces should then be wiped down with 30% bleach followed by water wash.

4.3.5. Product Detection and Characterization

The following is a series of assays we have successfully used in the detection and characterization of mIL-12 expressed from this transient expression system. Keep in mind that the strategy used for detecting your expressed protein-of-interest depends on the available tools (e.g., antibodies/Western analysis, functional ELISAs, enzyme assays) and how selective these tools are for your protein. We recommend processing leaf material from control plants infiltrated with *A. tumefaciens* carrying an "empty" vector (e.g., pBIB-Kan) in parallel with all procedures to ensure the presumed product of interest is indeed a product of the transgene and not an artifact or endogenous product.

4.3.5.1. Protein Extraction from Leaves

In choosing the extraction procedure for initial evaluation of your transiently expressed protein of interest, it is important to consider what is known about the overall stability of your protein. Issues including protease susceptibility, pH sensitivity, protein solubility, product stability in a crude protein extract and extract storage conditions (e.g., freeze/thaw impact, room temperature stable) need to be considered. We recommend that fresh protein extracts are used in your initial assessments until the stability of your protein produced in this transient expression system can be addressed.

1. Quickly transfer leaf sample from –80°C storage to a cold mortar (prechill mortar/pestle with liquid nitrogen) and grind into a fine powder (*see* **Note 12**). It is important to keep the leaf tissue frozen during this entire process by periodically adding liquid nitrogen to the mortar.
2. Transfer the ground plant tissue (e.g., from 0.5 to 1 g fresh weight of leaf) to a microcentrifuge tube(s). Add two volumes of protein extraction buffer (e.g., TA protein extraction buffer or PBS) at a ratio of 2 mL of buffer per gram of tissue and vortex to mix (*see* **Note 13**). Maintain sample extracts on ice until all tissue samples to be screened have been processed. For scmIL-12, 0.5 g of leaf tissue provides sufficient material for initial evaluations of mIL-12 protein. For protein purification and more in depth analyses, this process can be scale up (see **Note 14**).
3. Centrifuge extracts (~13,000*g*; 5 min; room temperature) to remove nonsolubilized particulate and transfer the supernatant to a new microcentrifuge tube.
4. Determine total soluble protein (TSP) of each sample using a standard Bradford-based assay (*see* **Note 15**).

5. Aliquot the samples before storage at –20°C. In the case of mIL-12, freezing did not adversely impact yields or bioactivity.

4.3.5.2. Quantification of mIL-12 by ELISA

An ELISA can be a valuable assessment tool for the detection and quantitation of transiently expressed transgene products. In the case of mIL-12, a heterodimer-specific mIL-12p70 ELISA (R&D Systems) was adapted that could specifically detect mIL-12 in crude leaf extracts. Briefly, Immunlon 4 HBX plates, coated overnight at room temperature with rat anti-mIL12 p70 monoclonal capture antibody (1 μg/mL), are blocked with PBS-T/1% BSA and serial dilutions of standard recombinant mIL-12, plant-derived mIL-12 samples, and non-transgenic leaf extract are loaded. Following a wash step, captured mIL-12 p70 is detected with biotinylated goat anti-mIL-12 antibody (100 ng/mL) and streptavidin HRP/TMB substrate solution according to protocols of the Microwell Peroxidase Substrate System. Following 15-min incubation, ELISA plates are read at $OD_{450\ nm}$ (*see* **Note 16**).

For mIL-12, we typically obtain conformational intact p70 products at levels of 3–7% of the TSP (*see* **Note 17**) with levels as high as 20% were produced in young leaves. In contrast, the average yield of this same mIL-12 construct in our best-expressing stable transgenic tobacco (*N. tabacum*) plant line is 1% TSP in leaves and 1.8% in derived hairy root cultures *(26)*.

4.3.5.3. SDS-PAGE and Western Analysis

Western immunoblot analyses are useful for verifying the molecular size and integrity of the transgene product. In the initial assessment of transiently expressed mIL-12, 5–10 μg of TSP was loaded and resolved on a nonreducing, 10% SDS-polyacrylamide gel and subjected to SDS-PAGE using a mini-format horizontal gel electrophoresis unit. Separated proteins were electroblotted to nitrocellulase membrane in transblot buffer using an electrophoretic gel transfer unit. The protein blot was rinsed briefly with 1× PBS and blocked overnight at 4°C in PBS-T/3% BSA. For immunoblot detection of the mIL-12, the primary antibody, goat antimouse mIL-12, was used at 1:500 and an alkaline phosphatase-conjugated, antigoat IgG served as the secondary detection antibody (1:3,000). Detection on immunoblots was carried out using the CDP-Star and Nitroblock Enhancer II system in accordance with manufacturers' procedures. A Western blot comparing plant-synthesized products from transient expression of mIL-12 constructs and pBIB-Kan empty vector control is shown in **Fig. 4.4.**

4.3.6. Coexpression of Two Independent Constructs

While the majority of our studies expressing IL-12 in plants have utilized the single chain coding sequence (scmIL-12), our lab has successfully demonstrated that the molecular machinery of the plant in the context of a transient assay can effectively assemble subunits expressed on independent constructs in different

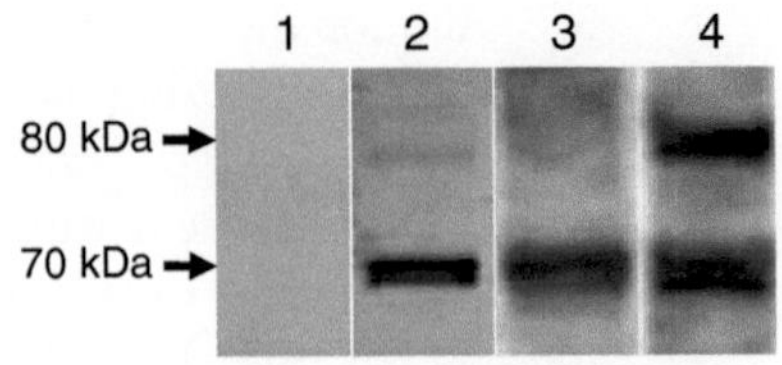

Fig. 4.4. Western immunoblot analysis of transiently expressed recombinant mIL-12. Proteins from crude leaf extracts were separated by nonreducing SDS-PAGE, transferred to membranes, and probed with an anti-mIL-12p70 antibody. Plant leaf extract infiltrated with empty pBIB-Kan vector served as the negative control in this analysis (*lane* 1) and animal cell-derived mIL-12 (*lane* 2) as a positive p70 control. Plant-derived mIL-12 p70 is detected in protein extracts from leaves infiltrated with *Agrobacterium* cultures containing the single chain IL-12 construct (scmIL-12; *lane* 3) and from leaves coinfiltrated with a 1:1 mixture of two *Agrobacterium* cultures individually encoding the mIL-12p40 or mIL-12p35 subunits (*lane* 4). Detection of p70 product in *lane* 4 confirms assembled p35 and p40 subunits that associate through disulfide interactions to form the p70 heterodimer. The additional 80 kDa band expressed when subunit constructs are coinfiltrated corresponds in size to a p40 homodimer known to assemble under these expression parameters. *mIL-12* mouse interleukin-12, *scmIL-12* single chain mouse interleukin-12.

A. tumefaciens strains to produce functional multimeric proteins. To test the utility of the plant transient expression system for coexpression and assembly of heterodimeric mIL-12, two individual *A. tumefaciens* strains, one harboring a p40 expression construct and the other coding for the p35 subunit of mIL-12, were mixed before the infiltration step. As shown in Fig. 4.4., the p70 IL-12 heterodimeric protein was successfully expressed and assembled with transient coexpression that is comparable to scmIL-12 expression. As shown in animal cells, subunit coexpression in plants also yielded p40 homodimers (p80), a competitive inhibitor of the mIL-12 in vivo, further supporting our choice of scmIL-12 for bioproduction. We have also utilized transient coexpression for assembly of light-chain/heavy-chain antibody components. This transient system circumvents some of the hurdles associated with stable expression of independent subunit constructs that includes no need for multiple antibiotic selection and better matched subunit expression levels for driving stochiometric assembly of the protein.

4.4. Notes

1. Alternative strains of *A. tumefaciens* have been used for transient expression including LBA4404, EHA105, AGL1, and AGLO *(12)*. For scmIL-12, no significant difference in yield

was observed using LBA4404 and AGLO. These bacterial strains may utilize different selectable markers to maintain resident Ti-derived plasmids (i.e., independent of the binary vector bearing the "gene of interest"); identify the appropriate antibiotic requirement for the *A. tumefaciens* strain used.

2. The binary plasmid carrying the "gene of interest" can be introduced into *A. tumefaciens* by electroporation methods as well as by the freeze/thaw method.

3. The YEP media should maintain continuous selection via antibiotics. For the *A. tumefaciens* strain LBA4404 described here, streptomycin (30 μg/mL) provides for Ti-plasmid vector selection and kanamycin (100 μg/mL) provides for selection of our pBIB-Kan-derived vectors. It should be noted that the pBIB-Kan vector also contains a neomycin phosphotransferase II (*nptII*; kanamycin resistance) plant selectable marker gene within the T-DNA (**Fig. 4.1.**). The plant selectable marker is not utilized in the transient system but we typically retain this feature as it facilitates mobilizing optimized constructs into stable transgenic plant lines if desired.

4. At harvest, bacterial cultures should have an $OD_{600\ nm}$ of >2.0. It is important to observe the formation of precipitates during culturing of *A. tumefaciens* strain LBA4404. After centrifugation, the pellet of LBA4404 should have a slight pink hue.

5. We have tested three different induction media reported in the literature. The IM described herein provided high expression of our target gene and is relatively simple to prepare. In addition, we evaluated the impact of increased acetosyringone concentration on the level of mIL-12 protein expression. No significant difference in recombinant protein expression levels was observed when acetosyringone concentration was elevated from 200 to 1,000 μM. Three media components appear to be most critical for successful transformation: (i) low pH (~pH 5.4), (ii) high sucrose levels (≥5%), and (iii) presence of acetosyringone (200 μM; *21).*

6. We tested the impact of inoculum density in the IM culture at the point of infiltration on recombinant mIL-12 protein expression levels. Infiltration with bacterial suspensions ranging from 0.1 to 2.0 $OD_{600\ nm}$ did not appear to significantly alter the amount of protein produced. We have observed this with other constructs and our observations were similar to those reported by Wydro et al. *(10).*

7. In our experience, the age (4–6 weeks from germination) and overall good health of the plants are critical in delivering the highest levels of protein for a given expression construct.

In general, older plants and leaves result in lower recovery of transgene products. Likewise, the level of transgene expression appears to correlate with the developmental stage of the leaf on a given plant.

8. Typically, we maintain vacuum until bubbles form on leaf surface and IM solution in the leaves starts to "bubble." It is very important to release the vacuum hose from the desiccator unit very rapidly to ensure successful infiltration. The exact time and vacuum pressure to obtain these conditions will vary lab-to-lab. Practice on a few dispensable plants to determine your vacuum conditions are resulting in uniformly darkened, water-soaked leaves.
9. We have demonstrated that using a single IM culture for infiltrating up to nine independent plants has no adverse effect on the amount of transgene product expressed. This is important in providing favorable cost advantages for large-scale industrial production of this transient technology. In our system, a 600 mL IM culture will provide sufficient bacterial inoculum for at least nine plants (additional culture is added to 400 mL beaker to maintain volume for leaf submersion in subsequent rounds of infiltration).
10. To perform the syringe infiltration method, it is important not to water plants 24 h before the infiltration procedure. More recently watered plants appear to not imbibe the IM solution as efficiently leading to lower transgene expression levels. Syringe infiltration has advantages when plant tissue is limited. The bacteria appears to remain localized to the site of infiltration (observed by leaf darkening as culture is infiltrated) enabling different strains/constructs to be infiltrated into different leaves of the same plant. However, there are several drawbacks with this method in comparison with the vacuum infiltration: (i) labor time for infiltrating leaves of 1 plant is ~30 min with the syringe method versus 3 min for vacuum, (ii) increased contamination of surrounding workspace versus a completely contained infiltration step with the vacuum method, and (iii) reduced consistency in yields since infiltration is less uniform.
11. While expression levels of transiently expressed transgenes are gene dependent, we typically see maximal target protein expression between 48 and 72 h postinfiltration. To date, only one construct tested in this system has required 96 h induction time for maximal target protein yields. We and others *(10)* have observed higher transgene product yields in younger leaves located in the top portion of the infiltrated plant compared to fully emerged leaves. The oldest lower leaves in general show less expression and are often

not harvested. For comparative analyses, we recommend using fully emerged, "middle tier" leaves.

12. A reproducible tissue grinding technique is important when utilizing this platform as a comparative screening tool for expressing recombinant proteins. Consistency in the grinding force used and in grinding time allotted to each sample can help in reducing unintended sample-to-sample variability. This is important when trying to compare the performance of (i) different expression constructs, (ii) various extraction buffers, and (iii) incubation time for protein expression.
13. We typically test multiple buffers (e.g., PBS or TA buffers) for initial extraction of a new transgene product. In the case of mIL-12, we observed no significant difference in mIL-12 recovery with PBS and TA buffers. We use PBS for scale-up mIL-12 production due to its better compatibility with downstream purification and shelf stability (i.e., TA buffer must be prepared fresh due to the storage sensitivity of ascorbic acid). If your protein is soluble and stable at lower pH, consider the use of an acidic pH extraction buffers (e.g., ≤pH 5.8) as they recover lower TSP thus providing an effective protein enrichment tool. In cases where product solubility may be an issue, extraction directly in 3X SDS protein loading buffer is often used to establish a "yield baseline".
14. For initial characterization of a previously untested construct, we typically process 1 g fresh weight of leaf tissue, which provides adequate sample for Western immunoblots, ELISAs, and total protein determinations. For larger scale preps (e.g., for protein purification and in vitro/in vivo assays), we typically process 20 g lots using the same ratio of tissue to buffer (2 mL/g fresh weight of tissue) and centrifuge the homogenate in 50-mL conical tubes (at 12–16,000*g*).
15. TSP yields vary depending on the extraction buffer and developmental stage of the leaf (see **Note 11**). For *N. benthamiana* with the Bradford assay noted, we typically yield 2–6 mg TSP/g fresh weight of leaf tissue.
16. In addition to providing an effective quantitation tool for IL-12 protein levels, this ELISA also confirms structural integrity of mIL-12 protein. This "conformational" mIL-12p70 ELISA uses a capture antibody that only recognizes correctly folded and/or assembled IL-12 p70 protein (R&D Systems).
17. In our experience, we would consider constructs supporting recombinant protein yields of 1% of TSP (in crude leaf extracts) to be good producers and construct yielding product at greater 5% to be outstanding.

Acknowledgments

Research supporting the optimization and utilization of this method was supported by a grant to Drs. Cramer and Dolan from the Arkansas Biosciences Institute funded by the Arkansas Tobacco Settlement Commission.

References

1. Fischer, R., Stoger, E., Schillberg, S., Christou, P., and Twyman, R. (2004) Plant-based production of biopharmaceuticals. *Curr. Opin. Plant Biol.* 7, 152–158.
2. Guillon, S., Tremouillaux-Guiller, J., Pati, P. K., Rideau, M., and Gantet, P. (2006) Hairy root research: Recent scenario and exciting prospects. *Curr. Opin. Plant Biol.* 9, 1–6.
3. Cramer, C. L., Boothe, J. G., and Oishi, K. K. (1999) Transgenic plants for therapeutic proteins: Linking upstream and downstream strategies. *Plant Biothecnol.* **240**, 95–118.
4. Garmory, H. S., Brown, K. A., and Titball, R. W. (2003) DNA vaccines: Improving expression of antigens. *Genet. Vaccines Ther.* 1, 2.
5. Voinnet, O., Rivas, S., Mestre, P., and Baulcome, D. (2003) An enhanced transient expression system in plants based on suppression of gene silencing by the p19 protein of tomato bushy stunt virus. *Plant J.* **33**, 949–956.
6. Knablein, J. (2005) Plant-based expression of biopharmaceuticals,in *Encyclopedia of Molecular Cell Biology and Molecular Medicine* (Meyers, R. A., ed.), Wiley-VCH Verlag GmbH & Co. KGaA, Weinheim, Germany, pp. 385–410.
7. Porta, C., Spall, V. E., Lin, T., Johnson, J. E., and Lomonossoff, G. P. (1996) The development of cowpea mosaic virus as a potential source of novel vaccines. *Intervirology* **39**, 79–84.
8. Stoger, E., Sack, M., Fischer, R., and Christou, P. (2002) Plantibodies: Applications, advantages and bottlenecks. *Curr. Opin. Biotechnol.* **13**, 161–166.
9. Sheludko, Y. V., Sindarovska, Y. R., Gerasymenko, I. M., Bannikova, M. A., and Kuchuk, N. V. (2006) Comparison of several *Nicotiana* species as hosts for high-scale *Agrobacterium*-mediated transient expression. *Biotechnol. Bioeng.* **96**, 608–614.
10. Ahmad, M. and Mirza, B. (2005) An efficient protocol for transient transformation of intact fruit and transgene expression in citrus. *Plant Mol. Biol. Rep.* **23**, 419a–419k.
11. Sindarovska, Y. R., Gerasymenko, I. M., Sheludko, Y. V., Komarnytskyy, I. K., Bannikova, M. A., and Kuchuk, N. V. (2005) Transgenic plants regenerated from hairy roots of *Nicotiana benthamiana*: A promising host for transient expression of foreign proteins. *Cytology and Genetics* **39**, 9–14.
12. Wydro, M., Kozubek, E., and Lehmann, P. (2006) Optimization of transient *Agrobacterium*-mediated gene expression system in leaves of *Nicotiana benthamiana*. *Acta. Biochim. Pol.* **53**, 289–298.
13. Thomas, C. L., Leh, V., Lederer, C., and Maule, A. J. (2003) Turnip crinkle virus coat protein mediates suppression of RNA silencing in *Nicotiana benthamiana*. *Virology* **306**, 33–41.
14. Marillonnet, S., Giritch, A., Gils, M., Kanzia, R., Klimyuk, V., and Gleba, Y. (2004) In planta engineering of viral RNA replicons: Efficient assembly by recombination of DNA modules delivered by *Agrobacterium*. *Proc. Natl. Acad. Sci. USA* **101**, 6852–6857.
15. Kapila, J., DeRycke, R., VanMontagu, M., and Angenon, G. (1997) An *Agrobacterium*-mediated transient gene expression system for intact leaves. *Plant Sci.* **122**, 101–108.
16. Kay, R., Chan, A., Daly, M., and McPherson, J. (1987) Duplication of CaMV 35S promoter sequences creates a strong enhancer for plant genes. *Science* 236, 1299–1303.
17. Carrington, J. C. and Freed, D. D. (1990) Cap-independent enhancement of translation by a plant potyvirus 5' nontranslational region. *J. Virol.* **64**, 1590–1597.
18. Becker, D. (1990) Binary vectors which allow the exchange of plant selectable markers and reporter genes. *Nucleic Acids Res.* **18**, 203.
19. Adorini, L. (1999) Interleukin-12, a key cytokine in Th1-mediated autoimmune diseases. *Cell. Mol. Life Sci.* **55**, 1610–1625.

20. Fieschi, C. and Casanova, J. (2003) Mini-review: The role of interleukin-12 in human infectious diseases: Only a faint signature. *Eur. J. Immunol.* **33**, 1461–1464.
21. Lieschke, G. J., Rao, P. K., Gately, M. K., and Mulligan, R. C. (1997) Bioactive murine and human interleukin-12 fusion proteins which retain antitumor activity in vivo. *Nat. Biotechnol.* **15**, 35–40.
22. Hellens, R., Mullineaux, P., and Klee, H. (2000) Technical focus: A guide to *Agrobacterium* binary Ti Vectors. *Trends Plant Sci.* **5**, 446–451.
23. Holsters, M., de Waele, D., Depicker, A., Messens, E., Van Montangu, M., and Schell, J. (1978) Transfection and transformation of *A. tumefaciens. Mol. Gen. Genet.* **163**, 181–187.
24. Murashige, T. and Skoog, F. (1962) A revised medium for rapid growth and bio-assays with tobacco tissue cultures. *Physiol. Plant.* **15**, 473–479.
25. Clough, S. J. and Bent, A. F. (1998) Floral dip: A simplified method for *Agrobacterium*-mediated transformation of *Arabidopsis thaliana. Plant J.* **16**, 735–743.
26. Liu, J., Dolan, M. C., Reidy, M. J., and Cramer, C. L. (2008) Expression of bioactive single-chain murine IL-12 in transgenic plants. *J. Interferon Cytokine Res.* **28**, 27–38.

Chapter 5

Production and Localization of Recombinant Pharmaceuticals in Transgenic Seeds

Thomas Rademacher, Elsa Arcalis, and Eva Stoger

Summary

Among the many plant-based production systems that have been developed for pharmaceutical proteins, seeds have the useful advantage of accumulating proteins in a relatively small volume, and recombinant proteins are very stable in dry seeds allowing long-term storage and facilitating distribution before processing.

To take full advantage of the natural ability of endosperm cells to store large amounts of protein in a protected subcellular environment, it is useful to target recombinant proteins to appropriate storage organelles. In this chapter, we describe the distinct types of protein storage organelles in the cereal endosperm and a protocol for the detection of recombinant proteins in these organelles by immunofluorescence and immunogold labelling.

The use of food and feed crops for the production of pharmaceutical proteins such as edible vaccines implies the need for strict separation of the transgenic seeds from the food and feed chain. For improved traceability visual markers may be co-expressed with the gene of interest in engineered seeds. DsRed is one example for a fluorescent protein that can be detected with high sensitivity using low tech equipment. We therefore describe the generation of transgenic maize plants expressing DsRed in a constitutive manner, and we point out the advantages of using this marker during the process of transformation and selection of plant tissue and later during breeding of transgenic lines into elite germplasm.

Key words: Cereals Protein bodies Storage organelles, Transgenic maize seeds, Endosperm, DsRed, Fluorescent protein, Immunofluorescence Immunogold, Microscopy.

5.1. Introduction

A large number of protein-based biopharmaceuticals, such as vaccines, antibodies and human replacement proteins for a broad range of potential applications, are currently in development

Loïc Faye and Véronique Gomord (eds.), *Methods in Molecular Biology, Recombinant Proteins From Plants, vol. 483*

DOI: 10.1007/978-1-59745-407-0_5

and therefore likely to create an increasing global demand in the foreseeable future. Conventionally, these molecules are produced using mammalian cell lines (for complex multi-subunit or glycosylated proteins) and bacterial cells (for simple, non-glycosylated proteins). However, the fermentation of cell cultures is expensive in terms of equipment, media and the need for skilled personnel, and there is limited opportunity for scale-up.

Transgenic plants have a number of distinct advantages over culture-based systems and offer a safe, convenient and cost-effective system for large-scale production *(1–4)*.

The main advantage is the anticipated cost savings, reflecting the large amount of biomass that can be produced in a short time with no need for specialized equipment or expensive media. Sowing, growing and harvesting can be carried out using traditional agricultural practices and unskilled labour. Moreover, scale-up can be achieved rapidly and inexpensively, simply by cultivating more land.

A variety of different plants can be used as expression hosts for pharmaceutical proteins. Among the production platforms that have been explored, seeds have the useful advantage of accumulating proteins in a relatively small volume, and recombinant proteins are very stable in dry seeds allowing long-term storage and facilitating distribution before processing. For example, antibodies or vaccines expressed in cereal seeds remain stable at ambient temperatures for years *(5)*. This is desirable if the material must be transported over a long distance or to remote areas.

Different seed crops, including cereals, grain legumes and oilseeds, have been explored by several companies *(6)* as production platforms for pharmaceutical and technical proteins, and the first two commercial plant-derived recombinant proteins, avidin and β-glucuronidase, were produced in maize seeds *(7,8)*.

Depending on the plant species different parts of the seed may be specialized for the accumulation of nutrients. While, for example, in legume seeds the cotyledons of the embryo are the main storage organs, this function is mainly assumed by the endosperm in cereals. The latter is a highly specialized, dedicated tissue that is consumed by the embryo upon germination. It is plausible that the restriction of transgene expression to this tissue will prevent potentially detrimental effects of the recombinant protein on the host plant due to its absence in vegetative tissues. Several endosperm-specific promoters, many derived from seed storage protein genes, are available. Early examples include the maize *zmZ27* zein promoter, a maize *Waxy* promoter and the rice small subunit ADP-glucose pyrophosphorylase promoter, all of which are endosperm-specific in maize, and the rice glutelin 1 (*Gt1*) promoter, which is endosperm-specific in rice, wheat and maize *(9–16)*. Even more restricted expression can be achieved with aleurone-specific promoters from barley *(12,17)*. Promoters derived from pea and bean storage protein genes have been used successfully for seed-specific transgene expression in grain legumes *(18–20)*, and high expression levels have been observed with the USP promoter *(20)*. One of the highest levels of expression recorded to date (36% TSP) was achieved using the seed specific arcelin 5-I promoter from *Phaseolus vulgaris* to express a single chain antibody fragment in *A. thaliana (21)*.

To take full advantage of the unique morphological characteristics of seed cells the subcellular targeting of the recombinant protein needs to be carefully considered. The prevalent storage organelles and corresponding compounds (protein, starch and oil) differ between species causing obvious differences in cell morphology. For example, cells from cereal endosperm are filled with protein bodies (pbs) and starch grains, while the endosperm or embryo cells of tobacco, rape and safflower are dominated by oil bodies. SemBioSys Genetics, Inc., developed a platform strategy for oilseeds in which the target recombinant protein is expressed as a fusion with oleosin to simplify protein isolation. The fusion protein can be recovered from oil bodies using a simple extraction procedure and the recombinant protein is separated from its fusion partner by endoprotease digestion *(22)*.

Targeting recombinant proteins to starch granules could be very useful for the oral or mucosal delivery of vaccines *(9)*, but such strategies yet need to be explored to their full extent. For proteins requiring extensive folding or glycosylation, the most obvious intracellular destinations are the protein storage organelles, as these have developed to facilitate stable protein accumulation.

Protein transport within cereal endosperm cells is complicated by the abundance of functionally distinct types of protein storage organelles *(23–26)*. Unlike dicotyledonous plants, cereals synthesize different classes of storage proteins with different solubilities and subcellular destinations. While most globulins are transported to protein storage vacuoles (PSVs) via the Golgi apparatus, prolamines of maize and rice remain within the ER and aggregate into separate pbs by direct enlargement of the ER network *(24)*. Consequently, cereal endosperm cells contain two types of protein storage organelles: ER-derived (pre-Golgi) prolamin bodies, and post-Golgi PSVs.

This specialized architecture of cereal endosperm cells has been shown to influence the intracellular route of a given recombinant protein *(27,28)*, and studies of recombinant glycoproteins in different tissues and plant species have indicated that the subcellular fate of a protein is dependent on its intrinsic properties and state of assembly, but also on tissue and species-specific factors *(27–30)*.

Various N-terminal signal peptides have been used successfully to target proteins to the endomembrane system of plant cells, including some of human origin *(31)*, heterologous plant signal peptides such as the barley α-amylase signal sequence *(32)*, and various leader peptides from seed storage proteins. While, in absence of further targeting information, secretion of the protein to the apoplast is generally expected, endosperm cells appear to exhibit a tissue-specific tendency towards intracellular protein deposition.

For example, human lysozyme expressed in rice endosperm with a functional signal peptide accumulated in PSVs, accompanied by distortion of pb morphology *(10)*. Similar observations were made for recombinant phytase produced in rice and wheat endosperm *(27,28)*, an IgG antibody and cytomegalovirus glycoprotein B expressed in tobacco seeds *(30,33)*, and bacterial amylase and bovine β-casein expressed in bean and soybean seeds *(18,34)*.

In combination with an N-terminal signal peptide a C-terminal KDEL sequence can be added to retrieve proteins from the *cis*-Golgi to the ER *(35)*. This is a widely used strategy to increase the accumulation of a recombinant protein in plant cells *(36)* and to avoid plant-specific N-glycan modification in the Golgi *(37)*.

In transgenic rice endosperm cells, recombinant proteins containing an ER-retention signal accumulated mainly *(31)* or exclusively *(38)* in ER-derived pbs.

In wheat, KDEL-tagged recombinant human serum albumin was targeted to the ER lumen in leaf cells, but deposited in storage protein aggregates within the vacuole of endosperm cells. It is most likely that the recombinant protein followed the same route as the bulk of endogenous glutenins and prolamins, which in wheat aggregate and bud off the ER, but later become incorporated into the vacuole by an autophagy-like process *(28)*. Petruccelli et al. *(30)* observed the unexpected partial secretion and vacuolar delivery of a KDEL-tagged antibody in tobacco seeds, and in *Arabidopsis* seeds a recombinant antibody fragment including a KDEL sequence was delivered to the periplamic space and caused aberrant localization of ER chaperones and endogenous seed storage proteins *(39)*. These studies indicate that species-specific properties of seed cells may lead to unexpected deposition patterns of recombinant proteins. It is therefore sensible to include studies of recombinant protein localization in the analysis of transgenic plants. An example of such study is illustrated in **Fig. 5.1** for transgenic wheat endosperm, and a procedure for the preparation of seed tissue and for immunodetection of recombinant proteins in cereal endosperm is included in this chapter.

Figure 5.1. (continued) Sections were incubated overnight with a rabbit polyclonal anti-phytase serum, followed by goat anti-rabbit Alexa Fluor 594 as a secondary antibody. Strong labelling is observed within the pbs, whereas no significant labelling is visible in the apoplast (arrows) or in any other compartment of the cell. (**e–f**) Localization of HSA-KDEL (**e**) and phytase (**f**) in ultrathin sections. Numerous gold particles are distributed within the prolamin bodies (pr), whereas no significant labelling is visible within the triticin inclusion bodies (arrowheads).

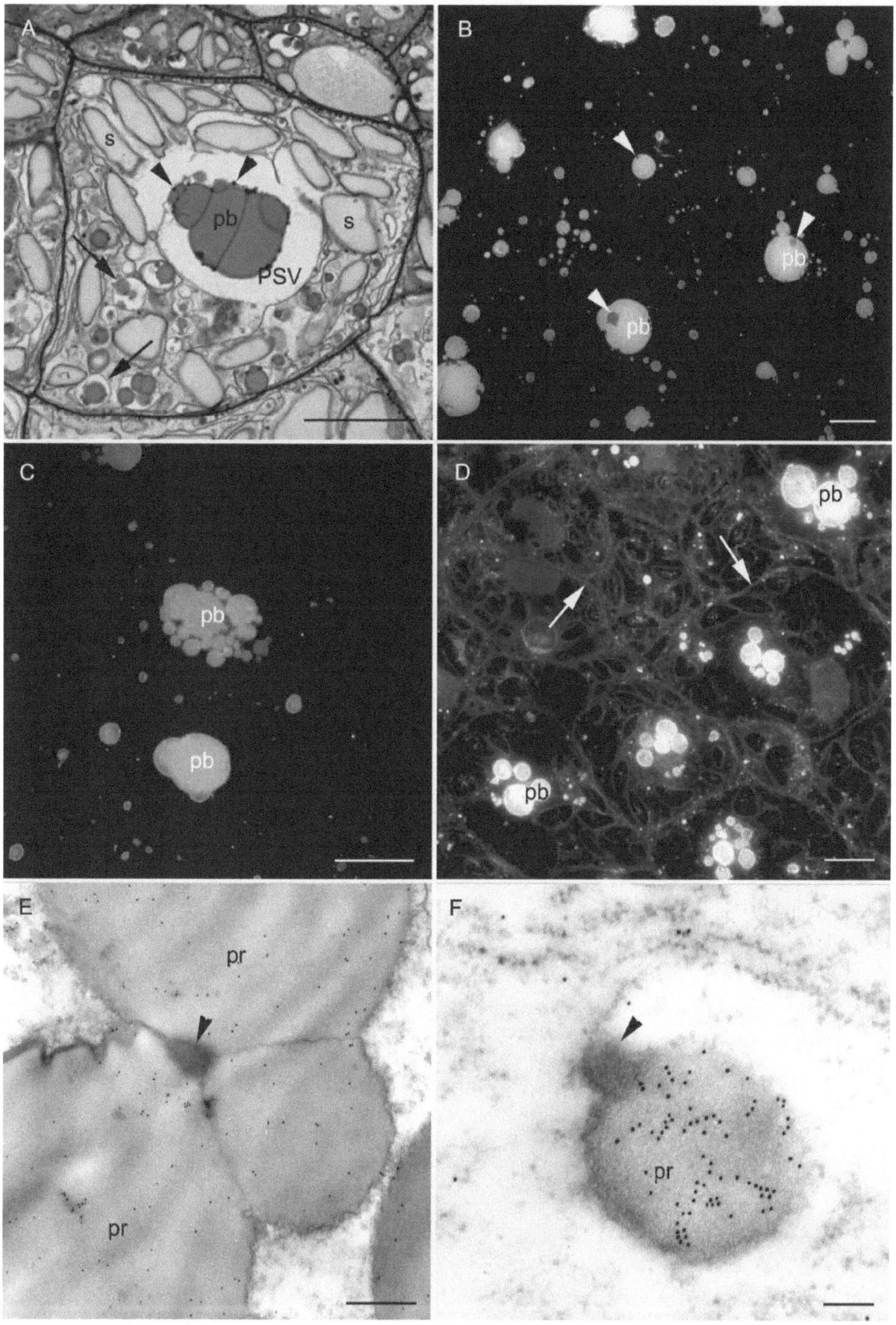

Figure 5.1. Immunolocalization of two recombinant proteins (phytase and human serum albumin) in wheat endosperm. (**a**) Semi-thin section, stained with toluidine blue. Developing wheat endosperm cells contain a large central protein storage vacuole that includes a protein body (pb). A large number of pre-vacuolar compartments containing blue pbs are also visible (arrows). Triticin inclusion bodies in the periphery of the prolamin pbs (arrowheads) and the spindle-shaped starch grains (s) are indicated. (**b**) Fluorescence immunolocalization of wheat gliadins in a semi-thin section. Sections were incubated O/N with a rabbit polyclonal anti-wheat gliadin serum; as a secondary antibody, a goat anti-rabbit Alexa Fluor® 594 serum was used. Prolamin bodies show specific labelling, while the triticin inclusion bodies (arrowheads) appear non-labelled. (**c**) Fluorescence immunolocalization of recombinant HSA-KDEL in a semi-thin section. Sections were incubated overnight with a rabbit polyclonal anti-HSA serum, followed by a goat anti-rabbit Alexa Fluor 594 serum as a secondary antibody. Significant labelling is observed in the pb within the central PSV, as well as in the smaller pbs distributed within the cytoplasm. (**d**) Fluorescence immunolocalization of recombinant phytase in a semi-thin section.

Several of the seed crops that appear attractive for the production of pharmaceutical proteins and vaccines are cereal and legume species that are widely cultivated for food and feed purpose. While the availability of simple and inexpensive food processing techniques for such species is an advantage, additional safety measures must be taken to prevent cross-pollination and inadvertent mixing of transgenic seeds with food or feed crop. The crop must therefore be grown in isolation from breeding materials to avoid genetic and mechanical mixing. An additional strategy is the development of phenotypic markers for identity preservation, for example, green or purple fruit colour in tomato as a label for transgenic lines expressing pharmaceuticals. Alternatively, fluorescent marker proteins such as green fluorescent protein (GFP) or red fluorescent protein (DsRed) may be used as traceable markers for the macroscopic detection of transgenic pollen and seed (**Fig. 5.2**). In this way, the presence of a linked transgene can be monitored visually *(40)* to facilitate and accelerate breeding programs into elite germplasm, and transgene movement can be traced without molecular techniques. Below we describe the generation of transgenic maize expressing DsRed in a constitutive manner, and we provide details on the sensitive macroscopic detection of transgenic tissue using low tech equipment.

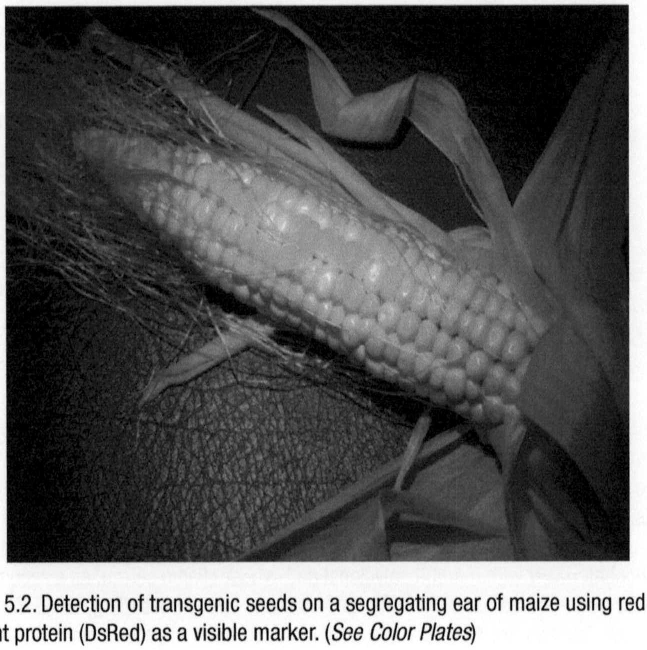

Figure 5.2. Detection of transgenic seeds on a segregating ear of maize using red fluorescent protein (DsRed) as a visible marker. (*See Color Plates*)

5.2. Materials

5.2.1. Plant Material

1. Maize (*Zea mays* L.). The transformation protocol for maize (**Subheading 5.3.1**) is optimized for embryogenic callus from immature embryos of cv. HiII (A188XB73). Typically, plants are maintained in a climate-controlled growth chamber at 28°C/24°C day/night temperature with a 14-h photoperiod. Mature seeds can be collected after 4–5 months. Immature seeds for immunodetection can be collected ~15–25 days after pollination.

5.2.2. Maize Tissue Culture

1. Plant culture room with illumination system.
2. Laminar flow hood.
3. Fine forceps and scalpel.
4. Petri dishes.
5. Culture pots and tubes.
6. Parafilm.
7. Sterile, double-distilled water.
8. 70% (v/v) ethanol.
9. 20% (w/v) sodium hypochlorite solution.
10. Tween 20.

5.2.3. Growth Media and Additives

5.2.3.1. Checklist

1. Basic MS (Murashige and Skoog) salts (*see* **Note 1**).
2. Basic N6 salts (*see* **Note 1**).
3. Minimal organics (*see* **Table 5.1**).
4. Sucrose.
5. 2,4-Dichlorophenoxyacetic acid (2,4-D).
6. Proline.
7. 6-Benzylaminopurine (6-BAP).
8. Gelrite.
9. Phosphinothricin (PPT), supplied as glufosinate-ammonium powder (*see* **Note 2**).
10. Mannitol.
11. Sorbitol.
12. $AgNO_3$

5.2.3.2. Stock Solutions

1. PPT stock: 10 mg/ml in H_2O.
2. 2,4-D stock: 2 mg/ml in DMSO.
3. $AgNO_3$ stock: 2 mg/ml in H_2O.

Table 5.1
Organic media components. All stocks should be stored at 4°C. The organic compounds should be added after cooling below 50°C

Organic components (mg/l)

Thiamine HCl	0.1
Pyridoxine HCl	0.5
Glycine	2
Myo-inositol	100
Nicotinic acid	0.5
Casein hydroxylate	1,000

Make up as 100× stock in sterile distilled water. Filter sterilize.

4. 6-BAP stock: 10 mg/ml in 1 N NaOH.
5. Minimal organics (*see* **Table 5.1**).

5.2.3.3. Media Composition

All media should be freshly prepared with ultrapure water (>18 MΩ/cm) and autoclaved or filter-sterilized (0.2 µm). Heat-sensitive components (selective agents, vitamins) must be added after cooling to below 50°C.

Callus Induction and Growth:

1. Basic N6 medium for callus induction and growth: Make up N6 salts from a ready powder. Add 30 g sucrose, 2 mg 2,4-D and 3 g gelrite per litre of medium. Before adding the gelrite, adjust pH to 5.8 with 1 N KOH. After autoclaving, add organic media components and 0.425 ml of $AgNO_3$ stock solution.
2. CIM (callus induction medium): basic medium plus 0.69 g proline (added before autoclaving).
3. CIB (bombardment medium): CIM plus 0.2 M mannitol and 0.2 M sorbitol.
4. CIS (selection medium): basic medium plus 3 mg/l PPT.
5. 50 mg/l hygromycin B.

Plant Regeneration:

1. Basic MS medium for plant regeneration and maintenance: Make up MS salts from a ready powder. Add 30 g sucrose and 3 g gelrite per litre of medium. Before adding the gelrite, adjust pH to 5.8 with 1 N KOH. After autoclaving, add organic media components and 0.425 ml of $AgNO_3$ stock solution.

2. MSR1 (regeneration medium): basic medium supplemented with 2 mg BAP and 3 mg/l PPT.
3. MSR2 (maintenance medium): basic medium supplemented with 3 mg/l PPT.

5.2.4. Maize Transformation

1. Stock solution of plasmid DNA carrying the transgene of interest (*see* **Notes 3–5**) and the DsRed gene (*see* **Note 6**).
2. Stock solution of plasmid DNA carrying the selectable marker gene (*see* **Note 7**).
3. Particle accelerator gun and accessories.
4. Gold particles (0.7–0.9 um; Bio-Rad).
5. 100 mM spermidine.
6. 30% Polyethylene glycol (PEG) (MW 8,000).
7. 100% ethanol.
8. Tris-EDTA (TE) buffer.
9. 2.5 M $CaCl_2$.
10. Microfuge.
11. Vortexer.
12. Sonicator.

5.2.5. DsRed Visualization

1. Epifluorescence microscope with filter cube for green excitation (*see* **Note 8**).
2. Stereomicroscope with a halogen cold light source and green excitation filter (*see* **Note 9**).
3. Red filter (*see* **Note 9**).

5.2.6. Immunocytochemistry

5.2.6.1. Equipment

1. Fluorescence microscope with appropriate filter sets (*see* **Note 10**).
2. Transmission electron microscope.
3. Low temperature embedding device with UV-source (*see* **Note 11**).
4. Ultramicrotome (*see* **Note 12**).
5. Fine forceps (number 5, non-magnetic) and scalpel.

5.2.6.2. Consumables and Reagents

1. 0.1 M Phosphate buffer (pH 7.4): 10.9 g/l Na_2HPO_4 (anhydrous), 3.2 g/l NaH_2PO_4 (anhydrous).
2. Fixative: 4% paraformaldehyde plus 0.2% glutaraldehyde in phosphate buffer, pH 7.4 (*see* **Note 13**).
3. Phosphate buffered Saline (PBS): 137 mM NaCl, 2.7 mM KCl, 8.1 mM Na_2HPO_4, 1.5 mM KH_2PO_4.
4. PBST: PBS plus 0.25% Tween 20.
5. LRWhite resin (*see* **Note 14**).

6. Benzoyl methyl ether.
7. Beem capsules (size 00).
8. Microscopy glass slides.
9. Gold grids (*see* **Note 15**).
10. Staining solution for LM: 1% toluidine blue in 1% aqueous borax w/v.
11. Staining solution for Transmission electron microscopy (TEM): 2% aqueous uranyl acetate (*see* **Note 16**).
12. Blocking solution: 5% BSA (Bovine Serum Albumin) in phosphate buffer, pH 7.4.

5.3. Methods

5.3.1. Transformation of Maize

The tissue culture and transformation protocol has been adapted from Armstrong and Green *(41)* and Armstrong et al. *(42)*. The manipulation of seeds, explants and callus should be carried out using sterilized tools and equipment in a laminar flow hood to avoid microbial contamination.

5.3.1.1. Induction of Embryogenic Maize Callus for Transformation

1. Collect immature ears (~13–16 days after pollination; embryos should measure 2–3 mm) from greenhouse-grown maize plants.
2. Soak the ears for 1 min in 70% (v/v) ethanol and rinse in distilled water.
3. Immerse ears in 500 ml 20% (w/v) sodium hypochlorite containing 0.5 ml Tween 20 for 15 min. Transfer work to the laminar flow hood.
4. Rinse seeds at least three times in sterile distilled water.
5. Excise the embryos using two pairs of forceps.
6. Place the embryos scutellum-uppermost on CIM.
7. Incubate the plates in darkness at 24°C for 2 weeks.
8. Transfer callus tissue to fresh medium every 3 weeks until fast growing callus is established (*see* **Note 17**).
9. Four to 6 h before bombardment, transfer embryos to bombardment medium (CIB; (*see* **Note 18**).

5.3.1.2. Preparation of DNA-Coated Gold Particles for Bombardment

1. Mix 2.5 mg of gold particles with 5 μg of the selectable marker plasmid and 10 μg of the plasmid carrying the DsRed gene and the transgene of interest. Add TE buffer to 100 μl Vortex for 30 s.
2. Add 100 μl of 100 mM spermidine and vortex for 30 s.
3. Add 100 μl of 30% PEG (MW 8,000) and vortex for 30 s.

4. Slowly add 100 µl of 2.5 M $CaCl_2$ with continuous vortexing. When all the $CaCl_2$ has been added, vortex for a further 10 min.
5. Centrifuge at 12,000×*g* for 30 s and discard supernatant.
6. Wash with 200 µl 100% ethanol and centrifuge at 12,000×*g* for 30 s.
7. Add 100 µl 100% ethanol and sonicate briefly to break up clumps of particles (*see* **Note 19**).
8. Spot 5–10 µl of the suspension onto the centre of a carrier disc and allow to air dry.

5.3.1.3. Bombardment Using Bio-Rad Helium Gun (PDS 1000/He)

1. Follow manufacturer's guidelines for loading carrier disc into the barrel of the helium gun and bombard plant material at 900–1,300 psi (*see* **Note 20**).
2. Bombard twice with a 4-h interval between bombardments (maintain the embryos in darkness at 24°C on CIB during the interval).

5.3.1.4. Selection and Regeneration of Transgenic Maize Plants

1. Transfer bombarded callus to CIM medium 12–16 h after bombardment. Incubate for 2 days in darkness at 24°C.
2. For the selection of transformed embryogenic callus, transfer tissue to CIS medium and incubate for 2 weeks in darkness at 24°C.
3. Observe tissue under green light and identify red fluorescent sectors. Subculture the red fluorescent callus at 2-week intervals on fresh CIMS medium in darkness at 24°C (*see* **Note 21**).
4. At the end of the third round of selection, transfer embryogenic callus to MSR1 medium for regeneration. Incubate plates at 24°C under low light (100 mE) for the first 10 days and stronger light (130 mE) for the subsequent 10 days, with an 18-h photoperiod (*see* **Note 22**).
5. Transfer elongated shoots to tissue culture tubes containing MSR2 medium (*see* **Note 23**).
6. When transgenic maize plantlets are rooted and ~10 cm long, they can be transferred to soil (*see* **Note 24**).
7. Transgenic plants may be self- or cross-pollinated with non-transgenic elite germplasm (*see* **Notes 25** and **26**).
8. Check pollen and seeds for DsRed fluorescence to identify homozygous offspring and to assess the number of transgene loci.

5.3.2. Immunolocalization of Recombinant Proteins in Cereal Endosperm

We describe here a standard protocol for fixation and embedding of plant tissues in general and seed endosperm in particular. The method combines a mild fixation at low temperatures to optimize antibody-antigen recognition in the sample, but compromises the ultrastructural integrity of the tissue to a certain extent. In particular, the membranes may appear disturbed, as the lipids can be partially extracted during dehydration.

5.3.2.1. Fixation and Embedding

1. Immerse the tissue in a drop of fixative and cut small pieces, not bigger than 1 mm^3 (*see* **Notes 27** and **28**).
2. Fix the samples overnight at 4°C. The volume of fixative used should be enough to cover the samples.
3. Wash the samples with phosphate buffer (5 steps of 15 min each) at 4°C.
4. Dehydrate the samples in ethanol series (*see* **Note 29**):
   30 min in 50% ethanol at 4°C.
   30 min in 70% ethanol at –20°C.
   30 min in 96% ethanol at –20°C.
   30 min in 100% ethanol at –20°C.
5. Infiltrate the samples with LRWhite resin (*see* **Note 30**). The infiltration is performed at –20°C (*see* **Note 31**). Incubate samples:
   For 3 h in a mixture of resin/absolute ethanol 1:3 (v/v).
   For 3 h in a mixture of resin/absolute ethanol 1:1 (v/v).
   For 3 h in a mixture of resin/absolute ethanol 3:1 (v/v).
   For 3 h in a mixture of resin/absolute ethanol 1:3 (v/v).
6. Incubate overnight in pure resin.
7. Change to fresh resin and incubate for another 3 h.
8. Place infiltrated tissue samples in pre-cooled beem-capsules filled with resin (*see* **Note 32–34**). Polymerization of the resin takes ~24 h and is carried out under UV light at –20°C.

5.3.2.2. Sectioning

For light microscopy, semi-thin sections (1 μm) are obtained with an ultramicrotome and collected on glass slides (*see* **Note 35**). For electron microscopy, sections showing silver interferences (~90 nm) are collected on 400 mesh gold grids. Sections are then stained (*see* **Subheading 5.3.2.3**) or incubated with antisera for immunodetection.

5.3.2.3. Staining

For light microscopy, toluidine blue is used to stain proteins and nuclei in blue. Sections on a glass slide are covered with toluidine blue solution for 30 s and then rinsed with distilled water.

For electron microscopy (EM), samples are stained in aqueous uranyl acetate according to the following protocol (*see* **Note 36**):

1. Float grids on drops of staining solution with the side holding the section facing downwards for 10 min.
2. Wash grids 4–5 times by floating them on drops of double distilled water.

5.3.2.4. Immunolocalization

The method described is an indirect method, where the primary antibodies bind to the antigen. The antigen–antibody reaction is visualized using a secondary antibody coupled to a fluorophor for fluorescence microscopy or to a gold particle for electron microscopy.

1. Block the sections (either on glass slides for fluorescence microscopy or on gold grids for electron microscopy) with blocking buffer for 15 min at room temperature (RT).
2. Incubate sections with the appropriate antibody diluted in phosphate buffer overnight at 4°C (*see* **Notes 37–39**).
3. Wash three times (10 min each) with PBST.
4. Incubate sections with the secondary antibody diluted in phosphate buffer for 1–2 h at room temperature (*see* **Note 40**).
5. Wash twice (10 min each) with phosphate buffer.
6. Wash twice (5 min each) with distilled water.
7. Leave to air dry.

Sections can be directly observed under a fluorescence microscope (*see* **Note 41**), whereas the samples for electron microscopy must be stained with uranyl acetate as described in **Subheading 5.3.2.3**.

5.4. Notes

1. The ingredients for these basic media can be purchased as ready-mixed powders from many suppliers (e.g., Duchefa, NL; Gibco-BRL, USA; Sigma, GER).
2. The choice of selective agent depends on the selectable marker used for transformation. PPT selection is used with the markers *bar* or *pat*, from *Streptomyces hygroscopicus*, both of which encode PPT acetyltransferase.
3. A strong and seed-specific promoter such as the gt1 promoter *(16)* is useful for the expression of the gene of interest. In some cases, it may also be necessary to modify the coding region of the transgene to match the codon preference of cereals. Appropriate targeting signals need to be added to achieve high accumulation levels and the desired N-glycan structures. Most recombinant proteins accumulate to higher levels when targeted to the secretory pathway using a signal peptide. Further increases in yield can often be obtained by adding a C-terminal KDEL sequence which causes proteins to be retrieved from the Golgi apparatus and returned to the ER in the manner of a resident ER protein.

4. Care must be taken, because although some promoters may be described as 'seed specific', a low level of activity may be present in other tissues such as pollen.
5. The insertion of an intron sequence in the 5′untranslated region of the transgene is often useful to boost expression in monocotyledonous species.
6. Several DsRed genes are commercially available. Monomeric DsRed does not form aggregates and is therefore more likely to be tolerated by plant cells without side effects. DsRed may be targeted to various subcellular compartments (ER, apolast, caytosol, plastids). Targeting to plastids is useful to avoid interference with the recombinant protein of interest and to achieve bright fluorescence.
7. The selectable marker gene (*bar*) may be included on the same plasmid as the transgene of interest but this is not necessary. Co-transformation with two plasmids results in co-integration at the same locus in most plants. If the simultaneous delivery of two genes of interest is desired (e.g., genes encoding the heavy and light chain of an IgG antibody), a useful setup includes the delivery of the transgene encoding the heavy chain on the same plasmid with the selectable marker, and the transgene encoding the light chain together with the DsRed on a second plasmid. Callus tissue showing both herbizide resistance and red fluorescence is then selected to recover transgenic plants containing both antibody genes.
8. The excitation spectrum of DsRed has a maximum at 558 nm (green) and two minor shoulders at 526 nm (green) and 490 nm (blue). The emission maximum is at 583 nm (amber/orange). It is not necessary to buy a special DsRed-filter set for epifluorescence microscopes. Standard filter cubes with a green excitation range, for example, for Cy3 or TRITC, are compatible with DsRed. All these filter cubes contain a band pass suppression filter which removes the red fluorescence of chlorophyll (>650 nm). In addition, the excitation of chlorophyll fluorescence is very low with green light. DsRed can also be excited with blue light. This feature can be utilized to exclude autofluorescence artifacts that are usually visible with one excitation wavelength only.

   The longer wavelength of green excitation light is less damaging for the plant tissues than blue and especially UV light. In addition, green excitation light can easier penetrate plant tissues and is less absorbed by plant compounds than blue or UV light.
9. For macroscopic observation of plants, a portable cold light source with a green excitation filter and 1 m fibre optics (KL2500LCD, Schott, Germany) may be used in combination with a simple red filter (e.g., light red, Lee, UK).

Alternatively, a green LED lamp may be used with a 12V standard household bulb socket and a red filter.

10. GFP and tetramethylrhodamine isothiocyanate (TRITC) filter sets are a convenient choice since a wide range of suitable fluorescence labelled antibodies is available for these wavelength combinations. A 10× objective allows an overview of the section and a 60× immersion objective is ideal to resolve the intracellular structures.
11. Suitable equipment is commercially available (e.g., Leica AFS2). Alternatively, self-made setups may also be used (e.g., freezer with UV lamps mounted on the door).
12. Self-made glass knives are generally used to obtain semi-thin sections. A diamond knife is recommended for ultrathin sections.
13. A stock solution of 10% paraformaldehyde in phosphate buffer pH 7.4 can be stored at –20°C.
14. LRWhite is an acryl resin that is ideal for embedding samples for immunocytochemistry as it polymerizes under UV light at –20°C.
15. Grids are available in various shapes. We prefer 400 hexagonal mesh grids because they improve the stability of the section under the electron beam while providing a wide open area. Because of the low stability of LRWhite sections under the electron beam, it is recommended to coat the grids before use. Drop Formvar on a quiet, clean water surface to form a thin layer of greyish–silver appearance. Place the grids on top and collect them using a piece of parafilm so that the coated grids are then attached to the Parafilm, with the coated side facing upwards. After air drying the coated grids can be stored in a Petri dish.
16. The staining solution can be prepared in advance and stored in the fridge. To remove uranyl acetate precipitates, it is recommended to pass the solution through a 0.2 µm syringe filter before use.
17. The callus may be maintained and used for several months, but will gradually lose its regeneration capacity.
18. Osmoticum treatment has been shown to improve the efficiency of particle bombardment. *(43)*.
19. Discard the suspension if clumps persist. Try again with DNA of better quality.
20. 20. We use a Bio-Rad pneumatic helium gun, model PDS 1000/He. When considering investing in or hiring a particle gun, the manufacturers will advise on setting up and operating procedures. Bottled helium gas, carrier sheets, rupture discs and stopping sheets are available from the same supplier.

21. The use of DsRed for visual screening will help to reduce the number of escapes.
22. The time intervals may vary for individual experiments. If embryogenic sectors are not readily visible, another round of subculture on CIS medium may be required.
23. If roots do not readily develop, transfer to MSR2 medium with 10 g sucrose (rather than 30 g) is recommended.
24. In maize plants derived from tissue culture, male and female flowers may not always mature simultaneously, depending on the growth conditions. It is therefore advisable to grow parallel batches of non-transgenic plants as pollen donors.
25. Cv. HiII has been selected for optimal tissue culture performance, but plants show poor agricultural characteristics such as yield and pathogen resistance. Selected transgenic lines must therefore be crossed into elite germplasm. It has also been demonstrated that the content of recombinant protein in seeds of a given transgenic line can be increased over several plant generations by continuous selection of individuals showing high transgene expression *(32)*.
26. Seeds of transgenic lines can be stored for years, ideally at low temperatures and humidity. However, their germination capacity will decrease, and seed stocks must be replenished in regular intervals.
27. The optimal age of the endosperm is a stage at mid-maturation. Younger seeds tend to be too soft for proper handling, while in older seeds the intracellular structure is masked by abundant starch grains that disturb the shape of the pbs. It is also recommended to use one half of each seed to confirm the presence of the recombinant protein by immunoblot analysis.
28. Immerse the seed sample in the fixative and cut the tissue on a piece of dental wax. The wax will prevent squeezing and the cut surface will be immediately in contact with the fixative improving its preservation. From some seeds (e.g., tobacco seeds) the hard seed coat needs to be removed to improve the penetration of the fixative.
29. Samples can be left in 70% ethanol overnight at 4°C if desired.
30. For UV polymerization, a catalyst (benzoyl methyl ether, 0.5% v/v) must be added to the resin for polymerization under UV light. After addition of the catalyst the resin should be kept in the dark at 4°C.
31. Plant tissues, and seeds in particular, are difficult to infiltrate. Depending on the species, longer infiltration steps may be necessary. Samples can be left overnight at –20°C at any of the steps.

32. A small paper label can be embedded together with the sample for easy identification of the resulting blocks.
33. Put one drop of resin-catalyst mixture in the capsules before adding the tissue piece. Except for very small samples (e.g., tobacco seeds), it is not necessary to embed more than one piece of sample per capsule.
34. Capsules should be overfilled and tightly sealed, as the presence of oxygen disturbs the polymerization.
35. Sections are placed on a water drop and left to dry on a hot plate. That way sections will spread evenly and stick to the glass.
36. Staining with uranyl acetate should be carried out under a fume hood.
37. The optimal incubation time depends on the antibody used.
38. The optimal dilution should be empirically determined for a given antibody. Commonly used dilutions range between 1:100 and 1:500.
39. 39. Polyclonal sera tend to give stronger signals than monoclonal antibodies. The labelling intensity depends on the number of epitopes exposed on each particular section. Therefore, although monoclonal antibodies are more specific, they often result in lower signals.
40. For example, TRITC- or Fluorescein-5-isothiocyanat (FITC)-labelled antibodies are suitable for fluorescence microscopy. Nevertheless, we prefer Alexa Fluor®-labelled antibodies as they are more stable and do not fade easily. For electron microscopy, antibodies labelled with 10 nm gold are optimal for routine work. For dilution follow the manufacturer's instructions.A combination of different fluorophors (e.g., Alexa Fluor 488 and Alexa Fluor 594) or different size gold particle (10 and 15 nm gold) can be used for double labelling.
41. No mounting is needed. Sections may be observed without cover slip.

References

1. J.K.Ma, Drake, P.M., Christou, P. (2003) The production of recombinant pharmaceutical proteins in plants. *Nat. Rev. Genet.* 4, 794–805.
2. Ma, J.K., Chikwamba, R., Sparrow, P., Fischer, R., Mahoney, R., Twyman, R.M. (2005) Plant-derived pharmaceuticals – the road forward. *Trends Plant Sci.* 10, 580–585.
3. Whaley, K.J., Zeitlin, L. (2005) Preventing transmission: plant-derived microbicides and mucosal vaccines for reproductive health. *Vaccine* 23, 1819–1822.
4. Stoger, E., Sack, M., Nicholson, L., Fischer, R., Christou, P. (2005) Recent Progress in Plantibody Technology. *J. Curr. Pharm. Des.* 11, 2439–2457.

5. Stoger, E., Sack, M., Perrin, Y., Vaquero, C., Torres, E., Twyman, R.M., Christou, P., Fischer, R. (2002) Practical considerations for pharmaceutical antibody production in different crop systems. *Mol. Breed.* 9, 149–158.
6. Stoger, E., Ma, J.K.C., Fischer, R., Christou, P. (2005) Sowing the seeds of success: pharmaceutical proteins from plants. *Curr. Opin. Biotechnol.* 16, 167–173.
7. Hood, E.E., Witcher, D.R., Maddock, S., Meyer, T., Baszczynski, C., Bailey, M., Flynn, P., Register, J., Marshall, L., Bond, D. *et al.*. (1997) Commercial production of avidin from transgenic maize: characterization of transformant, production, processing, extraction and purification. *Mol. Breed.* 3, 291–306.
8. Witcher, D.R., Hood, E.E., Peterson, D., Bailey, M., Bond, D., Kusnadi, A., Evangelista, R., Nikolov, Z., Wooge, C., Mehigh, R. *et al.*. (1998) Commercial production of b-glucuronidase (GUS): a model system for the production of proteins in plants. *Mol. Breed.* 4, 301–312.
9. Chikwamba, R.K., Scott, M.P., Mejia, L.B., Mason, H.S., Wang, K. (2003) Localization of a bacterial protein in starch granules of transgenic maize kernels. *Proc. Natl. Acad. Sci. USA* 100, 11127–11132.
10. Yang, D.C., Guo, F.L., Liu, B., Huang, N., Watkins, S.C. (2003) Expression and localization of human lysozyme in the endosperm of transgenic rice. *Planta* 216, 597–603.
11. Rascon-Cruz, Q., Sinagawa-Garcia, S., Osuna-Castro, J.A., Bohorova, N., Paredes-Lopez, O. (2004) Accumulation, assembly, and digestibility of amarantin expressed in transgenic tropical maize. *Theor. Appl. Genet.* 108, 335–342.
12. Horvath, H., Huang, J., Wong, O., Kohl, E., Okita, T., Kannangara, C.G., von Wettstein, D. (2000) The production of recombinant proteins in transgenic barley grains. *Proc. Natl. Acad. Sci. USA* 97, 1914–1919.
13. Alvarez, M.L., Guelman, S., Halford, N.G., Regiardo, M., Ryabuskina, N., Stein, J., Vallejos, R.H. (2000) Silencing of HMW glutelins in transgenic wheat expressing extraHMW subunits. *Theor. Appl. Genet.* 100, 319–327.
14. Lamacchia, C., Shewry, P.R., Di Fonzo, N., Forsyth, J.L., Harris, N., Lazzeri, P.A., Napier, J.A., Halford, N.G., Barcelo, P. (2001) Endosperm-specific activity of a storage protein gene promoter in transgenic wheat seed. *J. Exp. Bot.* 52, 243–250.
15. Qu, L.Q., Takaiwa, F. (2004) Tissue specific expression and quantitative potential evaluation of seed storage component gene promoters in transgenic rice. *Plant Biotechnol. J.* 2, 113–125.
16. Zheng, Z., Kawagoe, Y., Xiao, S., Li, Z., Okita, T., Hau, T.L., Lin, A., Murai, N. (1993) 5'¢ distal and proximal cis-acting regulator elements are required for developmental control of a rice seed storage protein glutelin gene. *Plant J.* 4, 357–366.
17. Hwang, Y.S., Nichol, S., Nandi, S., Jernstedt, J.A., Huang, N. (2001) Aleurone and embryo-specific expression of the b-glucuronidase gene controlled by the barley Chi26 and Ltp1 promoters in transgenic rice. *Plant Cell Rep.* 20, 647–654.
18. Philip, R., Darnowski, D.W., Maughan, P.J., Vodkin, L.O. (2001) Processing and localization of bovine b-casein expressed in transgenic soybean seeds under control of a soybean lectin expression cassette. *Plant Sci.* 161, 323–335.
19. Perrin, Y., Vaquero, C., Gerrard, I., Sack, M., Drossard, J., Stoger, E., Christou, P., Fischer, R. (2000) Transgenic pea seeds as bioreactors for the production of a single-chain Fv fragment (scFV) antibody used in cancer diagnosis and therapy. *Mol. Breed.* 6, 345–352.
20. Saalbach, I., Giersberg, M., Conrad, U. (2001) High-level expression of a single-chain Fv fragment (scFv) antibody in transgenic pea seeds. *J. Plant Physiol.* 158, 529–533.
21. De Jaeger, G., Scheffer, S., Jacobs, A., Zambre, M., Zobell, O., Goossens, A. *et al.*. (2002) Boosting heterologous protein production in transgenic dicotyledonous seeds using Phaseolus vulgaris regulatory sequences. *Nat. Biotechnol.* 20, 1265–1268.
22. Moloney, M., Boothe, J., Van Rooijen, G. (2003) Oil bodies and associated proteins as affinity matrices. January 21, 2003; US Patent 6509453.
23. Okita, T.W., Rogers, J.C. (1996) Compartmentation of proteins in the endomembrane system of plant cells. *Annu. Rev. Plant Physiol. Plant Mol. Biol.* 47, 327–350.
24. Shewry, P.R., Halford, N.G. (2002) Cereal seed storage proteins: structure, properties and role in grain utilization. *J. Exp. Bot.* 53, 947–958.
25. Vitale, A., Hinz, G. (2005) Sorting of proteins to storage vacuoles: how many mechanisms? *Trends Plant Sci.* 10, 316–323.
26. Robinson, D.G., Oliviusson, P., Hinz, G. (2005) Protein sorting to the vacuoles of plants: a critical appraisal. *Traffic* 6, 615–625.

27. Drakakaki, G., Marcel, S., Arcalis, E., Altmann, F., Gonzalez-Melendi, P., Fischer, R., Christou, P., Stoger, E. (2006) The intracellular fate of a recombinant protein is tissue-dependent. *Plant Physiol.* 141, 578–586.
28. Arcalis, E., Marcel, S., Altmann, F., Kolarich, D., Drakakaki, G., Fischer, R., Christou, P., Stoger, E. (2004) Unexpected deposition patterns of recombinant proteins in post-ER compartments of wheat endosperm. *Plant Physiol.* 136, 3457–3466.
29. Nicholson, L., Gonzalez-Melendi, P., van Dolleweerd, C., Tuck, H., Perrin, Y., Ma, J.K.C., Fischer, R., Christou, P., Stoger, E. (2005) A recombinant multimeric immunoglobulin expressed in rice shows assembly-dependent subcellular localization in endosperm cells. *Plant Biotechnol. J.* 3, 115–127.
30. Petruccelli, S., Otegui, M.S., Lareu, F., Tran Dinh, O., Fitchette, A.C., Circosta, A., Rumbo, M., Bardor, M., Carcamo, R., Gomord, V., Beachy, R.N. (2006) A KDEL-tagged monoclonal antibody is efficiently retained in the endoplasmic reticulum in leaves, but is both partially secreted and sorted to protein storage vacuoles in seeds. *Plant Biotechnol. J.* 4, 511–527.
31. Torres, E., Gonzalez-Melendi, P., Stoger, E., Shaw, P., Twyman, R.M., Nicholson, L., Vaquero, C., Fischer, R., Christou, P., Perrin, Y. (2001) Native and artificial reticuloplasmins co-accumulate in distinct domains of the endoplasmic reticulum and in postendoplasmic reticulum compartments. *Plant Physiol.* 127, 1212–1223.
32. Hood, E.E., Bailey, M.R., Beifuss, K., Magallanes-Lundback, M., Horn, M.E., Callaway, E., Drees, C., Delaney, D.E., Clough, R., Howard, J.A. (2003) Criteria for high-level expression of a fungal laccase gene in transgenic maize. *Plant Biotechnol. J.* 1, 129–140.
33. Wright, K.E., Prior, F., Sardana, R., Altosaar, I., Dudani, A.K., Ganz, P.R., Tackaberry, E.S. (2001) Sorting of glycoprotein B from human cytomegalovirus to protein storage vesicles in seeds of transgenic tobacco. *Transgenic Res.* 10, 177–181.
34. Czihal, A., Conrad, B., Buchner, P., Brevis, R., Farouk, A.A., Manteuffel, R., Adler, K., Wobus, U., Hofemeister, J., Baumlein, H. (1999) Gene farming in plants: expression of a heat stable Bacillus amylase in transgenic legume seeds. *J. Plant Physiol.* 155, 183–189.
35. Munro, S., Pelham, H.R. (1987) A C-terminal signal prevents secretion of luminal ER proteins. *Cell* 48, 899–907.
36. Conrad, U., Fiedler, U. (1998) Compartment-specific accumulation of recombinant immunoglobulins in plant cells: an essential tool for antibody production and immunomodulation of physiological functions and pathogen activity. *Plant Mol. Biol.* 38, 101–109.
37. Ko, K., Tekoah, Y., Rudd, P.M., Harvey, D.J., Dwek, R.A., Spitsin, S. et-al.. (2003) Function and glycosylation of plant-derived antiviral monoclonal antibody. *Proc. Natl. Acad. Sci. USA* 100, 8013–8018.
38. Takaiwa, F., Takagi, H., Hirose, S., Wakasa, Y. (2007) Endosperm tissue is good production platform for artificial recombinant proteins in transgenic rice. *Plant Biotechnol. J.* 5, 84–92.
39. Van Droogenbroeck, B., Cao, J., Stadlmann, J., Altmann, F., Colanesi, S., Hillmer, S., Robinson, D.G., Van Lerberge, E., Terryn, N., Van Montagu, M., Liang, M., Depicker, A., De Jaeger, G. (2007) Aberrant localization and underglycosylation of highly accumulating single-chain Fv-Fc antibodies in transgenic Arabidopsis seeds. *Proc. Natl. Acad. Sci. USA* 104, 1430–1435.
40. Harper, B.K., Mabon, S.A., Leffel, S.M., Halfhill, M.D., Richards, H.A., Moyer, K.A., Stewart, C.N. Jr (1999) Green fluorescent protein as a marker for expression of a second gene in transgenic plants. *Nat. Biotechnol.* 17, 1125–1129.
41. Armstrong, C.L., Green, C.E. (1985) Establishment and maintenance of friable, embryogenic maize callus and the involvement of L-proline. *Planta* 164, 207–214.
42. Armstrong, C.L., Green, C.M., Phillips, R.L. (1991) Development and availability of germplasm with Type II culture formation response. *Maize Genet. Coop. Newsletter* 65, 92–93.
43. Vain, P., McMullen, M.D., Finer, J.J. (1993). Osmotic treatment enhances particle bombardment-mediated transient and stable transformation of maize. *Plant Cell Rep.* 12, 84–88.

Chapter 6

Production of Antibody Fragments in *Arabidopsis* Seeds

Bart Van Droogenbroeck, Kirsten De Wilde, and Ann Depicker

Summary

Plants offer a number of attractive benefits over conventional mammalian or bacterial cell culture systems for the production of valuable pharmaceutical and industrial proteins. Currently, antibodies and their derived fragments represent the largest and most important group of biotechnological products in clinical trials. In particular, single-chain antibodies are an interesting class of biopharmaceuticals because they are able to overcome specific problems associated with full-length antibodies. Another valuable antibody format is the scFv-Fc: fusion of the Fc domain to a single-chain variable fragment restores antibody effector functions, allows purification, and mimicks, despite being a 'single-gene' product, the bivalent properties of a full-length IgG. Although many different plant-based production platforms have been evaluated for antibody production, seeds are especially attractive because, as natural storage organs, they provide an optimal biochemical environment for the accumulation and long-term storage of large amounts of functional proteins. This chapter describes how to achieve high-level seed-specific expression of antibody fragments, how to select the best transgenic lines, and how to evaluate the accumulation level in the seed stocks from the selected lines.

Key words: Molecular farming, Biopharmaceuticals, Seed-specific expression, Antibody fragments, Single-chain variable fragment (scFv), scFv-Fc antibody.

6.1. Introduction

Over the last two decades, many proof-of-concept studies have illustrated the potential of plant-based platforms for the production of valuable proteins *(1, 2)*. Despite technological developments on different levels, including transformation methods, control of gene expression, and protein targeting, recombinant protein accumulation levels, achieved in numerous plant-based

Loïc Faye and Véronique Gomord (eds.), *Methods in Molecular Biology, Recombinant Proteins From Plants, vol. 483*

DOI: 10.1007/978-1-59745-407-0_6

platforms under development, remain too low for commercial exploitation *(3)*. Two of the most important strategies to improve the recombinant protein yield in plants are based on (i) the development of novel expression cassettes and (ii) the amelioration of protein stability and accumulation through the use of signals that target the protein to particular intracellular compartments *(4)*. The high levels at which many seed storage proteins accumulate make their regulatory sequences (including promoter, signal peptide, and 3' flanking sequences) excellent tools to achieve high accumulation levels of heterologous proteins *(5)*. In addition, production of a recombinant protein in a seed-specific manner allows exploitation of all other advantages associated with seed tissues, such as long-term storage without loss of functionality. Therefore, it is not surprising that different seed crops are among the first to be utilized as commercial plant-based production platforms *(6)*. Specific subcellular targeting can also greatly influence product yield, presumably by affecting protein stability, folding, and exposure to proteases *(7)*. This is especially true for full-length antibodies and complex antibody fragments because post-translational modifications, required to obtain a fully functional recombinant protein, are associated with specific plant cell compartments *(8)*. Simpler antibody-derived formats, such as single-chain variable (scFv) fragments, are less demanding in terms of assembly and post-translational modifications, but most scFv fragments accumulate to higher levels when targeted to the endomembrane system, with an additional yield increase when the protein is retained within the endoplasmatic reticulum (ER) through an H/KDEL C-terminal tag *(9, 10)*.

Taking this information into account, different seed-specific expression cassettes have been developed based on the regulatory signals from the seed storage proteins genes *β-phaseolin* and *arcelin 5-I* of common bean (*Phaseolus vulgaris*), in combination with a signal peptide from the *Arabidopsis thaliana* seed storage protein 2S albumin and an KDEL tag, for entry and retention in the ER, respectively. With these seed storage promoter constructs, different antibody formats have been expressed in *Arabidopsis*. Accumulation of a model scFv reached exceptionally high levels, no less than 36.5% of the total soluble protein (TSP), without loss of scFv antigen-binding activity and affinity *(11)*. To our knowledge, these levels are among the highest reported for a recombinant protein produced in stable transgenic plants. More recently, the high accumulation levels obtained with these constructs were confirmed for the more complex scFv-Fc fusion antibody format *(12)*. This scFv-Fc format is very attractive because the expression of a single synthetic 'gene' results in a format mimicking the bivalent properties of IgG molecules. Thanks to the presence of the Fc domain, antigen-binding avidity and antibody effector functions are restored when compared to a scFv and also result in a prolonged serum half-life. In addition, scFv-Fc antibodies benefit from the Fc domain as a convenient affinity handle for purification and immunochemistry *(13, 14)*.

In this chapter, we will focus on the generation of high levels of seed-specific expression of antibody fragments, the selection of the best transgenic lines, and the evaluation of the accumulation level in the seed stocks from the selected lines.

6.2. Materials

6.2.1. Bacterial Strains

The cloning and subcloning of scFv and scFv-Fc constructs is carried out in *Escherichia coli* DH5α. *Agrobacterium* C58C1Rif^R is used for transfer of the T-DNA constructs to *Arabidopsis* with the floral dip transformation method (*see* **Subheading 6.3.1.2**) *(15)*.

6.2.2. Plant Material

For the experiments, transgenic plant lines expressing scFv fragments and scFv-Fc fusion proteins and the corresponding non-transgenic *A. thaliana* (L.) Heynh., ecotype Colombia 0 are used.

6.2.3. Special Laboratory Tools and Equipment

1. Life Science UV/Visible Spectrophotometer Beckman DU 530 (Beckman Coulter, Fullerton, CA).
2. Silwet L77 (Lehle Seeds, Round Rock, TX).
3. ARACON bases and sheets (Lehle Seeds).
4. Saran Wrap (Dow).
5. Regular potting soil.
6. Liquid nitrogen.
7. Mortar and pestle.
8. Miracloth Calbiochem (Merck).
9. Grinding device MM200 (Retsch, Haan, Germany).
10. Centrifuge 5417R (Eppendorf).
11. Speed vac device Eppendorf Concentrator 5301 (Eppendorf).
12. Microfuge tubes of 1.5 and 2.0 mL.
13. 96-well microtiter plates.
14. Petri dishes.
15. VERSAmax tunable microplate reader (Molecular Devices, Sunnyvale, CA).

16. Softmax® Pro Software 3.0 (Molecular Devices).
17. Mini-Protean II™ Electrophoresis Unit for protein gel electrophoresis (Bio-Rad).
18. Cellophane.
19. ImageMaster® VDS Densitometer (Amersham Pharmacia Biotech, now part of GE Healthcare, Little Chalfont, UK).
20. ImageMaster VDS Software 3.0 (GE Healthcare).

6.2.4. Media, Solvent, Buffers, and Other Solutions

All buffers and solutions are made as aqueous solutions. Solutions for widely used procedures, such as sodium dodecyl sulphate-polyacrylamide gel electrophoresis (SDS-PAGE), are composed according to the established protocols *(16)*.

1. Luria Bertani (LB) medium: 10 g of tryptone, 5 g of yeast extract, 5 g of sodium chloride per litre.
2. K1 selective medium: 2.15 g/L half-strength Murashige and Skoog salts, 0.5 g/L 2-(*N*-morpholino)ethanesulfonic acid (MES) supplemented with 10 g/L sucrose. Bring this solution to pH 5.7 and complete with 7 g/L agar, 50 mg/L kanamycin, 50 mg/L nystatin, and 500 mg/L carbenicillin.
3. Seed extraction buffer: 50 m*M* Tris-HCl, pH 8.0, 200 m*M* NaCl, 5 m*M* EDTA, 0.1% Tween 20 (v/v). Keep at 4°C until use. Dissolve 1 tablet of Complete® protease inhibitor in 0.5 mL double-distilled water (ddH_2O). Just before use, add 20 μL of this protease inhibitor solution to every millilitre of extraction buffer that will be used (*see* **Note 1**).
4. Complete® protease inhibitor (Roche Diagnostics, Vilvoorde, Belgium).
5. Bovine serum albumin (BSA) (Sigma).
6. Reagents A and B (Bio-Rad).
7. Electrophoresis running buffer (1×): 25 m*M* Tris, 192 m*M* glycine, 0.1% (w/v) SDS, pH 8.3.
8. Protein sample buffer (5×): 5 mL of 0.5 M Tris-HCl, pH 6.8, 4 mL 100% glycerol, 0.5 g SDS, 5% (w/v) dithiothreitol (DTT), dissolved in 10 mL ddH_2O. Add 0.01% (w/v) bromophenol blue. Store at –20°C.
9. Protein gel staining solution: 0.1% (w/v) Coomassie® Brilliant Blue R-250, 40% (v/v) methanol, 10% (v/v) HOAc.
10. Protein gel destaining solution: 40% (v/v) methanol, 10% (v/v) HOAc.
11. Standard reagents for SDS-PAGE.

6.3. Methods

6.3.1. Generation of Transgenic Lines Expressing an Antibody Fragment

6.3.1.1. Assembly of a Seed-Specific Expression Cassette

To generate transgenic *Arabidopsis* plants with high-level accumulation of a recombinant protein of interest, a seed-specific expression cassette has to be created. In our cloning strategy, this requires the combination of fragments derived from different plasmids, resulting in a binary T-DNA vector of over 17 kb (*see* **Fig. 6.1**; *11*). Because of the size of both vector and the 3' end of the *arcelin 5-I* gene, a complex multistep cloning procedure is required. Since this falls outside the scope of the chapter, this is not described in detail. Briefly, to direct the antibody fragment of interest to the ER and keep it there (*see* **Subheading 6.1**), a signal peptide has to be added at the amino-terminal and the tetrapeptide KDEL at the carboxy-terminal ends. Finally, a specific tag, such as a *c-myc* tag *(17)*, can be included for detection. A 21-amino-acid-long signal peptide is used that is derived from the *Arabidopsis* seed storage protein 2S2 albumin *(18)*.

After sequence confirmation of the construct in *E. coli DH5α*, this binary vector is transferred by electroporation to an *Agrobacterium* strain *(19)*. Below, we describe in detail (i) the generation of *Arabidopsis* transformants with the floral dip method, and (ii) the selection of the primary transformants.

6.3.1.2. Plant Transformation

Transgenic *Arabidopsis* plants are obtained by *Agrobacterium*-mediated transformation with the simple floral dip method *(15)*.

1. Bring 1 mL of LB medium in a 50-mL Falcon tube; inoculate with the *Agrobacterium* strain that contains the binary vector. Grow for 8–9 h at 28°C in a rotary shaker at 230 rpm.
2. Add 10 mL of LB medium and incubate overnight at 28°C in a rotary shaker at 230 rpm.

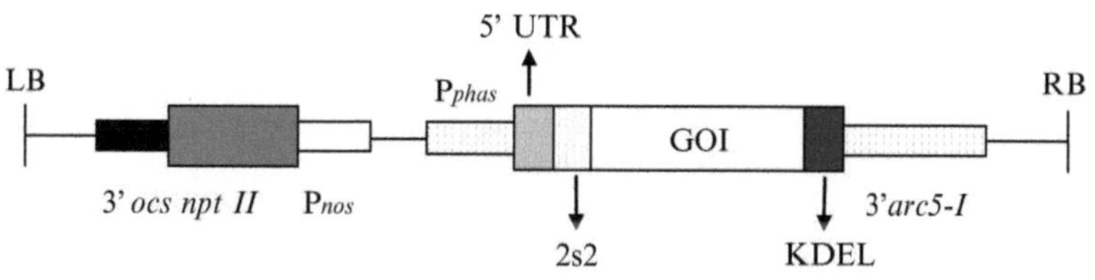

Fig. 6.1. Schematic outline of the T-DNA region of the binary seed-specific vector pP*phas* (not to scale). For a more detailed explanation, *see* **Subheading 6.3.1.1**. Selection marker consisting of the P*nos*, nopaline synthase gene promoter; *npt* II, coding sequence of the neomycin phosphotransferase II gene; 3′ *ocs*, 3′ end of the octopine synthase gene. Seed-specific expression cassette created from P*phas*, β-phaseolin gene promoter (–1 to –1470; GenBank accession no. J01263); 5′*UTR*, 5′ untranslated region of *arc5-I* gene (13 bp; part of GenBank accession no. Z5020); 3′ *arc5-I*, 3′-flanking regulatory sequences of the *arc5-I* genomic clone (4001 bp; part of GenBank accession no. Z50202); 2S2 SP, signal peptide of the *Arabidopsis* 2S2 albumin seed storage protein gene; GOI, sequence of the gene of interest encoding a single-chain variable fragment (*scFv*) or *scFv-Fc* antibody fragment; c-myc tag; C-terminal myc tag; KDEL, endoplasmic reticulum (*ER*) retention signal; RB and LB, T-DNA right and left border, respectively.

3. The next morning, prepare a 1/10 dilution of the sample, and check the optical density (OD) of the suspension culture. Therefore, put 900 µL of LB medium in a cuvette and add 100 µL of the culture (*see* **Note 2**). Measure at a wavelength of 600 nm. As a control sample, use 1 mL of LB. The OD value must be somewhere in-between 1.5 and 2.0; if the value is lower, further incubate the culture at 28°C and measure each 30 min.
4. Mix 20 µL of a 0.05% (v/v) Silwet and 40ml of ddH_2O containing a 10% (w/v) sucrose solution. Add this mixture to the 10-mL culture (*see* **Note 3**).
5. Gently shake the culture and dip the plant in the suspension for 10 s (*see* **Note 4**). Typically five plants (the so-called T_0 plants) per construct are dipped to obtain a sufficiently large number of independent transformants.
6. Return the plants to normal growing conditions in the greenhouse, that is, 21°C on a 16-h light/8-h dark cycle, with 55% relative humidity.
7. Further grow the plants until most of the siliques are dry, stop watering the plants to hasten maturation.
8. Harvest the T_1 seeds when siliques are completely dried, which is normally around 8 weeks after floral dip. Store the seeds at room temperature.

6.3.1.3. Selection of Transgenic Lines

To select the primary transformants, the T_1 seeds harvested on the dipped plants (T_0 plants) have to be sown on a selective medium, containing the appropriate antibiotic. For demonstration purposes, kanamycin is used as a selective antibiotic, so that the transgenic lines containing the T-DNA described in **Fig. 6.1** can be selected.

1. Pack 1,000 seeds (~25 mg) of every dipped plant in Miracloth filter paper and put in a –70°C freezer for 2 days (*see* **Note 5**). Surface-sterilize seeds and allow them to germinate on selective K1 medium *(20)*.
2. Incubate plates in a growth chamber under the following conditions: 21°C on a 16-h light/8-h dark cycle. Transformed plants are easy to identify after 10–14 days of growth on the selective medium.
3. Transfer T_1 plants that are large enough after 3–4 weeks to soil. Allow them first to grow in a growth chamber and transfer them, when growing well, after a few days to the greenhouse

under exactly the same conditions, that is, 21°C on a 16-h light/8-h dark cycle and with 55% relative humidity.

4. Harvest T_2 segregating seeds, after self-pollination, on the T_1 plants as described above (*see* **Subheading 6.3.1.2, steps 7–8**).

6.3.2 Selection of Single-Locus Lines and Protein Extraction

To be able to generate genetically stable homozygous transgenic lines in the T_3 generation and beyond *(21)*, we first identify the single-locus T-DNA insertion lines in the T_2 population by segregation analysis.

6.3.2.1. Segregation Analysis

1. For the segregation analysis, sow 64 seeds of each T_2 segregating seed stock on selective MS medium as described above (*see* **Subheading 6.3.1.3, steps 1, 2**).
2. After 3 to 4 weeks, count the resistant (green) and sensitive (only pale cotyledons) seedlings. Use the ratio of these values, together with the expected values for a 3:1 single-locus segregation in the χ^2 statistical test to verify the 3:1 segregation ratio *(22)*.

6.3.2.2. Protein Extraction

Depending on the amount of seed samples to be extracted and the availability of a grinding device, such as a Retsch MM200, seeds can be ground manually (*see* **step 1a**) or mechanically (*see* **step 1b**). Both grinding procedures are described, although the mechanical grinding procedure is, of course, less labour intensive and time saving when a large amount of samples needs to be processed. A comparison of the protein yield obtained with both methods is illustrated in **Fig. 6.2.**

1a. For manual grinding, weigh 10 mg of *Arabidopsis* seeds in an 1.5-mL microfuge tube. Keep the tube on ice while crushing the seeds with a file.

1b. For mechanical grinding, weigh 10 mg of *Arabidopsis* seeds in an 2.0-mL microfuge tube. Add two metal balls with a diameter of 4 mm and cool the microfuge tubes in liquid nitrogen, as well as the holders. Place the filled holders in the Retsch MM 200 device and shake for 2 min at a mill frequency of 30/s (*see* **Note 6**).

2. Add 1 mL of hexane, vortex, and centrifuge at maximal speed (±20,000*g*) for 5 min at room temperature. Discard the supernatant.
3. Repeat **step 2**.
4. Put the samples in a speed vac device for 10 min or longer at room temperature, until they are completely dry.
5. Add 600 µL of extraction buffer, mix, and vortex for 1 min.
6. Centrifuge again at maximal speed (±20,000*g*) for 5 min at room temperature.

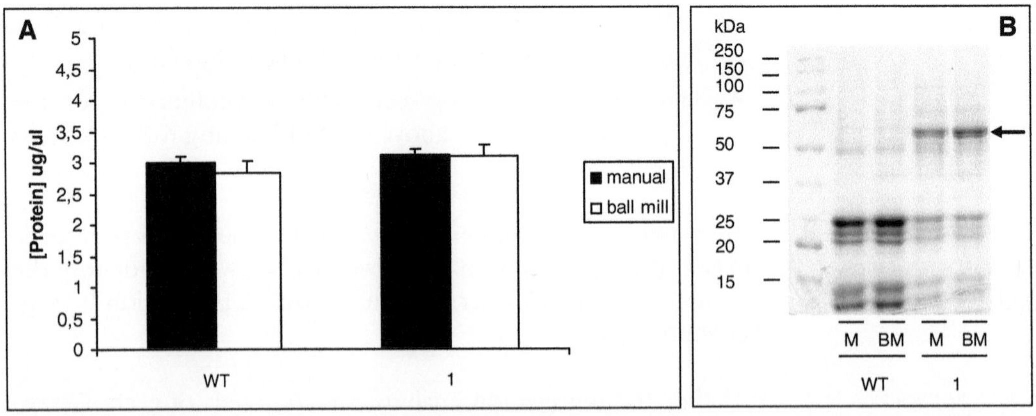

Fig. 6.2. Comparison of a manual and an automated mechanical seed grinding procedure. (**a**) No difference in the protein yield obtained with the two different methods. Each bar is the mean of three values ±SE. (**b**) Coomassie-stained, reducing sodium dodecyl sulphate-polyacrylamide gel electrophoresis (*SDS-PAGE*) gel on which 40 µg of total seed protein extract was loaded. In the transgenic line, fewer seed storage proteins (at the bottom of the gel) are present in the total seed protein extract because a large part of the biosynthetic capacity in these transgenic seed stocks is used to produce the recombinant protein. The arrow marks the height at which the doublet of single-chain variable fragment (*scFv-Fc*) monomer bands are visible in the seed extracts of the transgenic lines (the doublet of recombinant protein bands is due to differential glycosylation) (*12*). WT, wild-type non-transgenic *Arabidopsis* ecotype Col0; 1, transgenic *Arabidopsis* line expressing a model *scFv-Fc* fusion antibody (expected molecular mass ~55 kDa). M, manual grinding; BM, mechanical grinding with a Retsch ball mill device.

7. Collect 450 µL of supernatant in a new 1.5-mL microfuge and keep on ice.
8. Add another 400 µL of extraction buffer, mix to the pellet, and vortex for 1 min.
9. Centrifuge again at maximal speed (±20,000*g*) for 5 min at room temperature.
10. Take another 250 µL of the supernatant and add to the former 450 µL. Discard the pellet.
11. Add 150 µL of glycerol (100%) to each protein extract and store at –20°C.

6.3.3. Determination of the Protein Concentration in the Total Seed Extracts

To determine the protein concentration in *Arabidopsis* seed extracts, the Bio-Rad Dc Protein assay is used, which is based on the method of Lowry (*see* **Note** 7; *23*).

1. To carry out the protocol, follow the manufacturer's instructions. To determine the protein concentration in the samples, use BSA as standard. Therefore, prepare in duplicate four dilutions of a BSA protein standard: 1.6 mg/mL, 0.8 mg/mL, 0.4 mg/mL, 0.2 mg/mL of BSA in ddH_2O. Include ddH_2O in the assay as a 'blanco' sample in duplicate.
2. Prepare four dilutions of each protein extract, the dilution factor of the protein extract depending on the protein yield.

Use the following dilution series of every seed extract to obtain at least three measurements within the linear detection range: 115×, 230×, 460×, and 920×.

3. Pipette 5 μL of BSA standards, the blanco, and the dilutions of the samples into a clean, dry microtiter plate.
4. Add 25 μL of Reagent A and subsequently 200 μL of Reagent B into each well. Shake the plate for 5 s.
5. After 15 min, read absorbances that would be stable for ~1 h in a spectrophotometer at 750 nm. With the Softmax Pro software, the protein concentrations are calculated automatically.

6.3.4. Evaluation of the Accumulation Level of the Antibody Fragments in the Seed Stocks of the Selected Transgenic Lines

The variation in transgene expression in the lines obtained after transformation with the described seed promoter constructs is generally low *(5, 11)*. High-accumulating lines can normally be identified after analysis of the total seed protein extracts from a group as small as 20 independent transformants (*see* **Fig. 6.3A**) *(11, 12)*. In addition, protein gel blot analysis is not necessary to verify the accumulation levels because high accumulation can be obtained with these seed promoter constructs. Therefore, we will describe only how to evaluate the accumulation level of antibody fragments in the seed stocks of the selected single-locus transgenic lines by SDS-PAGE analysis and how to quantify them with Coomassie-stained SDS-PAGE gels (*see* **Fig. 6.3**).

1. After standard SDS-PAGE *(13)* with the Bio-Rad Mini-Protean II Electrophoresis Unit™, load the samples onto a 0.75-mm thick gel and allow them to migrate at 180 V until the bromophenol blue tracking dye reaches the bottom of the gel. For quantification, use a number of lanes to load different quantified amounts of the same, purified, recombinant protein (*see* **Fig. 6.3B**).
2. Stain the gel with the protein gel staining solution for 30 min in a Petri dish, under shaking.
3. Decolorize with the protein gel destaining solution for 1 h 30 min in a Petri dish, under shaking.
4. Incubate the gels in 10% glycerol to avoid shrinking and breaking (*see* **Step 6**).
5. Take a digital picture of the gels for quantification with a densitometre. Define lane and standard with the Softmax Pro software and quantify the selected bands according to the software manufacturer's instruction (*see* **Note 8**).
6. Place the Coomassie-stained SDS-PAGE gels between two wetted cellophane sheets at room temperature until they are completely dry.

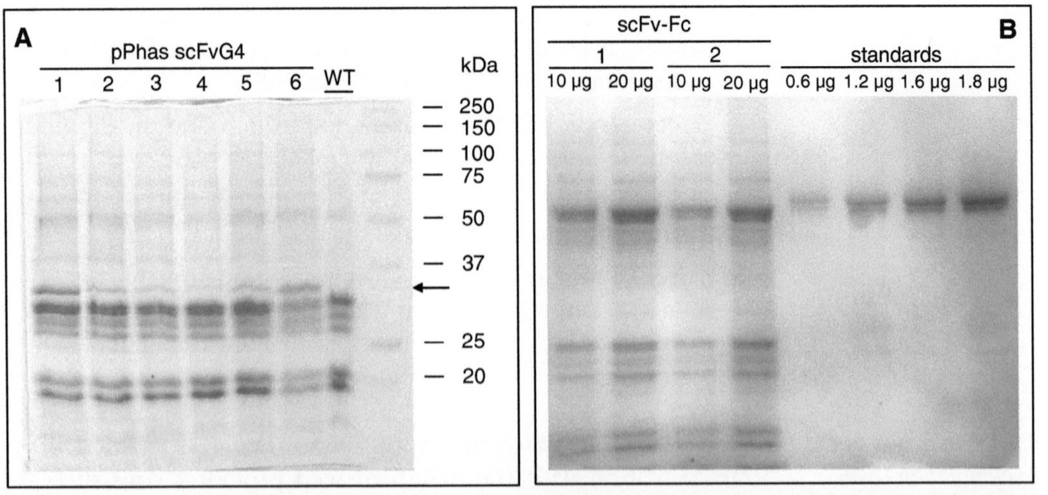

Fig. 6.3. Evaluation of the accumulation level of two different antibody fragments in total seed protein extracts of *Arabidopsis*. (**a**) Coomassie-stained, reducing sodium dodecyl sulphate-polyacrylamide gel electrophoresis (*SDS-PAGE*) gel illustrating the accumulation of the scFvG4 protein (expected molecular mass ~36 kDa). Forty micrograms of total seed protein extract from both wild-type Col 0 (WT) and six different pPhas-scFvG4 transgenic lines was loaded (1–6). The arrow indicates the height at which the single-chain variable fragment (*scFv*) bands are visible in the seed extracts of the transgenic lines. (**b**) Coomassie-stained, reducing *SDS-PAGE* gel used to estimate *scFv-Fc* accumulation levels in two different transgenic seed stocks (1 and 2), of which 10 and 20 µg of total seed protein extract was loaded. As standards, quantified amounts were used of protA purified *scFv-Fc* produced in *Pichia pastoris* (*12*). Based on this gel, the *scFv-Fc* accumulation level in lines 1 and 2 are estimated to be ~14% of total soluble protein (*TSP*) and 10% of *TSP*, respectively.

7. Estimate the accumulation visually on these dried gels by comparing band intensity of the recombinant proteins present in the total seed extracts from the transgenic lines, with those of the standards loaded on the same gel (*see* **Step 1** and **Fig. 6.3b**).

6.4. Notes

1. Although the dissolved protease inhibitor can be stored at –20°C, prepare the mix of extraction buffer and protease inhibitor solution only for the samples that will be extracted at the moment, because non-used mix cannot be stored.
2. Mix your samples well before measurement to be sure that the cells are completely resuspended.
3. The surfactant must be mixed thoroughly into the solution after addition to the bacterial suspension. The mix should be used immediately for the floral dip.

4. Only use plants with a flowering stem of ~10–15 cm and with immature flowers and flower buds (±7–8 weeks after sowing). At this stage, plants normally have many unopened flower buds. The primary inflorescence can also be cut off to encourage a flush of secondary inflorescences. These plants are ready again to be used for floral dip as soon as they have developed a high number of new immature flower bud clusters.
5. Given that the typical transformation rates obtained with the floral dip method normally fall in the range of 0.5–3.0%, sowing 1000 seeds theoretically results in 5–30 transformants *(20)*. In our hands, more than 20 independent transformants were mostly obtained when ~1,000 seeds were placed on selective medium for germination. The incubation step at −70°C, which can be omitted, was introduced into the protocol to reduce the survival of thrips eggs harvested together with the seeds during a severe infection. Finally, nystatin and carbenicillin in the Kl medium were added to avoid fungal infection and to kill *Agrobacterium*, respectively.
6. Different holders for the Retsch mill devices are commercially available that allow processing of twice five 1.5-mL microfuge tubes or crushing of twice 96 seed samples. Special tubes and metal balls with a smaller diameter have to be used with the 96-tube holders.
7. Recombinant protein accumulation levels are determined as the percentage of total extractable/soluble seed protein. The three most used methods to measure total protein concentrations are UV-A_{280}, the Lowry method *(23)*, and the Bradford method *(24)*. As these methods give substantially different values or widely varying results for the same protein, proteins are extracted on a large scale from *Arabidopsis* seeds *(5)*. After salt removal and subsequent lyophilization of the extracts, total extracted protein is measured by weighing. Comparison of data with those obtained by more convenient quantification methods indicated that results obtained with the Lowry method are the most reliable to estimate the protein concentration in total seed protein extract *(5)*. Note that the standard curve should be prepared for each assay and for optimal results, in the same buffer as the sample, that is, in extraction buffer mixes containing 20% glycerol.
8. It is very important to take the time needed to define lane widths, identification of the bands, etc. with a densitometer. To increase confidence in the obtained results, different amounts of total seed protein extract of the same transgenic seed stocks can be loaded and used to estimate accumulation levels. In addition, quantification can be repeated by different people, both for the visual and densitometer-based method.

Acknowledgments

The authors thank Sylvie De Buck for critical reading of the manuscript and Martine De Cock for help preparing it. This work was supported by grants from the Pharma–Planta Consortium, which is funded by the European Union through the Framework 6 research program. K.D.W was indebted to the 'Bijzonder Onderzoeksfonds' of the Ghent University for a predoctoral fellowship.

References

1. Ma, J.K.C., Barros, E., Bock, R., Christou, P., Dale, P.J., Dix, P.J., Fischer, R., Irwin, J., Mahoney, R., Pezzotti, M., Schillberg, S., Sparrow, P., Stoger, E., and Twyman, R.M. (2005) Molecular farming for new drugs and vaccines. *EMBO Rep.* **6**, 593–599.
2. Ma, J.K.C., Chikwamba, R., Sparrow, P., Fischer, R., Mahoney, R., and Twyman, R.M. (2005) Plant-derived pharmaceuticals—the road forward. *Trends Plant Sci.* **10**, 580–585.
3. Twyman, R.M., Stoger, E., Schillberg, S., Christou, P., and Fischer, R. (2003) Molecular farming in plants: host systems and expression technology. *Trends Biotechnol.* **21**, 570–578.
4. Menkhaus, T.J., Bai, Y., Zhang, C., Nikolov, Z.L., and Glatz, C.E. (2004) Considerations for the recovery of recombinant proteins from plants. *Biotechnol. Prog.* **20**, 1001–1014.
5. Goossens, A., Dillen, W., De Clercq, J., Van Montagu, M., and Angenon, G. (1999) The *arcelin-5* gene of *Phaseolus vulgaris* directs high seed-specific expression in transgenic *Phaseolus acutifolius* and *Arabidopsis* plants. *Plant Physiol.* **120**, 1095–1104.
6. Stoger, E., Ma, J.K.C., Fischer, R., and Christou, P. (2005) Sowing the seeds of success: pharmaceutical proteins from plants. *Curr. Opin. Biotechnol.* **16**, 167–173.
7. Stoger, E., Sack, M., Nicholson, L., Fischer, R., and Christou, P. (2005) Recent progress in plantibody technology. *Curr. Pharm. Des.* **11**, 2439–2457.
8. Gomord, V., Sourrouille, C., Fitchette, A.C., Bardor, M., Pagny, S., Lerouge, P., and Faye, L. (2004) Production and glycosylation of plant-made pharmaceuticals: the antibodies as a challenge. *Plant Biotechnol. J.* **2**, 83–100.
9. Schillberg, S., Emans, N., and Fischer, R. (2002) Antibody molecular farming in plants and plant cells. *Phytochem. Rev.* **1**, 45–54.
10. Schouten, A., Roosien, J., van Engelen, F.A., de Jong, G.A., Borst-Vrenssen, A.W., Zilverentant, J.F., Bosch, D., Stiekema, W.J., Gommers, F.J., Schots, A., and Bakker, J. (1998) The C-terminal KDEL sequence increases the expression level of a single-chain antibody designed to be targeted to both the cytosol and the secretory pathway in transgenic tobacco. *Plant Mol. Biol.* **30**, 781–793.
11. De Jaeger, G., Scheffer, S., Jacobs, A., Zambre, M., Zobel, O., Goossens, A., Depicker, A., and Angenon, G. (2002) Boosting heterologous protein production in transgenic dicotyledonous seeds using *Phaseolus vulgaris* regulatory sequences. *Nat. Biotechnol.* **20**, 1265–1268.
12. Van Droogenbroeck, B., Cao, J., Stadlmann, J., Altmann, F., Colanesi, S., Hillmer, S., Robinson, D.G., Van Lerberge, E., Terryn, N., Van Montagu, M., Liang, M., Depicker, A., and De Jaeger, G. (2007) Aberrant localization and underglycosylation of highly accumulating single-chain Fv-Fc antibodies in transgenic *Arabidopsis* seeds. *Proc. Natl. Acad. Sci. USA* **104**, 1430–1435.
13. Shu, L., Qi, C.-F., Schlom, J., and Kashmiri, S.V.S. (1993) Secretion of a single-gene-encoded immunoglobulin from myeloma cells. *Proc. Natl. Acad. Sci. USA* **90**, 7995–7999.
14. Powers, D.B., Amersdorfer, P., Poul, M.-A., Nielsen, U.B., Shalaby, M.R., Adams, G.P., Weiner, L.M., and Marks, J.D. (2001) Expression of single-chain Fv-Fc fusions in *Pichia pastoris*. *J. Immunol. Methods* **251**, 123–135.
15. Clough, S.J., and Bent, A.F. (1998) Floral dip: a simplified method for *Agrobacterium*-

mediated transformation of *Arabidopsis thaliana*. *Plant J.* **16**, 735–743.

16. Smith, B.J. (1984) SDS polyacrylamide gel electrophoresis of proteins. in *Methods in Molecular Biology*, vol. 1: *Proteins* (Walker, J.M., ed.). Humana Press, Clifton, NJ, pp. 165–178.
17. Terpe, K. (2003) Overview of tag protein fusions: from molecular and biochemical fundamentals to commercial systems. *Appl. Microbiol. Biotechnol.* **60**, 523–533.
18. Krebbers, E., Herdies, L., De Clercq, A., Seurinck, J., Leemans, J., Van Damme, J., Segura, M., Gheysen, G., Van Montagu, M., and Vandekerckhove, J. (1988) Determination of the processing sites of an Arabidopsis 2S albumin and characterization of the complete gene family. *Plant Physiol.* **87**, 859–866.
19. Koncz, C.L, and Schell, J. (1986) The promoter of T_L-DNA gene *5* controls the tissue-specific expression of chimeric genes carried by a novel type of *Agrobacterium* binary vector. *Mol. Gen. Genet.* **204**, 383–396.
20. Clough, S.J. (2004) Plant transformation—*Agrobacterium*-mediated germ line transformation. in *Methods in Molecular Biology*, vol. 286: *Transgenic plants: Methods and Protocols* (Peña, L., ed.). Humana Press, Clifton, NY, pp. 91–101.
21. De Wilde, C., Van Houdt, H., De Buck, S., Angenon, G., De Jaeger, G., and Depicker, A. (2000) Plants as bioreactors for protein production: avoiding the problem of transgene silencing. *Plant Mol. Biol.* **43**, 347–359.
22. Chernoff, H., and Lehmann, E.L. (1954) The use of maximum likelihood estimates in $\chi 2$ tests for goodness-of-fit. *Ann. Math. Stat.* **25**, 579–586.
23. Lowry, O.H., Rosebrough, N.J., Farr, A.L., and Randall, R.J. (1951) Protein measurement with the Folin phenol reagents. *J. Biol. Chem.* **193**, 265–275.
24. Bradford, M.M. (1976) A rapid and sensitive method for the quantitation of microgram quantities of protein utilizing the principle of protein-dye binding. *Anal. Biochem.* **72**, 248–254.

Chapter 7

Production of Plantibodies in *Nicotiana* Plants

Marta Ayala, Jorge Gavilondo, Meilyn Rodríguez, Alejandro Fuentes, Gil Enríquez, Lincidio Pérez, José Cremata, and Merardo Pujol

Summary

Because of the wide use and high demand in medicine, monoclonal antibodies are among the main recombinant pharmaceuticals at present, although present limitations of the productive platforms for monoclonal antibodies are driving the improvement of the large-scale technologies and the development of alternative expression systems. This has drawn the attention on plants as expression system for monoclonal antibodies and related derivatives, owning the capacity of plants to properly express and process eukaryotic proteins with biological activity resembling that of the natural proteins. In this chapter, the procedures from the isolation of the monoclonal antibody genes to the biochemical and biological characterization of the plant-expressed monoclonal antibody are described.

Key words: Monoclonal antibody, Recombinant, Plant expression, *Nicotiana*, Transgenic.

7.1. Introduction

Biologicals are playing a significant role among pharmaceuticals, accounting at present for more than half of discovery-stage drug candidates *(1)*. Regarding monoclonal antibodies in particular, three of them were among the top ten biotechnology drug sellers in 2006 *(2)*. However, monoclonal antibody-based immunotherapy requires grams amounts of the compound per patient, and the worldwide demand is calculated in hundred of kilograms. Annual demands for at least five antibodies are greater than 100 kg *(3)*. This poses a great challenge for present large-scale exploited production technologies for monoclonal antibodies *(4)*, pushing for the expansion of capacities to triplicate the presently available,

Loïc Faye and Véronique Gomord (eds.), *Methods in Molecular Biology, Recombinant Proteins From Plants, vol. 483*

DOI: 10.1007/978-1-59745-407-0_7

although the expansion process should take several years until certification of the new facilities, creating uncertainty on the short-term possibilities to properly respond to demands increasing.

In this situation, besides being among the first pharmaceuticals proteins expressed in plants by the end of the eighties, monoclonal antibodies and related derivatives has continued the center of attention of the academic and industrial sector for their potential expression in plants *(5)*, due to the capacity of plants to perform eukaryotic posttranslational modifications (i.e., glycosylation, folding, and multimeric assembly), as well as the low risk of transmission of human pathogens and endotoxins *(6)*.

This chapter describes the procedures from the isolation of the genes corresponding to the chains of monoclonal antibodies, the transient and stable expression in plant cells and transgenic plants, as well as the biochemical and biological characterization of the plant-expressed monoclonal antibodies.

7.2. Materials

7.2.1. Isolation of Heavy and Light Chain Immunoglobulin Regions Using PCR

- 10^6 hybridoma or transfectoma cells that produce the desired antibody. In the latter case, if the antibody plasmid is available, skip the two first procedures and move directly to polymerase chain reaction (PCR) using the plasmid as template for genes isolation.

7.2.1.1. RNA Extraction

- RNA extraction reagent (e.g., TriPure™; Boehringer-Mannheim, Mannheim, Germany).
- Diethyl pyrocarbonate (DEPC)-treated water. Add 0.1 mL of DEPC to 100 mL. Shake vigorously to dissolve DEPC. Incubate the solution at 37°C for at least 12 h and then autoclave for 15 min, to inactive remaining DEPC.
- Isopropanol.
- 75% ethanol prepared in DEPC-treated distilled water.
- Chloroform (free of all additives such as isoamyl alcohol).
- Microcentrifuge.
- Table top centrifuge for cultured cells.
- UV spectrophotometer.

7.2.1.2. cDNA Preparation

- ProSTAR™ First-Strand RT-PCR Kit (Stratagene, La Jolla, CA). For cDNA production the kit contains: oligo (dT) primer, 0.1 μg/μL; MMLV-RT reverse transcriptase, 50 U/ μL; 10× Reverse transcriptase buffer; 40 U/μL RNasin; 100 mM solutions of dATP, dCTP, dGTP, dTTP; DEPC-treated water.
- Total RNA as template.

- Heating block.
- Microcentrifuge.

7.2.1.3. PCR Amplification of Immunoglobulin Heavy and Light Chains

- ProSTAR First-Strand RT-PCR Kit (Stratagene). For PCR, the kit contains: Taq polymerase (5 U/μL); 10× Taq DNA polymerase buffer, supplemented with $MgCl_2$; 100 mM solutions of dATP, dCTP, dGTP, dTTP (25 mM each).
- DNA plasmid, or cDNA, with the template of interest.
- Stock solution of 5′ and 3′ PCR oligonucleotide primers at 20 μM. Oligonucleotide primers are custom-synthesized by different companies (*see* **Notes 1–6**).
- Sterile 0.5 ml microcentrifuge PCR tubes.
- Mineral oil.
- Sterile distilled water.
- Thermo-cycling equipment.
- Agarose.
- DNA electrophoresis casting and running apparatus.
- QIAquick Gel Extraction Kit (QIAgen GmbH, Hilden, Germany).

7.2.1.4. Cloning of Heavy and Light Immunoglobulin Chains into a Constitutive Plant Expression Vector

- Wizard® Plus Minipreps DNA purification Systems (Promega, Madison, WI).
- Wizard DNA Clean-Up System (Promega).
- Plasmid designed (named Intermediary vector) (pHES74).
- Heavy and light chains (Products of PCR) (CHKDEL and CLKDEL).
- Restriction Endonucleases and buffers; (MulTI-CORE buffer, NcoI and EcoRI enzymes).
- T4 DNA Ligase enzyme and buffer.
- Competent cells of *Escherichia coli*.
- Agarose.
- Sterile dH_2O.
- LB medium.
- Selection.antibiotic.

7.2.2. Preparation of Binary Vectors Containing Expression Cassettes for Monoclonal Antibodies

- Wizard Plus Minipreps DNA purification Systems (Promega).
- Wizard DNA Clean-Up System (Promega).
- Plasmids with the heavy and light chains expression cassettes (pHES-CHKDEL and pHES-CLKDEL plasmids).
- Binary vector (pCAMBIA2300).
- Restriction endonucleases and buffers (E and H buffer, PstI and HindIII enzymes).

- T4 DNA ligase enzyme and buffer.
- Competent cells of *E. coli* and *Agrobacterium*.
- Agarose.
- Sterile dH_2O.
- LB medium.
- Selection antibiotic.

7.2.3. Assessing Vector Viability by Transient Expression

7.2.3.1. NT-1 Cell Suspension

NT-1L medium (shake-flask liquid medium): Prepare a Murashige and Skoog (MS) medium *(7)*. The MS medium can be purchased from Sigma-Aldrich, (Saint Louis, MO), Gibco BRL, (Carlsbad, CA), and Duchefa (Holland); (*see* **Note** 7). Add 1 mL of 2,4-dichlorophenoxyacetic acid solution (2,4 D 2 mg/mL dissolved in ethanol); adjust pH to 5.8, and autoclave for 20 min at 120°C, under 1 kgf/cm^2.

NT-1S medium (for Petri dishes solid medium): same as the NT-1L medium containing Phytagel (Sigma-Aldrich, catalogue number P8169) 2 g/L for solidification (Phytagel must be add before autoclaving the medium). Autoclave during 20 min at 120°C, under 1 kgf/cm^2.

7.2.4. Agrobacterium Tumefaciens

LB solid medium (for Petri dishes bacterial colony growth): dissolve in 950 mL Tryptone (Oxoid, England) 10 g, NaCL 10 g, Yeast Extract (Oxoid, England) 5 g. Adjust pH to 7, and add Agar Technical #3 (Oxoid, England) 12 g. Autoclave for 20 min at 120°C, under 1 kgf/cm^2.

YEB liquid medium (shake-flask liquid medium): dissolve in 95 mL "Lab-Lemco" Powder (Oxoid, England) 0.4 g, sucrose 0.5 g, Bacteriological Peptone (Oxoid, England) 0.5 g, Yeast Extract (Oxoid) 0.1 g, $MgSO_4 \cdot 7H_2O$ 0.05 g. Adjust pH to 7.2 with NaOH before autoclaving. Autoclave during 20 min at 120°C, under 1 kgf/cm^2.

MSLSM medium (medium for *Agrobacterium* NT-1 inoculation): medium containing MS salts, MS vitamins, and sucrose 20 g/L (see above the NT-1L medium) and adjusted to pH 5.5.

MSLM solution (solution for *Agrobacterium* leave infiltration): solution containing MS salts, MS vitamins, and adjusted to pH 5.5.

7.2.5. Agrobacterium-mediated Stable Transformation of Nicotiana

- 2.5% sodium hypochlorite (commercial grade bleaching)
- Tween 20
- Sterile double distillated water
- Rifampicin
- Cefotaxim (or Claforan)
- Kanamycin
- Carbenicillin

- Phosphinothricin
- Acetosyringone
- BAP
- Phytoagar
- MS salts and vitamins
- Sterile Petri dishes
- Scalpels

7.2.6. Screening of Transgenic Nicotiana Clones for Monoclonal Antibody Expression

7.2.6.1. Polymerase Chain Reaction

- Genomic DNA from transgenic and nontransgenic plants.
- Deionized water (dH_2O).
- Oligonucleotide primers (down and upstream of transgene).
- DNA polymerase and reaction buffer.
- $MgCl_2$ (25 mM).
- dNTP mix (10 mM).

7.2.6.2. Enzyme-Linked Immunosorbent Assay

- Extracts of total soluble proteins (TSP) from transgenic and nontransgenic plants (*see* **Note 8**).
- Polyvinylchloride (PVC) microtiter plate.
- Capture antibody and enzyme-antibody conjugate (Sigma).
- Substrate.
- Phosphate-buffered saline (PBS).
- Enzyme-linked immunosorbent assay (ELISA) plate reader.

7.2.6.3. Immunobloting

- Extracts of TSP from transgenic and nontransgenic plants (*see* **Note 8**).
- Deionized water (dH_2O).
- Acryalmide solution 40%.
- APS, 10%.
- TEMED.
- Loading 2× buffer.
- Sodium dodecyl sulfate (SDS)-polyacrilamide gel running 1× buffer.
- Nitrocellulose or polyvinylidene fluoride (PVDF) membrane.
- Transfer buffer (25 mM Tris, 190 mM glycine, and 20% methanol).
- Blocking solution.
- TBS.
- Capture antibody and enzyme-antibody conjugate (Sigma).
- Substrate.

7.2.7. Assessment of Biological Activity by Florescence-Activated Cell Sorting

- RPMI 1640 medium (Sigma-Aldrich),
- Fetal calf serum (PAA, Laboratories, Australia),
- Trypsin (Sigma-Aldrich),
- PBS,
- BSA (bovine serum albumin),
- Sodium azide (Sigma-Adrich),
- Propridium iodide (Sigma-Aldrich),
- Paraformaldehyde (Sigma-Aldrich),
- FITC or other fluorescent dye conjugate antibodies (Sigma-Aldrich),
- Cells culture flasks (Costar, Lowell, MA),
- Rubber policeman (Costar),
- 1.5 mL Eppendorf tubes,
- Polystyrene 50 mL centrifuge tube (Costar),
- 96-well round-bottomed microtiter plates (Costar).

7.3. Methods

7.3.1. Isolation of Heavy and Light Chain Immunoglobulin Regions Using PCR

Established and reliable methods of gene cloning and genetic engineering of immunoglobulins, and the development of the transgenic plant technology have merged to result in the use of genetically modified plants as bioreactors for the production of antibodies.

Production of complete antibody molecules in transgenic plants has recently been done with different proposes, for study of the accumulation pattern, the assembly of molecules, plant physiology and pathology studies, and as bioreactors for purification of proteins for immunotherapy in humans and other uses.

Since its initial demonstration, the expression of functional antibodies in transgenic plants has been considered highly promising for large-scale production. It is important to determine the best form to obtain antibody in relative short time as well as to determine the overall accumulation pattern in antibody producing plants and functionality for its final purpose.

The plantibody production has improved. Antibody molecules in the first report on the expression of full-size antibodies in plants were obtained by crossing of plants expressing the immunoglobulin chain, but later the two chains were obtained in the same transgenic plant.

The recombinant immunoglobulin chains are generally targeted to the plant endoplasmic reticulum where folding, assembly and glycosylation takes place *(8)*. It has been shown that some of the properties of immunoglobulins depend on their glycosylation. Plant glycosylation differs from that found in the mammalian cells, and these differences between the complex glycans could have a direct relevance for the application of the expressed plantibody.

In the following section, the steps from isolating antibody heavy and light chains genes to cloning in plant transformation vector are described.

7.3.1.1. RNA Extraction

Owing to the time and care that must be taken in preparing RNase-free reagents and plastic ware, we find it is cost-effective to purchase a kit for preparing RNA *(9)*. If one chooses not to use a kit, a detailed protocol for mRNA purification can be found in the Maniatis cloning manual *(10)*.

1. Wash harvested 10^6 culture cells two times with PBS or serum-free culture medium using low-speed centrifugation (106 g; 5–10 min). Remove remaining supernatant as much as possible (*see* **Note 9**).
2. Add 1 mL of TriPure to the cell pellet; homogenize sample by vortex.
3. Incubate 5 min at 15–25°C to dissociate nucleoprotein complexes.
4. Add chloroform (0.2 mL per 1 mL TriPure), shake vigorously 15 s, and incubate 2–15 min at 15–25°C.
5. Centrifuge at 7,200 g, 15 min, at 2–8°C. Three phases can be seen.
6. Transfer the aqueous phase (colorless) to a fresh tube (*see* **Note 10**).
7. Precipitate with isopropanol (0.5 mL per 1 mL TriPure), mix by inversion.
8. Incubate 5–10 min at 15–25°C.
9. Centrifuge at 7,200 g, 10 min, at 2–8°C. Discard supernatant.
10. Wash pellet 1× with 75% EtOH (1 mL per 1 mL TriPure).
11. Centrifuge at 7,200 g, 5 min, at 2–8°C. Discard supernatant.
12. Air dry pellet.
13. Resuspend in DEPC-treated water (20–30 µL). Incubate 10–15 min at 55–60°C to resuspend (*see* **Note 11**).
14. Determine the amount of RNA by measuring a diluted sample at optical density $(OD)_{260}$, assuming that an OD_{260} of 1.0 equals of 40 µg/mL. The ratio of the absorbance at 260 nm and 280 nm should be between 1.5 and 2. Lower

ratios indicate possible contamination of the RNA with protein, which may interfere with subsequent PCR reactions.

15. Store RNA at –15°C to –25°C.

7.3.1.2. cDNA Preparation

1. Mix 5–10 µg of total RNA and adjust the volume to 38 µL with DEPC-treated water in a microfuge tube.
2. Add 3 µL of oligo (dT) primer (100 ng/µL). Mix gently (*see* **Notes 12** and **13**).
3. Incubate at 65°C for 5 min. Cool the reaction slowly at room temperature (~10 min) to allow primers to anneal to the RNA.
4. Add 5 µL 10× reverse transcriptase buffer, 2 µL 100 mM dNTP mixture, 1 µL of RNasin, 1 µL of MMLV-RT.
5. Mix the control and experimental reactions gently and incubate at 37°C for 1 h.
6. Store at –80°C if necessary.

7.3.1.3. PCR Amplification of Immunoglobulin Heavy and Light Chains

1. Separate heavy and light chains primary PCR amplification reactions is done in 100 µL reactions containing (*see* **Note 14**)
   10.0 µL of 10× buffer,
   0.8 µL of 100 mM dNTPs,
   2.5 µL each 5′ PCR primer and 3′ PCR primer (20 µM solution),
   1–5 µL of cDNA,
   X µL of sterile dH_2O to a final volume of 99.5 µL.
2. Place the amplification reactions in a thermal cycler. Heat each PCR amplification reaction to 94°C for 5 min and then immediately cool each reaction to 54°C for 5 min.
3. Remove the reactions from the thermal cycler. Briefly microcentrifuge both the control and experimental reactions and then add 0.5 µL of Taq polymerase (5 U/µL) to each reaction.
4. Briefly microcentrifuge the reactions again.
5. Carefully overlay each PCR amplification reaction with a drop or two of mineral oil to prevent evaporation of the reaction components during thermal cycling.
6. Return the PCR amplification reactions to the thermal cycler and program cycling parameters that are optimal for the oligonucleotide, for immunoglobulin amplification, we suggest, the following schedule: denaturizing to 94°C, 1 min, annealing to 55°C, 1 min, extension to 72°C, 1 min, 25 cycles, with five additional minutes of extension to the temperature already described in the last cycle.

7. Load 10 μL of each PCR amplification reaction taken from below the mineral oil layer into separate lanes of an agarose gel.
8. Analyze the PCR amplification products by electrophoresis using appropriate molecular-weight makers.
9. PCR amplification reaction yields a ~0.7-kb PCR product for light chain and a ~1.4-kb product for heavy chain when visualized on an ethidium bromide (EtBr)-stained 0.8% (w/v) agarose gel (*see* **Note 15**).
10. The DNA amplified fragments, with the expected size, were purified in low meeting point agarose gels (Sigma), using the QIAquick Gel Extraction Kit (QIAGEN, GmbH), and were cloned independently in the pHES74 vector *(11)*.

7.3.1.4. Cloning of Heavy and Light Immunoglobulin Chains into a Constitutive Plant Expression Vector

Once the appropriate cDNA has been isolated from a hybridoma cell line, it must be inserted into a plant vector designed for high-level expression. Generally, a strong constitutive promoter is chosen to maximize transcription. The cauliflower mosaic virus is widely used in dicotyledonous plants. The presence of an intron often increases the rate of transcription, so they are also commonly included. Protein synthesis can be optimized by making sure the translational start site conforms to the Kozak consensus for plants and using translational enhancers such as the 5′ leader sequence derived from tobacco mosaic virus (TMV) RNA (the omega sequence). A very important consideration in the expression construct design is the protein targeting, since recombinant antibodies are more stable in some intracellular compartments than others and this contributes to the overall yield. As well, the immunoglobulins must be targeted to the secretory pathway to undergo glycosylation and the appropriated targeting could also facilitate protein isolation and purification *(12, 13)*.

Choosing a protein-expression system depends on the needs of investigator and long-term goal of manufacturing the recombinant protein.

The heavy and light chains of recombinant antibody are independently cloned in intermediary plasmids. Cloning details of heavy and light chains fragments obtained by PCR is here exemplified in the **Figs. 7.1** and **7.2** using pHES74 as intermediary plasmid *(11)* (*see* **Notes 16** and **17**).

1. Purify plasmids, the products of PCR containing the heavy and light chains (CHKDEL and CLKDEL), as well as the intermediary vector (pHES74) using Wizard Plus Minipreps DNA purification systems (Promega) (*see* **Notes 18** and **19**).
2. Digest 1 μg of DNA both from heavy and light chains (CHKDEL and CLKDEL) as well as intermediary vector (pHES74), using the pair of endonucleases *Nco*I and *Eco*RI

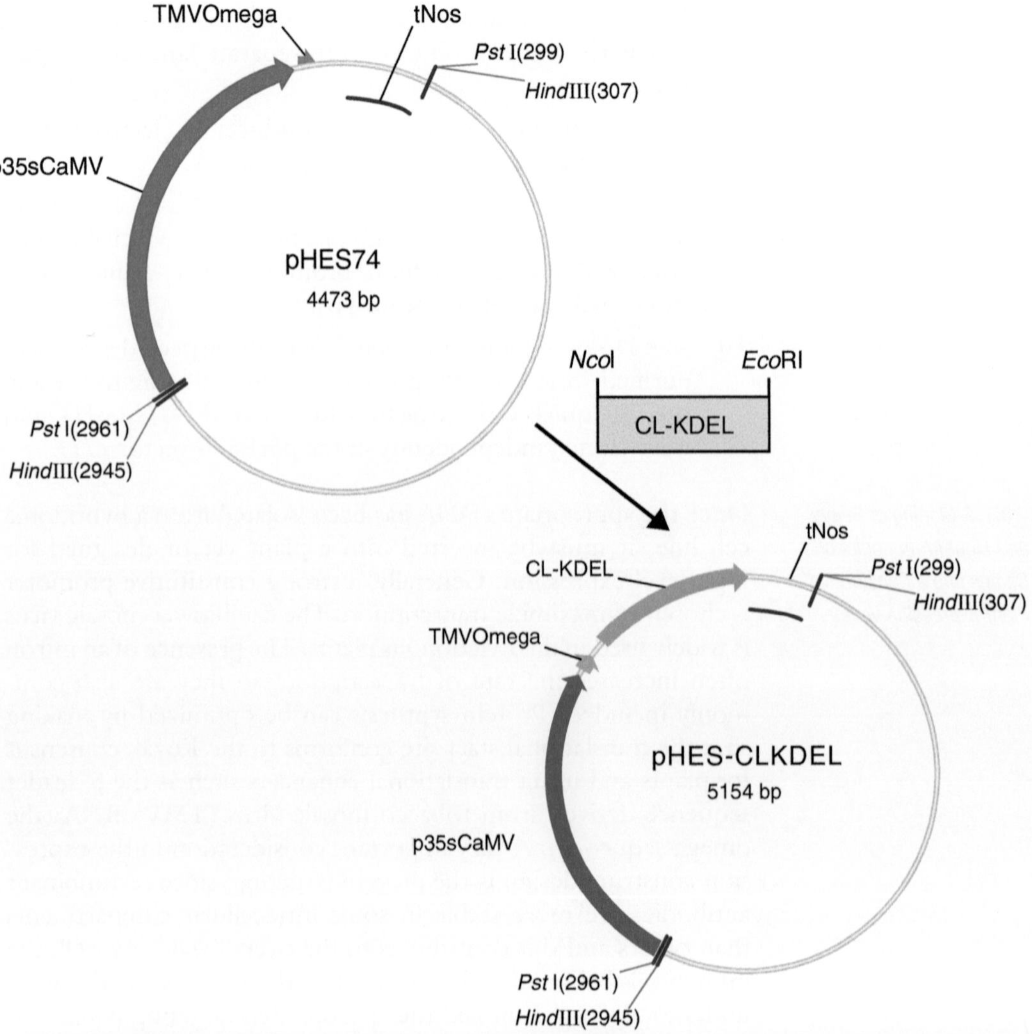

Fig. 7.1. A schematic overview of steps to clone the light chain antibody gene into the constitutive plant expression vector pHES74. *p35SCaMV* cauliflower mosaic virus promoter, *TMVOmega* 5′ untranslated region of tobacco mosaic virus (TMV), *CLKDEL* fragment of light chain antibody with additional amino acid sequence KDEL in carboxyl terminal, *tNOS* nopaline synthase terminator.

according to the appropriated cloning design and following manufacturer recommendations.

3. Confirm the products from digestion of both plasmids on an agarose gel (0.8% w/v).
4. Run a preparative agarose gel and elute the digested plasmids using Wizard DNA Clean-Up System (Promega), according to standard techniques.
5. To incorporate the bands from heavy and light chains into intermediary vector, calculate the appropriated vector–insert ratio, respectively, setup the next reaction:

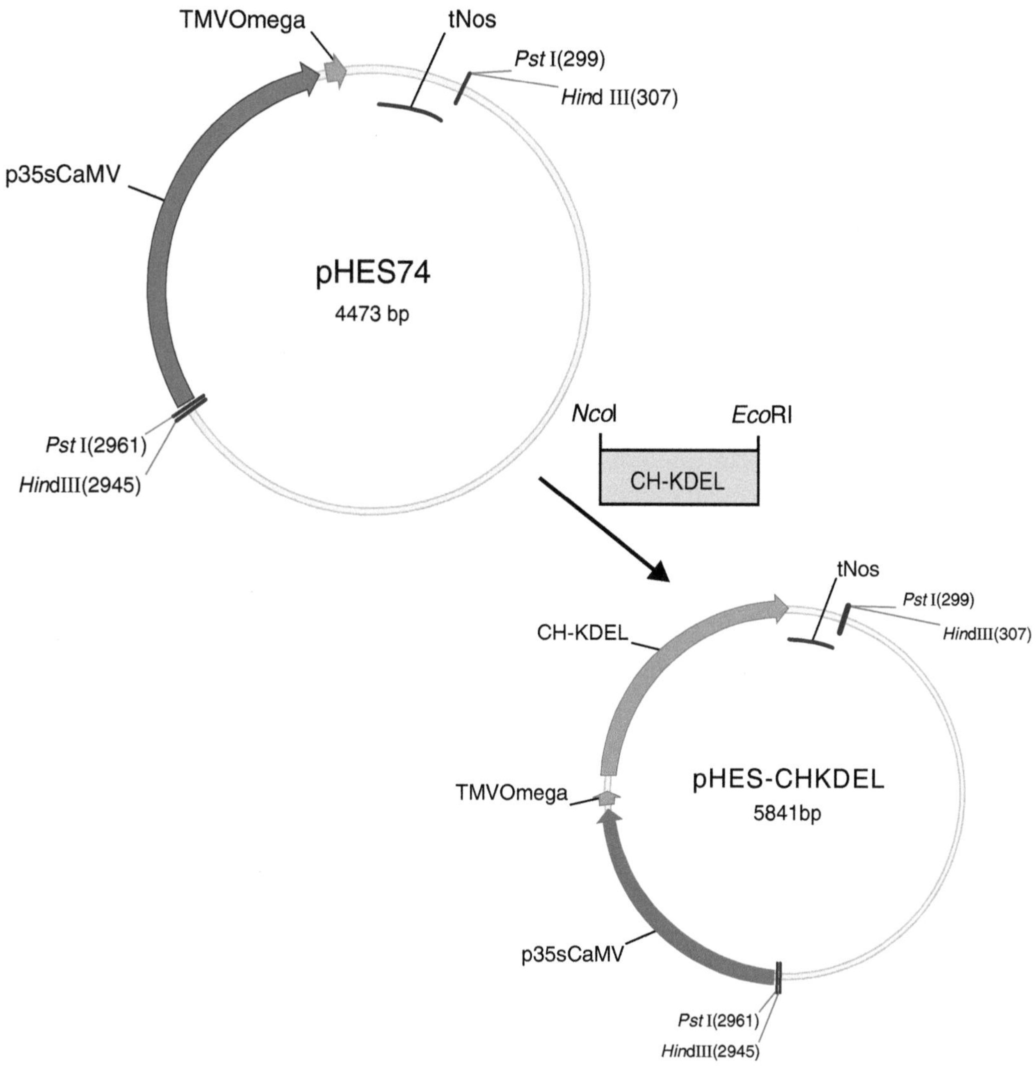

Fig. 7.2. Overview of steps to clone the heavy chain antibody gene into the constitutive plant expression vector pHES74. *p35SCaMV* cauliflower mosaic virus promoter, *TMVOmega* 5′ untranslated region of tobacco mosaic virus (TMV), *CHK-DEL* fragment of heavy chain antibody with additional amino acid sequence KDEL in carboxyl terminal, *tNOS* nopaline synthase terminator.

X ng of vector DNA,
X ng insert DNA,,
1 u (Weiss units) T4 DNA ligase,
1 μL ligase 10× buffer,
X μL of sterile dH_2O to a final volume of 10 μL.

6. Incubate the ligation reaction according to the guidelines for temperature and time.
7. Following the ligation reaction, transform the ligated DNA into competent cells of *E. coli* appropriated host strain (*see* **Note 20**).

8. Select positive clones on selection antibiotic-containing LB plates and confirm the successful ligation by digestion of DNA using restriction enzymes. This plasmid will be named "heavy or light chain intermediary vector" (pHES-CHKDEL and pHES-CLKDEL plasmids).

7.3.2. Preparation of Binary Vectors Containing Expression Cassettes for Mabs

The soil bacterium *A. tumefaciens* infects wounded parts of dicotyledonous and some monocotyledonous plants and causes undifferentiated, tumorous growth of plant tissue, the formation of so-called crown galls. Upon perception of wounding signals from the plants, phenolic substances, virulence genes are expressed and part of large, tumor-inducing (Ti) plasmid is transferred to the plant cells. This transferred DNA (T-DNA) is integrated into the plant genome and contains genes for synthesis of phytohormones, cytokines, and auxins, promoting a tumor-like proliferation and growth of infected cells *(14)*.

Therefore, this microbe acts as natural genetic engineer transforming a plant for specific needs. Usually, a Ti plasmid contains the oncogenes and ~35 *vir* genes as well as the T-DNA delimited by 25 pb long repeats, the left border (LB) and the right border (RB). These vectors are no longer oncogenic and allow easy manipulation of T-DNA, introduction of multiple cloning site, marker genes or selectivity marker between the two borders. The resulting binary vector Ti vectors are able to replicate in *E. coli* and in *Agrobacterium* simplifying the cloning procedure for introduction of foreign genes into T-DNA region. All gene manipulation steps can be performed on small Ti vectors in *E. coli* and only in a last step the Ti vector is introduced into *Agrobacteria* for subsequently plant transformation *(15)*. A variety of binary Ti vectors for *Agrobacterium*-based plant transformation are now available such as pCAMBIA2300 *(16)*.

The heavy and light chains coding regions are independently cloned in intermediary plasmids and later sequentially cloned into the binary vector (*see* **Notes 21–23**). We exemplified this method using pHES-CHKDEL and pHES-CLKDEL intermediary plasmids shown in **Fig. 7.3**.

1. Purify plasmids containing the heavy and light chains expression cassettes (pHES-CHKDEL and pHES-CLKDEL plasmids) as well as the binary vector (pCAMBIA2300) using Wizard Plus Minipreps DNA purification Systems (Promega).
2. Digest 1 μg of DNA both from plasmid with light chain expression cassette (pHES-CLKDEL) and from binary vector (pCAMBIA2300), using the first chosen endonuclease (*Pst*I) according the appropriated cloning design and following manufacturer recommendations.
3. Confirm the products from digestion of both plasmids on an agarose gel (0.8% w/v).

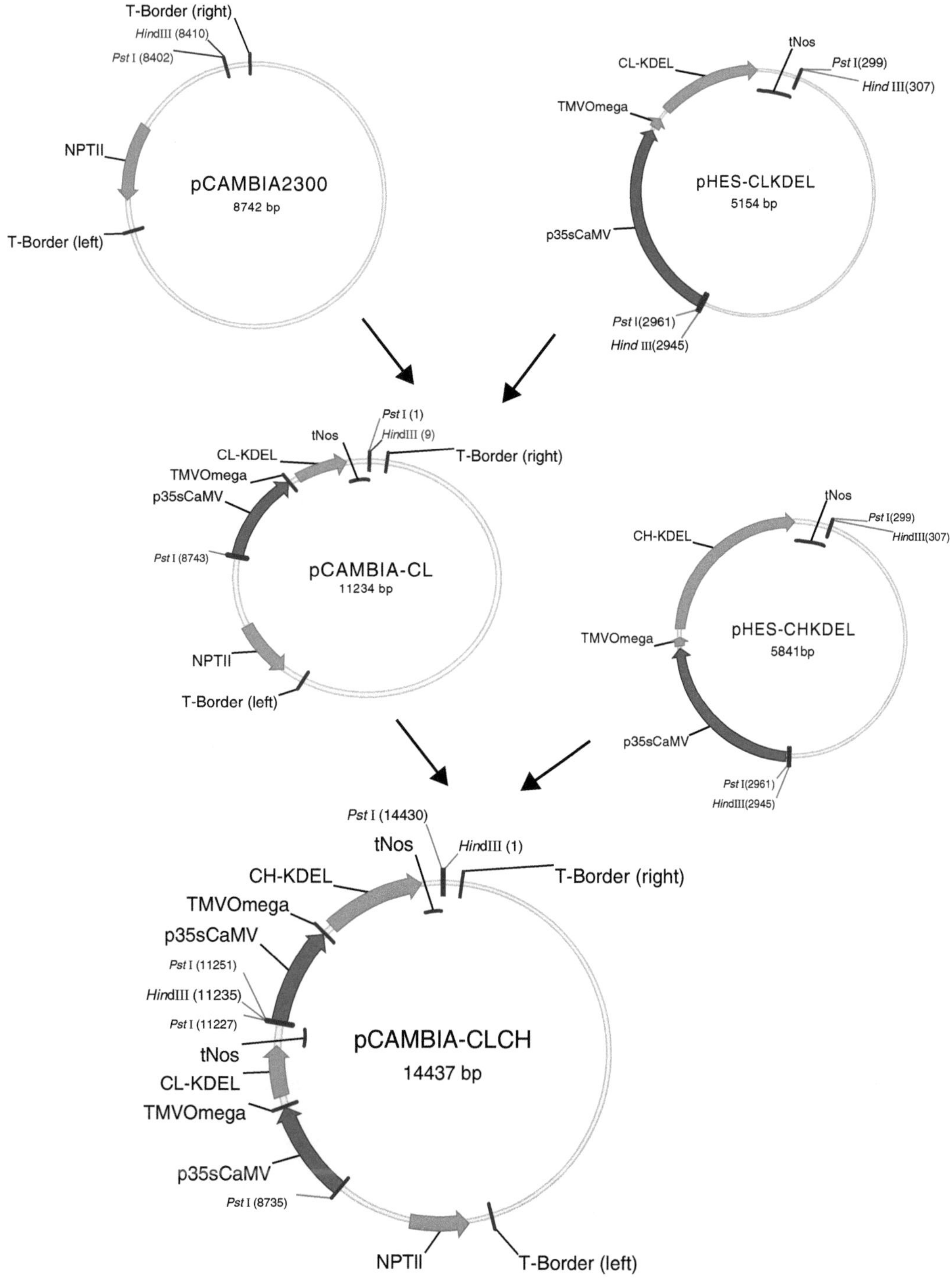

Fig. 7.3. Scheme showing the steps to construct the binary vector pCAMBIA2300. *p35SCaMV* cauliflower mosaic virus promoter, *TMVOmega* 5′ untranslated region of Tobacco Mosaic Virus (TMV), *CLKDEL and CHKDEL fragments* of heavy and light chain antibody with additional amino acid sequence KDEL in carboxyl terminal, *tNOS* nopaline synthase terminator, *NPTII* neomycin phosphotransferase II gene, *T-Borders* (left and right).

4. Run a preparative agarose gel and elute the digested plasmids using Wizard DNA Clean-Up System (Promega), according to standard techniques.
5. To incorporate the band from light chain expression cassette into binary vector, calculate the appropriated vector–insert ratio, set-up the next reaction.
   X ng of vector DNA,
   X ng insert DNA,
   1 u (Weiss units) T4 DNA Ligase,
   1 μL Ligase 10× Buffer,
   X μL of sterile dH_2O to a final volume of 10 μL.
6. Incubate the ligation reaction according to the guidelines for temperature and time.
7. Following the ligation reaction, transform the ligated DNA into competent cells of *E. coli* appropriated host strain.
8. Select positive clones on marker genes containing LB plates and confirm the successful ligation by digestion of DNA using restriction enzymes. This plasmid will be named "intermediary binary vector" (pCAMBIA-CLKDEL).
9. Once finished the cloning of the light chain cassette into the binary vector, the remaining steps describe the cloning of the heavy chain cassette.
10. Repeat the **steps 2–8**, but using heavy chain expression cassette (pHES-CHKDEL) and the "intermediary binary vector" (pCAMBIA-CLKDEL) plasmids.
11. Finally, the plasmid resultant (pCLCH) with two gene chains expression cassettes is used to transform competent cells of *Agrobacterium* appropriated host strain (*see* **Note 24**).
12. Select positive clone of *Agrobacterium* on selection antibiotic-containing LB plates and confirm the incorporation of final plasmid. This clone of bacteria was use to transform plants.

7.3.3. Assessing Vector Viability by Transient Expression

Transient expression analysis is of great application to access the functionality of vector constructions or particular recombinant gene in a rapid way. Transiently produced recombinant protein could be sufficient for chemical, physical, and even functional activity studies *(17–19)*. Among transient expression systems, the *A. tumefaciens* transfection methods became widely used because of several advantages. The most highlighted advantage is the lower costs of the transformation procedure comparing to so-called physical or direct methods for transient expression recover. In addition, the higher survival rate of cells subjected to transfection by *Agrobacterium* and, subsequently, the higher transformation efficiency made this method widely employed *(17–21)*. Other alternative of transient expression systems are based on viral vectors *(22)* which are often limited by the size of the gene to be expressed. Two most extensively used models for transient studies by *Agrobacterium* transfection have been *Nicotiana* plant leaves and tobacco cell suspension *(17, 20)*. They have in common the requirement for plant cells to enter in S-phase state for efficient *Agrobacterium* transformation. In the same time, the *Agrobacterium* cells could be applied in both cases after an equal pretreatment to achieve high efficiency of gene transfer.

7.3.3.1. NT-1 Cell Suspension

NT-1 tobacco cell line is characterized by several practical features: the possibility to propagate with low cost disposables, a few exigencies for *in vitro* manipulation and self-competence to be transformed by direct and indirect methods *(17, 23, 24)*. The high division capacity of this cell line can supports the transformable competence ensuring the higher transgene expression level and genome integration frequencies, which was observed in different plant species applying flow cytometry during the transformation process by *A. tumefaciens (25, 26)*. The NT-1 cell line can be maintained in the proliferation state for long time, refreshing the culture medium periodically, independently of the medium consistence.

The procedure here described comprises the cultivation and maintenance of plants and cells, the growth of *A. tumefaciens*, the inoculation of cultures to cells or plants, and the subsequent culture of cells or plants until expression is evaluated. The choice of cells or plants for transient expression should be decided according to the needs of the specific case.

7.3.3.2. A. Tumefaciens

A. tumefaciens is a soil bacterium causing of crown galls or tumors at or close to infection sites *(27)*. The tumor tissue provides the *Agrobacterium* with unusual aminoacids (opines). Genes required establishing tumors are transferred from *Agrobacterium* on a discrete portion of its DNA (T-DNA) into the nuclear genome of the host plant. The machinery necessary for this DNA transfer resides on a Ti plasmid. This Ti plasmid includes the T-DNA itself that by combining actions of the vir genes, clustered in the same plasmid, achieves the delivery of the T-DNA to the nucleus of the host plant cell *(27, 28)*. Removal of the genes within the T-DNA does not impede the ability of *Agrobacterium* to transfer this DNA, but does prevent the formation of tumors, and this *Agrobacterium* is called “disarmed.” The insertion of useful or genes of interest in the T-DNA provides a sophisticated tool for genetic engineering of plants. The development of binary Ti vectors and a range of disarmed *Agrobacterium* strains are two key advances that have made the *Agrobacterium* an almost universal tool for genetic transformation of plants *(16)*.

A. tumefaciens strain: LBA4404; disarmed Ti plasmid pAL 4404, Sm^r, chromosome of Ach5 (pTiAch5), marker gene rif, octopine type strain *(29)*. [Other *Agrobacterium* strain could be employed, e.g., C58C1 (pGV2260) *(14)*.]

Agroinoculation of Tobacco Cells from NT-1 Line

Suspension cells from NT-1 line were maintained by dilution 1/10 every 10 days from 1/2 total settled cell volume in the same fresh NT-1L medium, contained in 250 mL volume Erlenmeyer flask (see **Note 25**). The suspension growth is conditioned to shaking at 110 rpm in the darkness, under 25°C.

1. Suspension cells (3 mL; 0.3 mL settled cell volume) are collected 5 days after dilution (subculture) and distributed on the surface of a sterile Whatman N1 filter paper with the aid of a Buchner funnel, followed by incubation on NT-1S medium.
2. After incubation for 7 days (*see* **Note 26**) in darkness at 25°C, in the NT-1S medium/pretreatment period, the lawned cells are covered with *Agrobacterium* at concentration of 2 OD units (*see* **Note 27**).

   A. tumefaciens strain LBA4404 colony culture is maintained on LB solid growth medium, supplemented with rifampicin 50 mg/L, streptomycin 100 mg/L, and an antibiotic specific for binary vector selection. To obtain *Agrobacterium* suitable for transformation inoculate one bacterial colony to YEB liquid medium and growth under shaking (110 rpm) until the culture reaches 0.7–0.8 OD units. The bacterial suspension is sedimented by centrifugation at 1,870 g and 22°C. Resuspend the bacterial pellet in MSLSM medium, containing 75 µg/mL of acetosyringone (Sigma-Aldrich).
3. After 60-min infection period under laminar flow conditions plant cells are briefly drained and incubated in the sealed Petri culture plates for 5 days in the dark, at 25–26°C (*see* **Note 28**).
4. Collect the inoculated cells for transgene expression analysis (*see* **Note 29**).

Agroinfiltration of *Nicotiana* Leaves

Nicotiana tabacum and *Nicotiana benthamiana* plants are germinated and grown in multidishes pots, under artificial or natural 16-h light regime (5,000 lux intensity), with constant temperature of 24°C.

1. When the plants reach 4–6 leaves state infiltrate the leaves with *Agrobacterium* suspension (*see* Note 30). The infiltration is performed with the aim of 2-mL syringe unit into fully expanded leaves through small puncture on the abaxial site (*see* **Note 31**).
2. Keep the plants with infiltrated leaves in the previously described growth conditions for 3 days minimum (*see* **Note 32**).
3. Cut the infiltrated leaves from plants for transgene expression analysis.

7.3.4. *Agrobacterium*-Mediated Stable Transformation of *Nicotiana*

One of the major goals of modern plant genetic engineering is the introduction of genes for molecules of high-added values into plants with capacities to be scale up. In this direction, tobacco is very attractive for its large biomass production in addition to its simplicity for genetic transformation and conveniences for model studies. Plant regeneration and genetic transformation techniques have been developed for several types of tobacco tissues from different species *(Nicotiana* sp.) *(30)*. High frequency of shoot regeneration via direct organogenesis have been obtained on MS *(7)* medium supplemented with BAP using *in vitro* leaf discs or cotiledonary explants. These explants are susceptible for transformation by a wide range of *Agrobacterium* strains, and transgenic shoots can be obtained using kanamycin, hygromycin, phosphinothricin, and others selecting agents. High transformation frequencies are obtained and confirmed by rooting on selection medium.

The tobacco transformation methodologies described in this section apply for several kinds of explants and selecting agents.

1. Disinfect tobacco seeds in 2.5% sodium hypochlorite with four drops of Tween 20 for 15 min by occasional manual agitation (*see* **Note 33**).
2. Wash the seeds four to five times with sterile double distillated water by agitation and then sow on propagation medium. After 3 weeks the cotyledonal leaves should be ready to use.
3. **Steps 1** and **2** are not applicable if true leaves are the explants of choice (*see* **Note 34**).
4. The compositions of all culture media used in this work are described in **Table 7.1**; and the media are adjusted to pH 5.8, and autoclaved for 20 min at 120 °C, at 1 kgf/cm^2 (*see* **Note 35**).
5. Grow *Agrobacterium* from one colony inoculated in 5 mL of liquid LB medium, containing rifampicin 100 mg/L and kanamycin 30 mg/mL. The culture is incubated overnight at 28°C with 100 rpm shaking until the cell density reach OD_{620} 0.8–0.9 (*see* **Note 36**).
6. Harvest *A. tumefaciens* by centrifugation of 5 mL cell suspension at 1,870 g for 5 min.
7. Remove supernatant and resuspend the *Agrobacterium* cell pellet in 50 mL liquid MS medium. Transfer the solution to Petri dishes (about 25 mL per plate).
8. In Petri dishes, cut 0.7 cm disk with a scalpel and set the leaf upside down, adding a thin layer of MS salts and vitamins liquid medium (*see* **Note 37**).
9. Place 10 discs, upside down, per regeneration plate (15 × 100 mm).
10. Add *A. tumefaciens* suspension and let set there during cocultivation period (*see* **Note 38**).
11. Incubate plates at 22°C for 72 h, in dark conditions.

Table 7.1
Media for *Nicotiana* sp. tissue culture and genetic transformation (amount per liter)

Composition (stock)	Propagation	Cocultivation	Regeneration selection	Rooting
MS major salts (10×)	100 mL	100 mL	100 mL	100 mL
MS minor salts (100×)	1 mL	1 mL	10 mL	10 mL
Ferrous/NaEDTA (100×)	1 mL	1 mL	10 mL	10 mL
Sucrose	20 g	30 g	30 g	20 g
PhytoAgar	7 g	7 g	8 g	7 g
MS vitamin (100X)	10 mL	10 mL	10 mL	10 mL
Acetosyringone (AS)	–	0.04 g	–	–
BAP (1 mg/mL)	–	1 mL	1 mL	–
Carbenicillin (100 mg/mL)	–	–	5 mL	2.5 mL
Phosphinothricin (5 mg/mL)	–	–	1 mL	1 mL
Kanamycin (100 mg/mL)	–	–	1 mL	1 mL

Add double distillated water to complete 1 L of solution.

12. Transfer leaf discs to regeneration/selection plates, and subculture monthly to the same fresh medium in dim light at 25°C until shoots reach up 1 cm (*see* **Note 39**).
13. Remove shoots from leaves with forceps and culture them in MS basal medium with selective agent for rooting at 24°C and 16- h photoperiod with light intensity of 2,000 Lux.
14. After the plantlet is reaching the same high of the container, they can be transferred to soil for acclimatization and molecular and biological characterization.
15. Established seedlings can be transferred to greenhouse to set seeds (*see* **Note 40**).

7.3.5. Screening of Transgenic Nicotiana Clones for Monoclonal Antibody Expression

The ability to insert genes into the plant genome and the identification of functionally active antibodies is an essential step. Transgene expression is influenced by several factors that cannot be controlled precisely through construct design, which lead to variable transgene expression. Such factors include the position of transgene integration, the structure of transgene locus, gene copy number and the presence of truncated or rearranged transgene copies. Several strategies have been adopted in attempt to minimize variation in transgene expression. The transformation is followed by selection for cells with stably integrated copies of the target gene by following an antibiotic resistance gene that is introduced in the binary vector.

There are several commonly used transgenic plants testing protocols, including biological assays, as well as ELISA and Immunoblotings (Western blots) tests. PCR test is another assay frequently used to determine transgenesis, but it does not permit detection of the expression of antibody. The selection of one method or another depends on advantages, disadvantages, and the antibody activity from final product. Generally, the scientists include the various testing methods.

7.3.5.1. Polymerase Chain Reaction

The PCR method is more sensitive than the ELISA or Western blot methods and tests for the presence of the specific DNA sequence of the gene itself. The major advantage of PCR tests is sensitivity, that is, detection of DNA transgene at very low levels. Major disadvantages of the PCR protocol include length of time needed and cost, because the PCR tests also require more sophisticated equipment and greater expertise.

A typical amplification reaction includes the sample of target DNA, a thermostable DNA polymerase, two oligonucleotide primers, desoxynucleotide triphosphates (dNTPs), reaction buffer, magnesium, and optional additives. The components of the reaction are mixed and were placed in a thermal cycler, which is an automated instrument that takes the reaction through a series of different temperatures for varying amounts and times.

1. Isolate plant genomic DNA from leaves of transgenic and nontransgenic plants (*see* **Note 41**).
2. Mix the reactions components in the order listed below (*see* **Note 42**):
   X µL of sterile dH_2O to a final volume of 50 µL,
   1× reaction 10× buffer,
   0.2 mM dNTP mix,
   0.025 u/µL DNA polymerase,
   1.5 mM $MgCl_2$,
   1 µM from each oligonucleotide primers,
   10 ng/µL plant genomic DNA.
3. Place the tubes in a controlled temperature heat block.
4. The initial step is denaturing of DNA by heating to 95°C for 1 min (*see* **Notes 43** and **44**).
5. Reduces the temperature to melting temperature, to facilitate the stable association between denatured DNA and the primers, for 1 min (*see* **Note 45**).

6. Increase the temperature to ~74°C for 1 min, in the extension step.
7. Finally, repeat the denaturation/annealing/extension steps for 20–40 cycles.
8. Analyze the PCR products by agarose gel electrophoresis.

7.3.5.2. Enzyme-Linked Immunosorbent Assay

ELISAs combine the specificity of antibodies with the sensitivity of simple enzyme assays, by using antibodies or antigens coupled to an easily assayed enzyme that possesses a high turnover number. ELISAs can provide a useful measurement of antigen or antibody concentration. There are different versions of the ELISA method used antibody detection in plants, such as lateral flow strips and plate test. Advantages of the ELISA tests are speed, relative ease of use, and low cost *(31)*.

One of the most useful of the immunoassays is the two antibody "sandwich" ELISA. This assay is used to determine the antibody concentration in unknown samples. The sandwich ELISA requires two antibodies that bind to epitopes that do not overlap on the antigen. To utilize this assay, one antibody (the "capture" antibody) is purified and bound to a solid phase typically attached to the bottom of a plate well. Samples extracted from transgenic plants with plantibody is then added and allowed to complex with the bound antibody. Unbound products are then removed with a wash, and a second enzyme-antibody conjugate (the "detection" antibody) is allowed to bind to the plantibody, thus completing the "sandwich." The assay is then quantified by measuring the amount of antibody conjugate bound to the matrix, through the use of a colorimetric substrate. There are popular enzymes that convert a colorless substrate to a colored product, for example, alkaline phosphatase and peroxidase, and theirs should be stable, safe and inexpensive.

Major advantages of this technique are that the plantibody samples do not need to be purified before use, and that these assays are very specific. However, one disadvantage is that not all antibodies can be used.

1. Bind the capture antibody to the bottom of each well by adding ~50 μL of antibody solution to each well (20 μg/mL in PBS). PVC will bind ~100 ng/well (300 ng/cm^2) (*see* **Note 46**). The amount of antibody used will depend on the individual assay, but if maximal binding is required, use at least 1 μg/well. This is well above the capacity of the well, but the binding will occur more rapidly, and the binding solution can be saved and used again.
2. Incubate the plate overnight at 4°C to allow complete binding.
3. Wash the wells four times with PBS-0.1% Tween 20. A 500 mL squirt bottle is convenient. The antibody solution washes can be removed by flicking the plate over a suitable container.

4. Wash wells twice with PBS.
5. Add 50 μL of the samples from transgenic and nontransgenic plants to the wells. All dilutions should be done in extraction buffers. Incubate for at least 2 h at room temperature in a humid atmosphere (*see* **Note 47**).
6. Repeat the **step 2**, washing.
7. Add the enzyme-antibody conjugated following the working solutions described by the manufacturers.
8. Incubate for 1 h at room temperature in a humid atmosphere.
9. Wash with several changes of PBS.
10. Add substrate as indicated by manufacturer (*see* **Note 48**). After suggested incubation time has elapsed, optical densities at target wavelengths can be measured on an ELISA plate reader.
11. For quantitative results, compare signal of unknown samples against those of a standard curve. Standards must be run with each assay to ensure accuracy.

7.3.5.3. Immunobloting

Immunobloting procedures combine the resolution of gel electrophoresis with the specificity of antibody detection. Blotting can be used to ascertain a number of important characteristics of proteins and its permits to detect the antibody expression in transgenic plants. In this method, the proteins can be separated by size, electrical charge, and shape by applying the samples to a gel matrix in the presence of an electric current. Native polyacrylamide gel electrophoresis (PAGE) is the most popularly used under reducing and nonreducing conditions: presence or absence of a molar excess of the ionic detergent sodium dodecyl sulfate (SDS-PAGE) and reducing agent (DTT or β-mercaptoethanol), respectively. After the separation step, the proteins are transferred to solid support membranes (Western blotting) for detection with specific antibodies: one that directly binds the sample of interest (named primary antibody) and the second that binds to the antigen-binding primary antibody (named secondary antibody) that is conjugated to enzyme. The most common antibody label used in Western blotting is an enzyme such as alkaline phosphatase or horseradish peroxidase, which can be detected visually through the conversion of a colorimetric substrate to a colored precipitate at the site of antibody binding. Alternatively, chemiluminescent substrates may be employed.

These are the basic steps for immunoblotting procedures:

1. Prepare the samples, could be an extract of cells or tissues, in a gel electrophoresis sample buffer (*see* **Note 49**).
2. Run the SDS-PAGE electrophoresis gel according to the recommendations of the manufacturer of the gel apparatus.

3. Cut the membrane and filter paper (Whatman 3 MM, 6 sheets) to fit the gel exactly. Filter paper soaked in transfer buffer can be used to carefully remove the gel from the glass plates or plastic cassette, and to then transfer the gel to the membrane.
4. Remove all air bubbles between the gel and the membrane. This can be done easily by rolling a pasture pipette across the surface of the gel/membrane sandwich.
5. Place the membrane on the side of the positive electrode (anode).
6. Transfer the separated proteins using 1 A (constant current) for 1 h, or equivalent, in a wet transfer system, or at 0.7 A for 45 min in a semi-dry transfer system, with transfer buffer (*see* **Note 50**).
7. Place the membrane into blocking solution for 30 min or 1 h at room temperature (*see* **Notes 51** and **52**).
8. Wash the membrane several times with TBS.
9. Add the primary antibody and incubate for 2 h at room temperature with gentle agitation (*see* **Note 53**).
10. Repeat the washing step.
11. Add the secondary antibody, using the working dilution described by the manufacturers, and incubate for 1 h at room temperature with gentle agitation.
12. Repeat the washing step.
13. Add substrate as indicated by manufacturer.
14. Incubate for 30 min at room temperature in dark conditions with gentle agitation.
15. When the color has developed to desired intensity, stop the reaction by washing the membrane in dH_2O for several minutes.

7.3.6. Assessment of Biological Activity by Florescence-Activated Cell Sorting

Flow cytometry is a widely used method for analyzing expression of cell surface and intracellular molecules, characterizing and defining different cell types in heterogeneous cell populations, assessing the purity of isolated subpopulations, and analyzing cell size and volume. It allows simultaneous multiparameter analysis of single cells. It is predominantly used to measure fluorescence intensity produced by fluorescent-labeled antibodies detecting proteins or ligands that bind to specific cell-associated molecules, such as EGF receptor, T cells receptor and other.

In the flow cytometer, the sample is hydrodynamically focused to a tiny stream of single cells. Laser light is directed onto the sample as it flows through the chamber. There are a number of detectors to detect the light scattered from the cells/particles as they go through the beam. There is one in front of the light beam (forward scatter or FSC) and several sides on to it (side scatter or SSC). Fluorescent detectors are used for the detection of fluorochromes themselves.

Forward scatter correlates with the cell size and side scatter depends on the density of the particle/cell (i.e., number of cytoplasmic granules, membrane size), and in this manner cell populations can often be distinguished based on their difference in size and density *(32)*.

Fluorochromes used for detection/staining will emit light when excited by a laser with the corresponding excitation wavelength. These particles/cells can be detected individually and the data analyzed in computer program like WinMDi and other.

The flow cytometry can be used as a powerful method for analysis of biological activity of several molecules expressed in transgenic plants.

The method employed more frequently for this purpose is the indirect staining, in which the primary antibody is not fluorochrome labeled but is detected by a second fluorochrome-labeled antibody *(33)*. This second reagent may be an antibody with specificity for the first antibody. Other methods usually employed in florescence-activated cell sorting are the case of direct staining *(34)* and intracellular staining methods *(35)*.

7.3.6.1. Flow Cytometry Antibody Labeling

Direct Staining Protocol

General procedure

1. The cells are grown in RPMI 1640, supplemented with 10% fetal calf serum. For harvest anchored cells detached from the culture flasks using a rubber policeman or 2 mL of trypsin 0.25% for 10 min at 37°C (*see* **Note 54**).
2. Pellet the cells by centrifugation at 400*g* for 5 min.
3. Wash the cells with PBS and suspended in PBS, 1% sodium azide at a final concentration between 3×10^6 and 5×10^6 cells/mL (*see* **Notes 55** and **56**).
4. Add 1 mL of cell suspension to each tube.
5. Pellet the cells by centrifugation at 400 g for 5 min.
6. Resuspend the cells in 250 μL of PBS/BSA 3% containing 0.1–10 μg/mL of the primary labeled antibody, prepared following the procedure previously described (*see* **Note 57**).
7. Incubate for at least 30 min at 21–23°C in the dark.
8. Wash the cells 3× by centrifugation at 400 g for 5 min and resuspend them in 1–1.5 mL of PBS. Keep the cells in the dark on ice or at 4°C in a fridge until your scheduled time for analysis.
9. Analysis. For best results, analyze the cells on the flow cytometer as soon as possible (*see* **Note 58**).

Indirect Staining Protocol

Indirect labeling requires two incubation steps; the first with a primary antibody followed by a compatible secondary antibody. The secondary antibodies has the selected fluorescent dye (FITC, PE, Cy5, etc.) conjugated. Please, note that this is a general protocol and you may need to adapt it for your specific applications.

General procedure

1. The cells are grown in RPMI 1640, supplemented with 10% fetal calf serum. For harvest anchored cells detached from the culture flasks using a rubber policeman or 2 mL of trypsin 0.25% for 10 min at 37°C (*see* **Note 59**).
2. Pellet the cells by centrifugation at 400 g for 5 min.
3. Wash with PBS and suspended in PBS, 1% sodium azide at a final concentration between 3×10^6 and 5×10^6 cells/mL (*see* **Note 60**).
4. Add 1 mL of cell suspension to each tube.
5. Pellet the cells by centrifugation at 400 g for 5 min.
6. Resuspended the cells in 250 µL of PBS/BSA 3% containing 0.1–10 µg/mL of the primary antibody.
7. Incubate for 60 min at temperature from 21°C to 24°C, with occasional gentle agitation.
8. Wash the cells three times by add 1 mL of PBS and centrifugation at 400 g for 5 min.
9. Resuspended the cells in 250 µL of PBS/BSA 3% containing fluorochrome-labeled secondary antibody at the optimal dilution (according to the manufacturer's instructions).
10. Incubate for 30 min at temperature from 21°C to 24°C. This incubation must be done in the dark.
11. Wash the cells three times by add 1 mL of PBS and centrifugation at 400 g for 5 min. and resuspended them in 1.5 mL of PBS, 1% sodium azide.
12. Store the cell suspension immediately at 4°C in the dark until analysis (*see* **Note 61**).

7.3.6.2. Trouble Shooting Tips

1. No signal/weak fluorescence intensity
   - Signal not correctly compensated: Check positive single color control is set up correctly on flow cytometer and gated/compensated correctly to capture all the events.
   - Insufficient antibody present for detection: Increase amount/concentration of antibody.
   - Lasers not aligned: Ensure lasers on flow cytometer are aligned correctly by running flow check beads and adjusting alignment if necessary. If the lasers do not align correctly

or if drift occurs, you may need to consider having the machine serviced.

- Target protein not present/expressed at low level: Ensure tissue/cell type expresses target protein and that it is present in a high enough amount to detect.
- Offset too high/gain too low: Use the positive control to set up the flow cytometer correctly again, using the offset to ensure the fluorescent signal from cells is not being cut off, and increase the gain to increase the signal (within reason – care should be taken).
- Fluorochrome fluorescence has faded: Antibody may have been kept for too long or left out in the light. Fresh antibody will be required.

2. The primary antibody and the secondary antibody are not compatible: Use secondary antibody that was raised against the species in which the primary was raised (e.g., primary is raised in rabbit, use anti-rabbit secondary).
3. High fluorescence intensity
   - Antibody concentration too high: This will give high nonspecific binding or very high intensity of fluorescence. Reduce the amount of antibody added to each sample.
   - Excess antibody trapped: This can be a particular problem in intracellular staining where large fluorochrome molecules on the antibody can be trapped. Ensure adequate washing steps and include Tween or Triton in wash buffers.
   - Inadequate blocking: Add 1–3% blocking agent with antibody as well as a blocking step.
4. High background/high percentage of positive cells
   - Gain set too high/offset too low: Use the positive control to set up the flow cytometer correctly again, using the offset to reduce background from small particles and reduce the gain to decrease the signal.
   - Excess antibody: Decrease the antibody concentration. You can also add detergent to the wash buffers to ensure washing away of excess antibody.
5. Low event rate
   - Low number of cells/mL: Run 1×10^6 cells/mL. Ensure cells are mixed well (but gently).
   - Cells clumped, blocking tubing: Ensure a homologous single cell suspension by pipetting gently several times before staining. Ensure you mix again before running. In extreme cases, cells can be sieved or filtered to remove clumps (30 μL nylon mesh).

7.4. Notes

1. The forward and reverse primers (FOR and REV) used for the antibody light and heavy chain PCR amplification were designed according to the sequences reported for the FR1 domain of the variable light (V_L) and heavy (V_H) region of the antibody *(36, 37)* and the Ck and CH3 region sequence available from Kabat el al. (1991) *(38)*, respectively.
2. The reaction sometimes needs optimization, which is done by varying the relative concentration of the templates.
3. Both excess salt and excess of primers in the first-strand cDNA synthesis reactions will inhibit Taq DNA polymerase, therefore, use 1–5 μL of the first-strand cDNA synthesis reaction in any subsequent PCR amplification reaction.
4. For best results, PCR primers should be designed with similar melting temperatures ranging from 55°C to 80°C. The use of primers with melting temperatures within this range reduces false priming and ensures complete denaturation of unextended primers at 94–95°C.
5. The primers were designed to incorporate a 5′ terminal *Nco* I and 3′ terminal *EcoR* I restriction sites in the amplified DNA fragments. The backward (REV) primers contain a KDEL encoding sequence.
6. You should design your specific primers, or use primers described in the literature *(36–38)*.
7. If no commercial MS salts and vitamins are available you might follow the composition described in the next order. In 940 mL of distilled water add MS salts: 10 mL of solution containing KNO_3 190 g/L, NH_4NO_3 165 g/L; 10 mL of solution containing $MgSO_4{\cdot}7H_2O$ 37 g/L, $MnSO_4{\cdot}4H_2O$ 2.23 g/L, $ZnSO_4{\cdot}7H_2O$ 0.86 g/L, $CuSO_4{\cdot}5H_2O$ 2.5 mg/L; 10 mL of solution containing $CaCL_2{\cdot}2H_2O$ 44 g/L, KI 83 mg/L, $CoCL_2{\cdot}6H_2O$ 2.5 mg/L; 10 mL of solution containing KH_2PO_4 17 g/L, H_3BO_3 0.62 g/L, $NaMoO_4{\cdot}2H_2O$ 0.025 g/L; 10 mL of solution containing $FeSO_4$ 1.51 g/L, $NaEDTA{\cdot}2H_2O$ 3.724 g/L. Add 1 mL of MS vitamins solution *(20)* containing glycine 2 mg/mL, myo-Inositol 100 mg/mL, nicotinic acid 0.5 mg/mL, pyridoxine HCL 0.5 mg/L, thiamine HCL 0.1 mg/L; and sucrose 30 g/L.

   Mixture of $FeSO_4$ and NaEDTA should be performed after dissolving the $FeSO_4$ in 400 mL of distilled water by heating. The NaEDTA should be dissolved in 400 mL of distilled water and added to warm $FeSO_4$ solution to obtain a yellow-green solution. The resulted mixture should be kept in darkness. All MS salt solution should be kept in 4°C up to 6 months. The vitamin solution should be stored at −20°C.

8. The method to obtain the TSP from transgenic and non-transgenic plants depends on the characteristics of the antibody expressed. The most common is ground the plants leaves in liquid nitrogen to a fine powder and mixing with extraction buffer.
9. Use sterile disposable plastic ware and pipettes/tips reserved for RNA work only. Wear gloves all the time and take appropriate precautions to ensure RNase-free environment.
10. Carefully remove the upper aqueous phase for subsequent RNA isolation (**step 6**), making sure to avoid the interphase/organic phase.
11. To avoid incomplete solubilization of the final RNA pellet, do not let RNA pellet dry completely. In such case, increase the incubation time to 30 min at 55°C to solubilize RNA in DEPC-treated distilled water.
12. Because of the extreme sensitivity of the following PCR step, caution must be taken to avoid contaminating RNA and cDNA reactions with exogenous DNA. Use gloves, disposable pipette tips and plastic ware, and work in laboratory clean designated areas.
13. Some protocols suggest setting up two cDNA reactions using specific heavy or light chain cDNA primers that anneal 3′of the variable region of the target mRNA. We prefer to use a oligo dT, since only one cDNA reaction needs to be done, and one does not need a set of different cDNA primers if working with cells from different species. The quality of the subsequent PCR is not affected by the use of the oligo dT primer and in our experience it appears to work better.
14. Reactions are set up in a laminar flow hood to prevent contamination by exogenous DNA. Control reactions, which are identical to the standard reaction, but without added cDNA, are set up for each primer set.
15. If the yields of the PCR reactions are unsatisfactory, check the integrity of the RNA preparation by running a formaldehyde-agarose gel *(10)*.
16. The design of intermediary vector will depend on all of the factors previously described.
17. The immunoglobulin chain encoding DNA was cloned into a constitutive plant expression vector, pHES74, which contains the signal sequence of the sweet potato sporamin storage protein.
18. We recommend verifying the sequence of heavy and light chain genes products from PCR.

19. Some DNA sequences, when inserted into particular vectors, can lower the copy number of the plasmid. However, for high plasmid yield, you should choose the appropriate growth conditions and media type, strain selection, and the volume of culture.
20. The selection of *E. coli* strain will depend on the compatibility with the recombinant plasmid.
21. The selection of binary vector will depend on all of the factors previously described.
22. In our experience we recommend for the best results, to clone first the shortest chain expression cassette and after the highest, into binary vector.
23. In case plasmid DNA yield is a problem, different host strains and growth conditions should be tested.
24. The selection of *Agrobacterium* strain will depend on the compatibility with the recombinant plasmid.
25. Ten-days growth period between each subculture is sufficient to achieve one-half settle volume from the suspension. However if the physical conditions change the growth could be slower or faster, and consequently the growth period could change. It is important to perform the subculture at the one-half settle volume state to keep the suspension in homogenous cell types.
26. The pretreatment period induce high homogenous proliferation state of the sample and could vary according to physical conditions of the culture.
27. The cell lawn should be covered by *Agrobacterium* suspension to achieve high transformation rate.
28. Conditions for efficient transformation should fit the requirements for *Agrobacterium* virulence induction and tobacco cell proliferation. In the present method, the temperature for optimal tobacco cell division determined the temperature for cocultivation with *Agrobacterium*.
29. The inoculated cells could be harvested before the fifth cocultivation day for early assessment of transgene expression.
30. The leaves should be fully expanded. Usually, in the four leaves plant state they are located at the bottom of the plant. Other leaves can be infiltrated if they are fully expanded.
31. The infiltration of the bacterial suspension should occupy all leave interspace for high recombinant protein yield.
32. Keep plants under the optimal culture conditions above described for optimal transgene expression.
33. All the procedures from described in this section must be performed in a laminar flow hood to prevent contamination

of cultures. Before the sterilization process, seeds are maintained in distillated water for 1 h and then are treated with commercial ethanol for 30 s. The method using cotyledonal explants from germinates normally yield higher regeneration–transformation efficiency.

34. This method is useful also for transformation when the explants are haploid lines and cannot produce seeds.
35. For preparing the medium, the use of commercial salts and vitamins mixed is very convenient to obtain high and reproducible results. These products can be purchased from Sigma and Duchefa (*see* details in transient transformation section).
36. For the culture of LBA 4404 strain, the addition of 0.3 g/L $MgSO_4{\cdot}7H_2O$ to YEB medium avoids cell aggregation in the final culture (*see* details in ref. *(14)*. Culture media components for *A. tumefaciens* are detailed in transient transformation section.
37. For the transformation experiments, the cotyledonal explants are taken 20 days after seed sowing in MS medium, incubated at 16-h photoperiod and a temperature of 25°C. Cut the explants under sterile conditions into small pieces (about 2–3 mm in length). The addition of acetosyringone could improve the frequency of transformation.
38. The *Agrobacterium* suspension is drooped (10–30 μL) on explants, depending of tissue area.
39. To remove *Agrobacterium* after coculture period using PPT as selection agent is very convenient to use the carbenicillin instead of cefotaxime. We have noted an over growth of *Agrobacterium* on infested explants just after a week of culture in media supplemented with 500 mg/L cefotaxime and 5 mg/L PPT.
40. During the transfer, wash carefully the gel from the root tissue with tap water.
41. One important step in the detection of antibody transgene is the quality of plant genomic DNA.
42. Is very important to take great care to minimize the potential for cross-contamination between samples. Use individuals pipette tips for all additions, being careful not to cross-contaminate the samples
43. These steps of PCR should be optimized for each template and primer combination.
44. The reaction should be overlayered with nuclease-free mineral oil to prevent condensation and evaporation.
45. Numerous formulas exist to determinate the melting temperature of oligonucleotide primers. We recommend use 5°C below any calculated melting temperature.

46. For most applications, a PVC microtiter plate is best; however, consult manufacturer guidelines to determine the most appropriate type of plate for protein binding.
47. We recommend modifying the incubation time and temperature with samples from transgenic and nontransgenic plants, if your antibody expression was very low in plants. In our experience, the coating step should be 1 to 4 h at 37°C and the testing samples overnight at 4°C.
48. Some enzyme substrates are considered hazardous, due to potential carcinogenicity. Handle with care and refer to Material Safety Data Sheets for proper handling precautions.
49. It is important that gloves are worn at all times while handling the membrane to prevent cross-contamination.
50. For longer transfer times, it is recommended that electroblotting be performed at 4°C to prevent overheating and buffer decomposition.
51. To ensure the transfer is complete, the membrane can be stained with Ponceau's Concentrate.
52. The most popular membranes blocking agents are BSA and skim milk powder.
53. The antibodies' working dilutions, temperature, and incubation time required to give optimal results will vary depending the expression levels of antibody in transgenic plants.
54. Trypsin is neutralized with 10 mL of RPMI 1640 medium with 10% fetal calf serum.
55. Sodium azide can be used to prevent the modulation and internalization of surface antigens, which can produce a loss of fluorescence intensity. Do not add sodium azide to buffers if you are concerned with recovering cell function, for example, if cells are to be collected for functional assays.
56. Cells are usually stained in Eppendorf tubes. However, they can be stained in any container for which you have an appropriate centrifuge, for example, test tubes, polystyrene round-bottom 12 × 75 mm Falcon tubes, and 96-well round-bottomed microtiter plates. In general, cells should be spun down hard enough that the supernatant fluid can be removed with little loss of cells, but not so hard that the cells are difficult to resuspend.
57. Propridium iodide can also be added at this point for dead cell exclusion.
58. For extended storage (until 48 h), resuspended cells in cold PBS with 4% paraformaldehyde to prevent deterioration.
59. Trypsin is neutralized with 10 mL of medium RPMI 1640 with 10% fetal calf serum.

60. It is always useful to check the viability of the cells. It should be around 95% and not less than 90%.

61. For best results, analyze the cells on the flow cytometer as soon as possible. We recommend analysis on the same day. For extended storage as well as for greater flexibility in planning time on the cytometer, resuspended cells in cool PBS with 4% paraformaldehyde to prevent deterioration.

References

1. Lawrence, S. (2006) Biotech blockbusters consolidate markets. Nat. Biotechnol. 24(12), 1466.
2. Lawrence, S. (2007) Billion dollar babies – Biotech drugs as blockbusters. Nat. Biotechnol. 25(4), 380–382.
3. Sannes, L., Branca, M. (2006) The evolving market for monoclonal antibodies: Facing new opportunities and pitfalls. PharmaWeek http:// www.pharmaweek.com/Exclusive_Content/2_23.asp
4. Hood, E., Woodard, S.L., Horn, M.E. (2002) Monoclonal antibody manufacturing in transgenic plants – Myths and realities. Curr. Opin. Biotechnol. 13, 630–635.
5. Pujol, M., Ayala, M., Borroto, C. (2005) State of plant-made pharmaceuticals sector 2005: Setting healthy roots. Biotecnología Aplicada 22(1), 59–60.
6. Raskin, I., Ribnicky, D.M., Komarnytsky, S., Ilic, N., Poulev, A., Borisjuk, N., Brinker, A., Moreno, D.A., Ripoll, C., Yakoby, N., O'Neal, J.M., Cornwell, T., Pastor, I., Fridlender, B. (2002) Plants and human health in the twenty-first century. Trends Biotechnol. 20(12), 522–530.
7. Murashige, T., Skoog, F. (1962) A revised medium for rapid growth and bioassays with tobacco tissue culture. Physiol. Plant 15, 473–497.
8. Rodríguez, M., Ramírez, N., Ayala, M., Freyre, F., Pérez, L., Triguero, A., Mateo, C., Selman-Housein, G., Gavilondo, J. V., Pujol, M. (2005) Transient expression in tobacco leaves of an aglycosilated recombinant antibody against the epidermal growth factor receptor. Biotechnol. Bioeng. 89, 188–194.
9. Chomczynski, P. (1993) A reagent for the single-step simultaneous isolation of RNA, DNA and proteins from cell and tissue samples. Biotechniques 15, 532–537.
10. Sambrook, J., Fritsch, E.F., Maniatis, T. (1989) Molecular Cloning. A Laboratory Manual. 2nd ed. ColdSpring Harbor Laboratory, Cold Spring Harbor, N.Y.
11. López, A., Zaldúa, Z., Pimentel, E., García, M., García, R., Mena, J., Morán, R., Selman, G. (1996) Modification of sporamin gene from sweet potato with a synthetic DNA fragment. Nucleotide sequence and expression in *Escherichia coli*. Biotecnología Aplicada 13, 265–270.
12. Schillberg, S., Emans, N., Fischer, R. (2002) Antibody molecular farming in plants and plant cells. Photochem. Rev. 1, 45–54.
13. Fischer, R., Twyman, R.M., Schillberg, S. (2003) Production de antibodies in plants and use for global health. Vaccine 21, 820–825.
14. Deblaere, R., Bytebier, B., De Greve, H., Deboeck, F., Schell, J., Van Montagu, M., Leemans, J. (1985) Efficient octopine Ti plasmid-derived vectors for *Agrobacterium*-mediated gene transfer to plants. Nucl. Acids Res. 13, 4777–4788.
15. Hellens, R.P., Mullineaux, P.M., Klee, H. (2000) Technical Focus: A guide to *Agrobacterium* binary Ti vectors. Trends Plant Sci. 5, 446–451.
16. Hajdukiewicz, P., Svab, Z., Maliga, P. (1994) The small, versatile pPZP family of *Agrobacterium* binary vectors for plant transformation. Plant Mol. Biol. 25, 989–994.
17. Fuentes, A., Ramos, P., Ayra, C., Rodríguez, M., Ramírez, N., Pujol, M. (2004) Development of a highly efficient system for assessing recombinant gene expression in plant cell suspensions via *Agrobacterium tumefaciens* transformation. Biotechnol. Appl. Biochem. 39, 355–361.
18. Yang, H., Morita, A., Matsubayashi, Y., Nakamura, K., Sakagami, Y. (2000) A rapid and efficient system of *Agrobacterium* infection-

mediated transient gene expression in rice Oc cells and its application for analysis of the expression of preprophytosulfokine, a precursor of phytosulfokine-a, encoded by OsPSK gene. Plant Cell Physiol. 41, 814–816.

19. Dafhnis, F., Trujillo, L., Arrieta, J., Fuentes, A., Ramos, P., Quintero, O., García, B.M., Hernández, L. (2000) Recombinant expression of active *Penicillium minioluteum* endodextranase in alternative hosts. J. Biochem. Mol. Biol. Biophys. 4, 1–7.
20. Vaquero, C., Sack, M., Chandler, J., Drossard, J., Schuster, F., Monecke, M., Schillberg, S., Fischer, R. (1999) Transient expression of a tumor-specific single-chain fragment and a chimeric antibody in tobacco leaves. Proc. Natl. Acad. Sci. U S A. 96, 11128–11133.
21. Yang, Y., Li, R., Qi, M. (2000) In vivo analysis of plant promoters and transcription factors by agroinfiltration of tobacco leaves. Plant J. 22, 543–551.
22. Gleba, Y., Marillonnet, S., Klimynk, V. (2004) Engineering viral expression vectors for plants: The "full virus" and the "deconstructed virus" strategies. Curr. Opin. Plant Biol. 7, 182–188.
23. Sanford, J.C., Smith, F.D., Russell, J.A. (1993) Optimizing the biolistic process for different biological applications. Methods Enzymol. 217, 483–509.
24. Gynheung, A. (1985) High efficiency transformation of cultured tobacco cells. Plant Physiol. 79, 568–570.
25. Peña, L., Pérez, R., Cervera, M., Juárez, J., Navarro, L. (2004) Early events in *Agrobacterium*-mediated genetic transformation of citrus explants. Ann. Bot. 94, 67–74.
26. Villemont, E., Dubois, F., Sangwan, R., Vasseur, G., Bourgeois, Y., Sangwan-Norreel, B. (1997) Role of the host cell cycle in the *Agrobacterium*-mediated genetic transformation of Petunia: Evidence of an S-phase control mechanism for T-DNA transfer. Planta 201, 160–172.
27. Zupan, J., Muth, T., Draper, O., Zambrisky, P. (2000) The transfer of DNA from *Agrobacterium tumefaciens* into plants: A feast of fundamental insights. Plant J. 23, 11–28.
28. Sheng, O., Citovsky, V. (1996) *Agrobacterium*-plant cell DNA transport: Have virulence proteins will travel. Plant Cell 8, 1699–1710.
29. Hoekema, A. (1983) A binary plant vector strategy based on separation of vir and T-region of the *Agrobacterium tumafaciens* Ti-plasmid. Nature 303, 179–180.
30. Horsch, R.B., Fry, J.E., Hoffmann, N.L., Eichholtz, D., Rogers, S.G., Fraley, R.T. (1985) A simple and general-method for transferring genes into plants. Science 227, 1229–1231
31. Saiki, R.K., Scharf, S., Faloona, F., Mullis, K.B., Horn, G.T., Erlich, H.A., Arnheim, N. (1985) Enzymatic amplification of beta-globin genomic sequences and restriction site analysis for diagnosis of sickle cell anemia. Science 230, 1350.
32. Lyons, A.B., Parish,C.R. (1994) Determination of lymphocyte division by flow cytometry. J. Immunol. Methods 171, 131–7.
33. Perez, L., Ayala, M., Pimentel, G., Bell, H., Canáan-Haden, L., Bequet, M., Gonzalez, L.J., Miranda, M., Ravelo, R., Roque, L., Acevedo, B., Oliva, J.P.,Gavilondo, J.V. (2006) A multivalent recombinant antibody fragment specific for carcinoembryonic antigen. Biotechnol. Appl. Biochem. 43, 39–48.
34. Graves, S.W., Woods, T.A., Kim, H., Nolan, J.P. (2005) Direct fluorescent staining and analysis of proteins on microspheres using CBQCA. Cytometry Part A 65A, 50–8.
35. Moore, A.C., Gallimore, A., Draper, S.J., Watkins, K.R., Gilbert, S.C., Hill, A.V.S. (2005) Anti-CD25 antibody enhancement of vaccine-induced immunogenicity: Increased durable cellular immunity with reduced immunodominance. J. Immunol. 175, 7264–73.
36. Marks, J.D., Hoogenboom, H.R., Bonnert, T.P., McCafferty, J., Griffiths, A.D., Winter, G. (1991) By-passing immunization. Human antibodies from V-gene libraries displayed on phage. J. Mol. Biol. 222, 581–597.
37. Marks, J.D., Tristem, M., Karpas, A., Winter, G. (1991) Oligonucleotide primers for polymerase chain reaction amplification of human immunoglobulin variable genes and design of family-specific oligonucleotide probes. Eur. J. Immunol. 21, 985–991.
38. Kabat, E.A., Wu, T.T., Perry, H.M., Gottesman, K.S., Foeller, C. (1991) Sequences of Proteins of Immunological Interest. 5th ed. US Department of Health and Human Services, Bethesda.

Chapter 8

Physcomitrella patens: A Non-Vascular Plant for Recombinant Protein Production

David Liénard and Fabien Nogué

Summary

The moss Physcomitrella patens is a long-standing model for studying plant development, growth and cell differentiation in particular. Interest in this non-vascular plant arose following the discovery that homologous recombination is an efficient process. P. patens is, therefore, a tool of choice not only to study gene function but also for recombinant protein production. This system has many attributes that are advantageous for molecular farming: protein production in cell suspension, the possibility of generating targeted knockout mutants for glycoengineering and quantitative optimization for protein production. In terms of technical advances, P. patens is one of the most up-to-date plant expression systems and is a promising alternative to animal cell factories for the production of therapeutic proteins with either simple or highly complex structures.

Key words: Physcomitrella Patens, Moss, Homologous recombination, Cre/lox, Recombinant protein production.

8.1. Introduction

In the current context of increasing demand for therapeutic protein production, Physcomitrella patens is emerging as a unique system for molecular farming, which is cheap and efficient. Proteins produced by moss expression systems are post-translationally modified (by glycosylation, formation of disulphide bonds ...) and are manufactured in an animal product-free environment (free from bacterial toxins, mammalian viruses and other pathogens). This photoautotrophic plant can be easily grown as a suspension culture under strictly controlled conditions *(1, 2)*.

Loïc Faye and Véronique Gomord (eds.), *Methods in Molecular Biology, Recombinant Proteins From Plants, vol. 483*

DOI: 10.1007/978-1-59745-407-0_8

The P. patens life cycle is divided into two stages: the main haploid stage (the gametophyte constituted of the protonema and the gametophore), and a diploid stage represented by the sporogone, which develops on the gametophyte. The entire development cycle takes ten to twelve weeks in vitro under controlled temperature and light conditions and using simple mineral solutions *(3, 4)*. One of the interesting features of P. patens is that it is possible to keep it at the haploid stage in either liquid culture (agitated vials or bioreactors) or solid (Petri dish) medium. Vigorous growth of haploid protonema is maintained for long periods, by periodically inoculating new media with small pieces of growing protonema tissue (fragmented with an Ultraturax) *(5)*. As it grows fully differentiated, somaclonal variation, which is a typical problem of higher plant cell cultures, is avoided. Thus, with this system, molecularly stable cell lines are obtained which maintain the required level of recombinant protein production for long periods of cultivation. Furthermore, recombinant protein expression can be increased by using signal peptides to target secretion into the culture medium from which subsequent low-cost purification is easy due to the simplicity of the growth medium (water and minerals). Expression levels can also be improved by using appropriate molecular tools like 5¢ regulatory sequences such as promoters.

The P. patens N-glycosylation pathway was shown to be very similar to that of higher plants *(6)*. Thus, in the case of glycosylated protein production, this means that glycoepitopes such as the Lewis a, the bisecting β1,2 xylose and the core α1,3 fucose residues will be present on proteins *(7)*. These three kinds of residues are not found on mammal proteins and are immunogenic in humans if administered parenterally *(8)*. Thus, P. patens was engineered to produce a strain that does not add β1,2 xylose or α1,3 fucose, but instead produces plant-made pharmaceuticals bearing a core heptasaccharide identical to that of a human IgG. This was possible due to the property of homologous recombination in P. patens. For most eukaryotes, genetic transformation leads to the random integration of the introduced gene into the host genome (a process known as illegitimate integration). However, the presence in the exogenic DNA of sequences identical to host genome sequences leads to targeted integration by homologous recombination. This "gene targeting" was initially discovered in the yeast Saccharomyces cerevisae. The frequency of successful targeted integration is quite different between species and among sequences *(9)*. In angiosperms, this technique is unsuitable for practical use in plant biology because efficient targeting occurs at very low frequencies. P. patens seems to be the only plant in which targeted integrations with frequencies close to those observed in S. cerevisae are found *(3, 10, 11)*. Any kind of plasmid vector can be used since transformation is carried out using a PEG-based method rather than Agrobacterium intermediates. Two types of target plasmid vectors have been developed: replacement or insertional plasmid vectors. Replacement vectors contain two sequences which are homologous to the target gene. These two sequences are separated by, for example, a resistance gene. In this case, the endogenous sequence is replaced by a non-functional sequence following a double crossover event to produce a knockout mutant. Insertional vectors contain one single sequence which is homologous to the target gene. In this case, the entire vector is integrated into the host genome by a simple crossover event. This second technique is often used to study the effect of a single base change in the sequence. More recently, precise gene manipulation in this moss was further facilitated by the Cre/lox system – an invaluable gene expression control tool *(12)*. This system is used to delete undesired DNA sequences, such as resistance genes, and to modify chromosome architecture *(13)*.

One of the most recent examples of the potential for molecular farming in mosses was the production of a recombinant human erythropoietin. This gene encoding this protein was expressed in a knockout strain lacking the enzymes which add β1,2 xylose or α1,3 fucose. This was the first successful production of a highly glycosylated human protein, intended for drug treatment, in a plant system *(14)*. In this chapter, we provide details of the basic protocols for vegetative propagation and generation of transformed moss without resistance genes.

8.2. Materials

8.2.1. Plant Material

P. patens (Gransden wild-type) are grown axenically as suspension-cultured cells or in Petri dishes (90- or 45-mm diameter) containing liquid culture medium called $PPNH_4$ medium *(15)* solidified with 0.8% (w/v) agar. The Petri dishes are wrapped with "3M Micropore" tape. They are kept in a growth chamber at 23–25°C with a 16/8-h light/dark cycle (light about 60–75 $\mu mol\ m^{-2}\ s^{-1}$). Plants are subcultured every 7 days by inoculating plates or liquid medium with a freshly fragmented filter suspension or directly by transfer of some protonema onto plates.

P. patens can be stored as a protonema suspension and kept in sterile water at 4°C for 1 year.

8.2.2. Laboratory Equipment and Reagents

1. Plant Culture Room with illumination system
2. An orbital shaker
3. Sieves of 140 µm, 80 µm and 40 µm
4. "3M Micropore ™" tape Ref. 1530-0

5. Cellophane disk of 80-mm diameter (Ref. 13849, A. A. Packaging Limited, UK)
6. General tissue culture and plant culture materials
7. A homogenizer

8.2.3. Growth Media and Buffers

Media and buffers are made up as aqueous preparations except macro-elements which are freshly made. The media for tissue culture are used after sterilization using standard procedures.

8.2.3.1. Stock Solutions

1. Macroelements final concentration: 3.37 mM $Ca(NO_3)_2$, 1.01 mM $MgSO_4$, and 4.50 mM $FeSO_4$
2. Microelements 1,000× buffer: 0.22 mM $CuSO_4$, 0.19 mM $ZnSO_4$, 0.99 mM H_3BO_3, 1.97 mM $MnCl_2$, 0.23 mM $CoCl_2$, 0.17 mM KI and 0.10 mM Na_2MoO_4. This buffer must be autoclaved or filter-sterilized (0.2 μm) and then kept at 4°C
3. Phosphate 1,000× buffer: 1.84 M KH_2PO_4 (pH 7 with KOH)
4. NH_4 tartrate: 2.72 10^{-3} M

8.2.3.2. Media Composition

1. Liquid $PPNH_4$ medium is prepared with freshly made macro-elements, 1 mL of microelements 1,000× per litre of medium, 1 mL of phosphate 1,000× buffer per litre of medium and NH_4 tartrate.
2. Solid $PPNH_4$ medium is prepared as liquid with 8% (w/v) agar.

8.2.4. Moss Transformation

1. MMM medium: 0.48 M mannitol, 15 mM $MgCl_2$ and 0.1% MES. pH to 5.6 with KOH. Filter-sterilized (0.2 μm) and stored at 4°C.
2. PEG solution: 0.48 M mannitol, 0.10 M $Ca(NO_3)_2$ and 40% PEG 4000 (pH to 8.0 with 10 mM Tris). This solution is heated at 45°C to dissolve PEG. This medium must be filter-sterilized (0.2 μm) and stored at –20°C.
3. $PPNH_4$ medium with mannitol: liquid $PPNH_4$ medium with 0.48 M mannitol. This medium must be autoclaved and can be stored at room temperature.
4. Top layer: 0.48 M mannitol and 1.4% agar. This medium must be autoclaved and can be stored at room temperature.

8.2.5. Genomic DNA Extraction for Knockout Screening

Extraction buffer: 220 mM Tris-HCl (pH 7.5), 250 mM NaCl, 25 mM EDTA and 0.5% SDS

8.2.6. Genomic DNA Extraction for Southern Blotting

1. 2XCTAB: 2XCTAB (w/v), 100 mM Tris-HCl (pH 8.0), 1.4 M NaCl, 1% PVP
2. 10% CTAB: 10% CTAB and 0.7 M NaCl
3. CTAB precipitation buffer: 1% CTAB, 50 mM Tris-HCl (pH 8.0), 10 mM EDTA (pH 8.0)
4. HSTE: 10 mM Tris-HCl (pH 8.0), 1 mM EDTA (pH 8.0), 1 M NaCl

8.3. Methods

8.3.1. Stable Transformation of Physcomitrella Patens Moss

8.3.1.1. Amplification of Protonema Tissue for Protoplast Isolation

1. Collect 1 Petri dish worth of protonema tissue. The protonema should be less than 2 weeks old so either previously grown on solid $PPNH_4$ medium or freshly prepared on $PPNH_4$ from a culture stock at 4°C.
2. Transfer to a sterile tube with 10 mL sterile water.
3. Fragment (using homogenizer) and filter through 140 μm sieve.
4. Resuspend in sterile water.
5. Inoculate 2–3 mL of the freshly fragmented protonema per Petri dish poured with solid $PPNH_4$ medium and overlaid with cellophane disk.
6. Seal the plates with Micropore tape.
7. Incubate at 24°C with a 16-h light/ 8-h dark light cycle.
8. Prepare 4 plates per transformation and collect protonema after 5–7 days (best yield after 6 days).

8.3.1.2. Protoplast Isolation

1. Collect protonema from the 4 plates grown for 6 days on $PPNH_4$; use a sterile spatula.
2. Transfer to a Petri dish filled with 10 mL 8.5% mannitol.
3. Add 10 mL sterile 2% (w/v) non-purified Driselase.
4. Incubate at room temperature for 30 min with occasional gentle mixing.
5. Filter the preparation through two sterile superposed sieves (80 μm and 40 μm); to sterilize the sieves place them on a 100 mL beaker, wrap in aluminium foil and autoclave at 120°C for 20 min.
6. Rinse the sieves with 5 mL 8.5% mannitol.
7. Transfer the preparation to a sterile 30 mL "corex" tube.
8. Harvest the protoplasts by low-speed centrifugation at 400g for 5 min at room temperature and discard supernatant.

9. Gently resuspend the pellet in 10 mL 8.5% mannitol.
10. Repeat centrifugation at 400g for 5 min at room temperature and discard supernatant.
11. Gently resuspend the pellet in 10 mL 8.5% mannitol.
12. Take an aliquot (15 μL) of the suspension (use a sterile yellow cut pipette tip) and count the protoplasts with a hematocytometer ("Malassez").

8.3.1.3. Counting Protoplasts

1. Cover the hematocytometer with a coverslip and gently inject the 15 μL of protoplast solution between the slide and the coverslip.
2. 2. Count two lines: X number of protoplasts × 5 (lines) × 10^3 × 10 mL solution = number of protoplasts.
3. Repeat centrifugation at 600g for 5 min at room temperature and discard supernatant.
4. Gently resuspend the pellet in MMM solution at a finale concentration of 1.2 10^6 protoplasts/mL (see **Note 1**).

8.3.1.4. Transformation of Protoplasts (see **Note 2**)

1. Dispense 10 μg of digested DNA (max. vol. 30 μL) into sterile 14-mL plastic tubes.
2. Add 300 μL of the protoplast suspension and mix gently.
3. Add 300 μL PEG solution and mix gently.
4. Heat shock for 5 min at 43°C in a water bath.
5. Leave for 10 min at room temperature.
6. Progressively add to each sample: 5 × 300 μL and then 5 × 1 mL of liquid $PPNH_4$ medium with mannitol without mixing.
7. Keep the transformed protoplasts overnight in darkness at 24°C.

8.3.1.5. Plating Out the Transformed Protoplasts (see **Note 3**)

1. Reduce the volume of each sample to 4 mL by taking out 3.1 mL of each supernatant and add 4 mL of Top layer.
2. Mix by gently inverting the tube twice.
3. Plate out 2 mL of protoplasts + agar per plate (4 plates total per sample) on solid $PPNH_4$ with mannitol overlaid with cellophane disk.
4. Seal the plates with Micropore tape.
5. Incubate in the growth chamber at 24°C with a 16-h light/8-h dark light cycle.

8.3.1.6. Selection of Transformants

1. Estimate the number of regenerated protoplasts after 4 days of growth by observation under the microscope: count 5 different fields and take the average = X number of regenerated protoplasts × 1,600 × 4 = number of regenerated

protoplasts/transformation 300 μL protoplasts = 360,000 protoplasts per transformation, at the beginning in classical experiments 30–50% of the protoplasts should regenerate).

2. After 6 days of growth, transfer the cellophane disk + embedded protoplasts to Petri dishes containing solid $PPNH_4$ medium supplemented with corresponding antibiotics (50 μg/mL G-418 or 25 μg/mL Hygromycin B ....) and seal the plates with Micropore tape.
3. Incubate in the growth chamber at 24°C with a 16-h light/8-h dark light cycle.

8.3.1.7. Selection of Stable Transformants

1. After 7–10 days on antibiotics (90% of the resistant colonies are unstable transformants), transfer each cellophane disk to Petri dishes containing solid $PPNH_4$**without selection** and seal the plates with Micropore tape.
2. Incubate in the growth chamber at 24°C with a 16-h light/ 8-h dark light cycle for 10 days.
3. Return to selective plates, transfer the cellophane disk to Petri dishes containing solid $PPNH_4$ supplemented with corresponding antibiotics and seal the plates with Micropore.
4. Incubation in the growth chamber at 24°C with 16-h of light and for at least 10 days

8.3.1.8. Knockout Screening

1. Take 5 mm of resistant colonies (make a backup of each colony by inoculating on $PPNH_4$ medium).
2. Grind up the colony in an Eppendorf tube filled with 200 μL extraction buffer with an electric pestle and a bit of clean sand.
3. After 5 min centrifugation, take the supernatant (200 μL) and precipitate DNA with isopropanol (200 μL, same volume) at room temperature.
4. Discard the supernatant and dry the pellet 20 min at room temperature
5. Resuspend each pellet in 50–100 μL sterile water (depending on the size of the DNA pellet).
6. Use 1 μL for PCR analysis with primers specific for the resistance cassette and primers specific for the targeted gene (outside the transformation construct).

8.3.2. Transient Transformation of Physcomitrella Patens Moss

Same protocol as that for stable transformation described above until §3.1.4.

1. Harvest the protoplasts by low-speed centrifugation at 600g for 5 min at room temperature and discard supernatant.

2. Gently resuspend the pellet in 2 mL of $PPNH_4$ liquid medium with mannitol.
3. Incubate in the growth chamber at 24°C with a 16-h light/8-h dark light cycle.

8.3.3. Generation of a Clean Deletion by Transient Cre Recombinase Expression

1. Transiently express (see **Subheading 8.3.2**) a constitutive Cre expression cassette *(16)* in antibiotic resistant protoplasts.
2. Keep the transformed protoplasts in the darkness at 24°C for 5 h, then add 2 mL of Top layer to the supernatant.
3. Mix by gently inverting the tube twice.
4. Plate out 100 μL of protoplasts + agar per plate on solid $PPNH_4$ with mannitol overlaid with cellophane disk.
5. Regenerate protoplasts at a low density on non-selective medium for 2 weeks.
6. Replicate fragments of single protoplast-derived colonies on selective (i.e. 50 μg/mL G-418 or 25 μg/mL Hygromycin B ....) and non-selective medium.
7. Determine loss of antibiotic resistance after an additional week of growth. Replicates growing on non-selective medium are used for subsequent amplification and molecular analysis.

8.3.4. DNA Extraction for Southern Blot

1. Collect material from protonema tissue grown for a week under standard conditions and freeze it in liquid nitrogen.
2. Cool down a standard coffee grinder by grinding 10 g of dry ice in it.
3. Mix in the coffee grinder bucket 5–10 g of plant material with 5–10 g of dry ice.
4. Dispense 2XCTAB, 1 mL per g of fresh weight, freshly supplemented with 20 mM of beta-mercaptoethanol in a 50 mL Falcon tube, and preheat the extraction buffer at 65°C in a water bath. Preheat the 10% CTAB solution at the same time.
5. Slowly add the ground material to the tube, mix to obtain a homogeneous slimy soup and incubate at 65°C for 15 min.
6. Add 2XCTAB to a final volume equal to 2 times the weight of the sample
7. Bring the tubes back to room temperature (water bath) and add 1 volume of $CHCl_3$/isoamyl alcohol (24:1).
8. Mix vigorously to obtain a stable emulsion, then centrifuge at 2,600g for 15 min
9. Transfer the supernatant (aqueous phase) to a clean centrifuge tube and measure its volume.
10. Add 1/10 of this volume of 10% CTAB solution and mix gently.

11. Add 1 volume of CHCl3/isoamyl alcohol, mix gently and centrifuge for 10 min at 10,000g in a swing-out rotor.
12. Transfer the clear supernatant to a clean centrifuge tube and measure its volume. Add one volume of CTAB precipitation buffer and leave for 30 min at RT. A milky precipitate forms immediately upon the addition of precipitation buffer. If this does not appear, it means that the yield of the extraction is low.
13. Centrifuge for 15 min at 10,000g in a swing-out rotor. The supernatant can be yellow but should be completely clear at this stage. Otherwise repeat centrifugation.
14. Discard the supernatant and resuspend the pellet in 400 µL HSTE. Incubate the tube at 50°C to help redissolve the pellet and if necessary mix by pipette up and down.
15. Transfer the solution to an Eppendorf, rinse the wall of the tube with 100 µL of HSTE and pool in the Eppendorf.
16. Precipitate DNA with 2.5 volume of absolute ethanol and leave for 30 min at –20°C.
17. Centrifuge 10 min at 10,000g in a microcentrifuge. Discard supernatant, wash the pellet with 500 µL EtOH 70%, dry the DNA and then resuspend the pellet in water.

8.4. Notes

1. Start the transformation of the protoplasts without delay.
2. Do not forget the negative control (protoplasts without DNA).
3. Water bath or incubator must be at 45°C.
4. Prepare enough digested DNA (10 µg per transformation), that was cleaned up by phenol/chloroform extraction, EtOH precipitated with sterile solutions and resuspended in sterile water.
5. Do not forget to prepare plates (Petri dishes) with solid $PPNH_4$ with mannitol overlaid with cellophane disk.
6. Melt Top layer and cool down at 50°C.

References

1. Fischer, R., Twyman, R. M., and Schillberg, S. (2003) Production of antibodies in plants and their use for global health. Vaccine 21, 820–825.
2. Koprivova, A., Stemmer, C., Altmann, F., Hoffmann, A., Kopriva, S., Gorr, G., Reski, R., and Decker, E. L. (2004) Targeted knockouts of Physcomitrella lacking plant-specific immunogenic N-glycans. Plant Biotechnol. J. 2, 517–523.
3. Schaefer, D., Zryd, J. P., Knight, C. D., and Cove, D. J. (1991) Stable transformation of the moss Physcomitrella patens. Mol. Gen. Genet. 226, 418–424.

4. Schaefer, D. G., and Zryd, J. P. (1997) Efficient gene targeting in the moss Physcomitrella patens. Plant J. 11, 1195–1206.
5. Cove, D. (2005) The moss Physcomitrella patens. Annu. Rev. Genet. 39, 339–358.
6. Vietor, R., Loutelier, B. C., Fitchette, A., Margerie, P., Gonneau, M., Faye, L., and Lerouge, P. (2003) Protein N-glycosylation is similar in the moss Physcomitrella patens and in higher plants. Planta 218, 269–275.
7. Gomord, V., and Faye, L. (2004) Posttranslational modification of therapeutic proteins in plants. Curr. Opin. Plant Biol. 7, 171–181.
8. Gomord, V., Chamberlain, P., Jefferis, R., and Faye, L. (2005) Biopharmaceutical production in plants: problems, solutions and opportunities. Trends Biotechnol. 23, 559–565.
9. Capecchi, M. R. (2005) Gene targeting in mice: functional analysis of the mammalian genome for the twenty-first century. Nat. Rev. Genet. 6, 507–512.
10. Schaefer, D. G. (2001) Gene targeting in Physcomitrella patens. Curr. Opin. Plant Biol. 4, 143–150.
11. Schaefer, D. G., and Zryd, J. P. (2001) The moss Physcomitrella patens, now and then. Plant Physiol. 127, 1430–1438.
12. Trouiller, B., Schaefer, D. G., Charlot, F., and Nogue, F. (2006) MSH2 is essential for the preservation of genome integrity and prevents homeologous recombination in the moss Physcomitrella patens. Nucleic Acids Res. 34, 232–242.
13. Sauer, B. (1993) Manipulation of transgenes by site-specific recombination: use of Cre recombinase. Methods Enzymol. 225, 890–900.
14. Weise, A., Altmann, F., Rodriguez-Franco, M., Sjoberg, E., Baumer, W., Launhardt, H., Kietzmann, M., and Gorr, G. (2007) High-level expression of secreted complex glycosylated recombinant human erythropoietin in the Physcomitrella delta-fuc-t & delta-xyl-t mutant. Plant Biotechnol. J. 5, 389–401.
15. Ashton, N. W., Cove, D. J., and Featherstone, D. R. (1979) Isolation and physiological analysis of mutants of the moss, Physcomitrella-patens, which over-produce gametophores. Planta 144, 437–442.
16. Albert, H., Dale, E. C., Lee, E., and Ow, D. W. (1995) Site-specific integration of DNA into wild-type and mutant lox sites placed in the plant genome. Plant J. 7, 649–659.

Chapter 9

Production of Recombinant Proteins in Suspension-Cultured Plant Cells

Carole Plasson, Rémy Michel, David Lienard, Claude Saint-Jore-Dupas, Christophe Sourrouille, Ghislaine Grenier de March, and Véronique Gomord

Summary

Plants have emerged in the past decade as a suitable alternative to the current production systems for recombinant pharmaceutical proteins and, today their potential for low-cost production of high quality, much safer and biologically active mammalian proteins is largely documented.

Among various plant expression systems being explored, genetically modified suspension-cultured plant cells offer a promising system for production of biopharmaceuticals. Indeed, when compared to other plant-based production platforms that have been explored, suspension-cultured plant cells have the advantage of being totally devoid of problems associated with the vagaries of weather, pest, soil and gene flow in the environment. Because of short growth cycles, the timescale needed for the production of recombinant proteins in plant cell culture can be counted in days or weeks after transformation compared to months needed for the production in transgenic plants. Moreover, recovery and purification of recombinant proteins from plant biomass is an expensive and technically challenging business that may amount to 80–94% of the final product cost. One additional advantage of plant cell culture is that the recombinant protein fused with a signal sequence can be expressed and secreted into the culture medium, and therefore recovered and purified in the absence of large quantities of contaminating proteins. Consequently, the downstream processing of proteins extracted from plant cell culture medium is less expensive, which may/does balance the higher costs of fermentation. When needed for clinical use, recombinant proteins are easily produced in suspension-cultured plant cells under certified, controllable and sterile conditions that offer improved safety and provide advantages for good manufacturing practices and regulatory compliance.

In this chapter, we present basic protocols for rapid generation of transgenic suspension-cultured cells of *Nicotiana tabacum*, *Oriza sativa* and *Arabidopis thaliana*. These systems are powerful tools for plant-made pharmaceuticals production in highly controlled conditions.

Key words: Suspension-cultured plant cells, *Nicotiana tabacum*, *Arabidopsis thaliana*, *Oriza sativa*, Plant-made pharmaceutical (PMP).

Loïc Faye and Véronique Gomord (eds.), *Methods in Molecular Biology, Recombinant Proteins From Plants, vol. 483*

DOI: 10.1007/978-1-59745-407-0_9

9.1. Introduction

Altogether, the biochemical, technical and economic limitations on existing prokaryote and eukaryote expression systems, the growing clinical demand for complex therapeutic proteins, and the lack of bioreactor capacity have created substantial interest in developing new expression systems for the production of therapeutic proteins. These are the reasons why plants have emerged in the past decade as a suitable alternative to the current production systems, and today their potential for low-cost production of high quality, much safer and biologically active mammalian proteins is largely documented *(1, 2, 3, 4, 5, 6, 7)*. One of the major advantages of using transgenic plants, over other production systems available for large-scale production such as yeasts or *Escherichia coli*, is the ability of plant cells to perform most post-translational modifications required for protein activity and suitable pharmacokinetics properties *(8, 9)*. This is illustrated from their capacity to produce functional forms of complex mammalian proteins, including, for instance, collagens *(10)*, haemoglobin *(11)* and antibodies *(1, 3)* such as secretory Immunoglobulin A (IgA) *(12)*.

Among various plant expression systems being investigated, genetically modified suspension-cultured plant cells offer a promising system for production of recombinant pharmaceutical proteins *(13, 14, 15)*. Until the 1980s, plant cell cultures were mainly used for the production of useful secondary metabolites of therapeutic value such as shikonin, ajmalicine or taxol *(16, 17, 18)*. The first report on the production of recombinant proteins with a therapeutic interest in plant cells concerns the human serum albumin in potato and in tobacco by Simons et al. *(19)*.

Suspension-cultured plant cells have several advantages over greenhouse or field-grown whole plants for biopharmaceutical production, this expression system being totally devoid of problems associated with contamination with animal excreta, pesticides, chemicals used as fertilizer or to remove weeds, and pests, cultivation variables and gene flow in the environment. Because of short growth cycles, the timescale needed for the production of recombinant proteins in plant cell cultures can be counted in days or weeks after transformation compared with months needed for the production in transgenic plants. In addition, plant cell cultures can be more easily cultivated in a controlled environment than whole plants and, thus, are more suitable candidates for production of recombinant proteins under certified, controllable and sterile conditions with straightforward/easy purification protocols offering improved safety *(20)*, which provides advantages for good manufacturing practice (GMP) and regulatory compliance.

Compared to microbial fermentation, animal or insect cell cultures, plant cell cultures also have several advantages. Microbial systems have a limited capacity for the accurate post-translational modification of eukaryotic proteins. As a consequence, bacterial fermentation often results in the production of insoluble protein aggregates, and substantial costs are involved to solubilize these aggregates and refold a native protein. In addition, eliminating the endotoxins produced during bacterial fermentation adds to the cost of protein production. Regarding mammalian and insect cells, they are more expensive and difficult to cultivate due to complex nutrient requirements and their susceptibility to mechanical stress in bioreactors. Therefore, the lower cultivation temperatures and inexpensive defined culture media (relatively simple solutions containing no exogenously supplemented proteins or sera) make plant cell culture systems competitive with the microbial and animal culture systems. Nevertheless, recovery and purification of recombinant proteins from plant biomass is an expensive and technically challenging business that may amount to 80–94% of the final product costs *(21, 22)*. To reduce the costs of protein purification, one additional advantage of plant cell culture is that the recombinant protein can be fused with a signal sequence that permits secretion in the culture medium. This strategy facilitates the recombinant protein purification in the absence of large quantities of contaminating proteins. In addition, the downstream processing of protein extracted from culture medium is less expensive and does balance, in part, the high costs of fermentation *(15)*. Since the first report in 1990, transgenic plant suspension-cultured cells have been used for the production of a number of biologically active recombinant pharmaceutical proteins including mammalian proteins and antibodies *(23, 24, 25, 26, 27, 28, 29, 30)*. In these applications, the constitutive Cauliflower Mosaic Virus 35S RNA (CaMV35S) or ubiquitin promoter and a signal sequence were used to promote the expression and the secretion, respectively, of recombinant proteins. However, the extracellular concentrations of these proteins were low, ranging from 10 μg/L to 600 μg/L. Alternative promoters and plant cell culture systems that could significantly enhance the productivity of recombinant proteins are being investigated to reduce the costs to offset the substantially higher cost of biomass production through cell culture or fermentation.

The transfer of foreign genes into plant cells is most often performed using the biological vector *Agrobacterium tumefaciens (31)*. The Agrobacterium transformation is based on the co-cultivation of plant cells with *A. tumefaciens* and the subsequent selection of transformed plant cells after spreading on a selective medium *(31)* (see **Note 9.4.1**). In this chapter, we present basic protocols for rapid generation and characterization of transgenic suspension-cultured cells from *Nicotiana tabacum*, *Oriza sativa* and *Arabidopsis thaliana* producing plant-made pharmaceuticals.

9.2. Materials

9.2.1. Plant Material

1. Tobacco suspension-cultured cells (*Nicotiana tabacum* cv BY-2) are grown in Murashige and Skoog (MS) medium (see **Subheading 9.2.5**), in the dark, at 25°C, on an orbital shaker (170 rpm). Calli are grown on solid medium containing 100 μg/mL kanamycin and 250 μg/mL cefotaxime.
2. Japonica and Basmati rice calli (Oryza sativa L.) are initiated from mature dehusked seeds and are grown in N6-CHU modified medium (see **Subheading 9.2.5**), in the dark, at 28°C. Rice suspension-cultured cells are maintained in N6M-L medium, in the dark, at 22°C on an orbital shaker (115 rpm).
3. *Arabidopsis thaliana* seeds (ecotype Col-1) are sown on solid MS medium supplemented with 0.5 g/L MES, 10 g/L sucrose and 8 g/L agar, on a 16 h-day photoperiod, at 22°C. *Arabidopsis* calli are grown under the same conditions but in solid B5 medium. Suspension-cultured cells are grown in MS medium supplemented with 0.5 mg/L NAA and 0.05 mg/L kinetin in the same conditions, on an orbital shaker (140 rpm).

9.2.2. Bacterial Strains

1. *Agrobacterium tumefaciens* strain LBA4404 is used for gene transfer. Before inoculation, bacteria are grown for 24 h at 28°C in YEB medium (*see* **Subheading 9.2.5**) containing 300 μg/mL streptomycin and 100 μg/mL spectinomycin.

9.2.3. cDNA Material

1. A green fluorescent protein (GFP) fusion, or β-glucuronidase (Gus) is used as model protein to illustrate the procedure. The coding sequence of GFP *(32)* or Gus *(33, 34)* was inserted in an expression cassette containing the CaMV 35S promoter and the 3′ end from nopaline synthase. This expression cassette was then ligated to the binary plant transformation vector, pBLTI121 described in *(35)*, which carries the neomycin phosphotransferase gene as a selectable marker for plant cell transformation.
2. Alternatively, the coding sequence of Derp1, a major allergen from house dust mite, was inserted in the same expression cassette and then introduced into pBLTI121.

9.2.4. Special Laboratory Tools and Materials

1. A tissue culture room, maintained at 22°C or at 25°C or at 28°C in the dark.
2. A tissue culture room, at 22°C, with a 16 h-day photoperiod for *Arabidopsis.*
3. An orbital shaker.
4. An incubator with an orbital shaker, set up at 28°C to grow *Agrobacterium tumefaciens.*
5. General tissue culture and plant culture materials.
6. Liquid nitrogen.
7. Spectrophotometer

9.2.5. Media and Buffers

All media and buffers are made up as aqueous preparations. The media for tissue culture are used after a sterilization step using standard procedures.

Tobacco:

1. YEB medium: 1.3% (w/v) nutrient broth (Difco, Detroit, MI), 0.1% (w/v) yeast extract (Difco), 0.5% (w/v) sucrose, 2 mM $MgS0_4$.
2. Solid YEB medium: YEB medium supplemented with 1.4% (w/v) Bactoagar (Difco), 300 µg/mL streptomycin (Sigma, St. Louis, MO), and 100 µg/mL spectinomycin (Sigma).
3. MS medium for tobacco: 4.3 mg/mL of MS mineral salt mixture (ICN, Costa Mesa, CA), 3% (w/v) sucrose, 0.05 mg/mL myo-inositol, 1 mg/mL thiamine-HCl, 0.2 µg/mL 2,4 dichlorophenolacetic acid, 2 g/L KH_2P0_4, pH 5.2.
4. Solid MS medium: MS medium (pH 5.8) supplemented with 0.8% (w/v) Bactoagar and 100 µg/mL kanamycin (Sigma) and 250 µg/mL cefotaxime (Sigma).

Rice:

1. N6-CHU modified medium (N6M-L): 3.95 g/L macro- and microelement and vitamin N6-CHU mixture (Duchefa Biochemie B.V. Haarlem, the Netherlands.), 30 g/L sucrose, Casamino acid 2 g/L, 2,4D 2 mg/L, pH 5.8.
2. Solid N6-CHU modified medium (N6M): 3.95 g/L macro- and microelement and vitamin N6-CHU mixture (Duchefa Biochemie B.V.), 30 g/L sucrose, 1 g/L Casamino acid, 2 mg/L 2,4D, 7 g/L Vitro agar (Kalys, Saint Ismier, France), pH 5.8.
3. Selection solid N6-CHU modified medium (N6M-CK): N6M medium supplemented with 50 mg/L kanamycin (Duchefa Biochemie B.V.) and 250 mg/L cefotaxime (Duchefa, Biochemie B.V.).

4. Co-culture medium (N6M-AS): N6M medium supplemented with acetosyringone 100 μM (stock in 70% EtOH).

Arabidopsis:

1. MS medium for *Arabidopsis*: Macroelements, 1650 mg/L NH_4NO_3, 440 mg/L $CaCl_2$ $2H_2O$, 370 mg/L $MgSO_4$ $7H_2O$, 1.9 g/L KNO_3, 170 mg/L KH_2PO_4; Microelements, 8.6 mg/L $ZnSO_4$ $7H_2O$, 6.2 mg/L H_3BO_3, 22.3 mg/L $MnSO_4$ $4H_2O$, 25 μg/L $CuSO_4$ $5H_2O$, 0.83 mg/L KI, 0.25 mg/L Na_2MoO_4 $2H_2O$, 25 μg/L $CoCl_2$ $6H_2O$, 37.23 mg/L Na_2EDTA, 27.95 mg/L $FeSO_4$ $7H_2O$; Vitamins, 0.5 mg/L nicotinic acid (Sigma), 0.5 mg/L pyridoxin-HCl (Sigma), 0.1 mg/L thiamin-HCl (Sigma); 2 mg/L glycin, 100 mg/L myo-inositol, pH 5.7.
2. B5 medium: 3.2 g/L Gamborg (including vitamins, Duchefa, Biochemie B.V.), 0.5 g/L MES (Sigma), 0.5 mg/L 2,4 dichlorophenolacetic acid (Sigma), 0.05 mg/L kinetin (Sigma), 20 g/L glucose, 8 g/L agar, pH 5.8.
3. Denaturation buffer: 62.5 mM Tris-HCI, pH 6.8, containing 5% (v/v) β-mercaptoethanol, 10%(v/v) glycerol, 1%(w/v)SDS.

9.2.6. Gel Electrophoresis and Blotting Systems

All electrophoresis gels and solutions are made up as aqueous preparations. The procedures for standard sodium dodecyl sulfate-polyacrylamide gel electrophoresis (SDS-PAGE) and immunoblotting analysis are also described in *(36, 37)*, respectively.

1. SDS-PAGE resolving gel: 15% (w/v) acrylamide (acrylamide: bis-acrylamide 30:0.8), 375 mM Tris-HCl, pH 8.8, 0.1% (w/v) SDS.
2. SDS-PAGE stacking gel: 5% (w/v) acrylamide (acrylamide: bis-acrylamide 30:1.5), 175 mM Tris-HCl, pH 6.8, 0.1% (w/v) SDS.
3. SDS-PAGE sample buffer: 62.5 mM Tris-HCI, pH 6.8, 5% (v/v) β-mercaptoethanol, 10% (v/v) glycerol, 2% (w/v) SDS.
4. SDS-PAGE running buffer: 25 mM Tris-HCI, 192 mM glycine, 0.1 (w/v) SDS.
5. Gel-staining solution: 0.25% (w/v) Coomassie brilliant blue R250 in 50% (v/v) methanol, 5% (v/v) acetic acid.
6. Gel-destaining solution: 25% (v/v) methanol, 9% (v/v) acetic acid.
7. Blotting-transfer buffer: 25 mM Tris, pH 8.3, 192 mM glycine, 10% (v/v) methanol.
8. Membrane-staining solution: 0.1% (w/v) Ponceau Red in 3% (v/v) trichloroacetic acid. Destain with Tris buffered saline (TBS) buffer.

9. Membrane saturation solution: 3% (w/v) gelatin in TBS bufferv (heat for dissolution).
10. TBS buffer: 20 mM Tris-HCI, pH 7.5, 500 mM NaCI.

9.3. Methods

9.3.1. Agrobacterium Tumefaciens *Transformation*

1. Transfect the pBLTI121-derived constructs into *Agrobacterium tumefaciens* (LBA 4404) according to *(38)* and select the transformants on YEB solid medium.
2. Initiate liquid cultures of *Agrobacterium tumefaciens* by inoculating 3 mL YEB medium supplemented with streptomycin and spectinomycin with a single colony picked up from stock plate. Incubate at 28°C on a rotary shaker (200 rpm).
3. After 24 h, inoculate 3 mL of fresh medium with 30 μL of the preculture and, incubate for 24 h at 28°C on a rotary shaker (200 rpm). The bacteria are then ready to be used to transform plant cells.

9.3.2. Generation of Recombinant Tobacco BY-2 Cell Lines

1. Grow tobacco BY-2 suspension-cultured cells in a 500-mL Erlenmeyer flask containing 150 mL of MS medium, at 25°C, in the dark, with continuous shaking.
2. Infect 1 mL of 3-day-old tobacco suspension-cultured cells with 50 μL of the *Agrobacterium tumefaciens* preculture in a small Petri dish and keep the co-culture in the dark at 25°C for 2–3 days.
3. Centrifuge the tobacco BY-2 cells for 3 min, at 50×g and wash them 3 times with fresh culture medium.
4. Analyze the cells for expression (see **Subheadings 9.3.5.1** and **9.3.6**) or plate them onto solid-MS medium containing kanamycin and cefotaxime to produce stable transgenic cells within a month.

9.3.3. Generation of Recombinant Rice Cell Lines

9.3.3.1. Initiation of Calli

1. Sterilize mature dehusked rice seeds for 1 min with 96% ethanol and then for 40 min with 2% sodium hypochlorite under shaking. The seeds are rinsed 5 times with sterile water and dried on a sterile wipe.
2. Plate the sterile seeds on solid N6M medium and incubate at 28°C for 4 weeks.
3. After 4 weeks, excise embryo-derived calli and transfer them onto fresh N6M medium.
4. Subculture calli onto fresh N6M medium every 4 weeks.
5. Use 7 to 10-day-old calli for transformation.

9.3.3.2. Transformation

1. Initiate liquid cultures of *Agrobacterium tumefaciens* LBA 4404 by inoculating 3 mL YEB medium supplemented with streptomycin and kanamycin with a single colony picked up from a stock plate. Incubate at 28°C on a rotary shaker (200 rpm).
2. After 24 h, inoculate 5 mL of YEB medium with 10 μL of bacteria suspension and incubate for 24 h until optical density (OD_{600}) reached 0.8–1), then pellet the culture (4000×g, 10 min) and resuspend the bacteria in 5 mL of fresh YEB medium supplemented with 100 μM acetosyringone and use to transform calli.
3. Soak small pieces of calli (1–2 mm) into the 5 mL bacteria suspension in a Petri dish (50-mm diameter) and leave for 15 min *(39, 40)*.
4. Eliminate the bacteria suspension with a pipette.
5. Dry the calli onto a sterile wipe, transfer onto fresh N6M-AS medium and leave the co-culture for 4 days at 22°C, in the dark.
6. Recover the small calli and rinse them 5 times with N6M-L medium supplemented with 250 mg/L cefotaxime in a Petri dish (50-mm diameter), (by adding the medium, incubating for 1 min and removing the medium with a pipette. After a final wash in N6M-L medium, dry the calli onto a sterile wipe. Finally, transfer them on N6M-CK selection medium, in the dark at 22°C for 2 weeks, and move to 28°C for 2 more weeks.
7. Sub-culture the resistant calli onto fresh N6M-CK medium every month.

9.3.3.3. Initiation of Suspension Cultures

1. Initiation of suspension culture is possible 2 months after callus induction.
2. Dilute a callus (30 mg) into 30 mL of N6M-L medium in a 250-mL Erlenmeyer flask and incubate at 22°C, in the dark, with continuous shaking (115 rpm), for 14 days.
3. Transfer 10 mL of this first suspension in 150 mL of N6M-L medium in a 500-mL Erlenmeyer flask using a broken serological pipette and incubate for 14 days in the same conditions as above (see **step 2**).
4. Maintain rice suspension-cultured cells in 150 mL of N6M-L medium in a 500-mL Erlenmeyer flask in the conditions described in **step 2** and passage the cells onto fresh medium as in **step 3** every 14 days.

9.3.4. Generation of Recombinant *Arabidopsis* Cell Lines from Roots

1. To induce root formation, transfer 7-day-old seedlings in conical flasks containing 50 mL of MS medium and grow for 7 days, on a 16 h-day photoperiod, at 22°C, on an orbital shaker (140 rpm).

2. Infect 5-mm root segments with 10 mL of Agrobacterium preculture in a Petri dish (90-mm diameter) for 10 min in the dark, at 22°C with low shaking.
3. Dry the roots on absorbing paper and co-incubate the roots with Agrobacterium on solid B5 medium, for 48 h in the dark, at 22°C.
4. Wash the roots 3–4 times with B5 medium in a Petri dish with low shaking, and lay them down on absorbing paper.
5. Select transgenic cells onto B5 medium complemented with 50 mg/L kanamycin. Calli appear after a month (16 h-day photoperiod, 22°C).
6. Keep suspension-cultured cells in 500 mL Erlenmeyer flasks in 100 mL of MS medium supplemented with 0.5 mg/L NAA, 0.05 mg/L kinetin and 30 g/L sucrose, on the 16 h-day photoperiod, at 22°C with orbital shaking (140 rpm).

9.3.5. Detection of the Recombinant Proteins in Cell Extracts by Western Blotting

Expression of model proteins in cells is monitored after 0, 1, 2, and 3 days of co-culture with A. tumefaciens, using standard SDS-PAGE and immunoblotting procedures.

9.3.5.1. Immunodetection of the Recombinant Proteins Three Days After Transformation

1. Collect the cells, resuspend them (see **Subheading 9.3.2**) in 100 μL of denaturation buffer and boil for 10 min to allow for complete denaturing of the proteins.
2. Load 20 μL of protein extracts and resolve the proteins using a 15% (w/v) SDS polyacrylamide gel *(36)*.
3. Transfer the gel onto nitrocellulose and monitor the transfer efficiency by staining reversibly the membrane with Ponceau Red *(37)*. The recombinant protein is then immunodetected using standard procedure *(41)*.
4. Saturate the blot with 3% gelatin prepared in TBS, for at least 1 h at room temperature.
5. Incubate the blot in TBS containing 1% gelatin and the immunoserum at a convenient dilution for 2 h at room temperature.
6. Wash the blot with TTBS buffer 4 times for 15 min each.
7. Incubate the blot in TBS containing 1% gelatin and the suitable conjugate, that is, a goat anti-rabbit IgG conjugate coupled to horseradish peroxidase (HRP) at a dilution of 1:2,000, or HRP conjugated goat antibodies directed at mouse polyvalent immunoglobulins at a dilution of 1:500, for 1 h at room temperature.
8. Wash the blot 4 times with TTBS for 15 min each and once in TBS before development.
9. Prepare extemporaneously the development mixture, load on the membrane and shake gently to optimize the development reaction.

10. Stop the reaction by discarding the development mixture and rinsing the blot several times with distilled water. Dry the membrane and store between two sheets of filter paper.

9.3.5.2. Detection of the Recombinant Proteins from Transgenic Calli

1. Transgenic kanamycin-resistant tobacco microcalli appear after incubation at 25°C, in the dark, for about 3 weeks on MS-agar medium containing 100 μg/mL kanamycin and 250 μg/mL cefotaxime. Transgenic calli present an active growth and a yellow colour, compared to untransformed ones that turn dark brown.
2. Add 1 volume of twice concentrated warm denaturation buffer to 1 volume of kanamycin-resistant callus, grind, boil for 3 min and centrifuge (10 min, 4500×g). Analyze the supernatant by SDS-PAGE and detect recombinant proteins using immunoblotting, as described in **Subheading 9.3.5.1**.
3. Select calli exhibiting the higher level of recombinant protein expression to establish suspension-cultured cells.

9.3.5.3. Detection of the Recombinant Proteins from Transgenic Suspension-Cultured Cells

1. For analysis of the recombinant protein expression level in transgenic suspension-cultured cells, propagate 2-weeks-old transformed calli in 150 mL MS liquid medium in a 500-mL Erlenmeyer flask. Subculture the cell lines weekly with a 2% inoculum of 7-day-old cells. Note for tobacco cells, 3-day-old cells are used for expression analysis whereas for rice cells, 10–15 days old cells are used.
2. Filter the cells. The filtrate corresponds to proteins accumulated in the extracellular medium. Incubate the cells (30 min, 4°C) in culture medium containing 0.5 M NaCl to collect the proteins bound to the cell wall. After filtration, pool the proteins to the extracellular proteins to constitute the extracellular medium. Finally, precipitate the protein using 12% trichloroacetic acid (4°C) (final concentration).
3. For the cells, proceed as for the calli (see **Subheading 9.3.5.2**) to obtain the intracellular medium.
4. Analyze the expression of recombinant protein by SDS-PAGE before and after purification.
5. Monitor the expression of recombinant protein by immunoblotting, as described previously (see **Fig. 9.1** as an example and **Note 9.4** for additional comments).

9.3.6. Detection of the Recombinant Proteins Using Confocal Microscopy

GFP expression in plants can be viewed at the macroscopic level with hand-held UV lamps providing that the expression levels are sufficiently high and the variant of GFP used is excited by UV light.

For microscopic examination of GFP expressing cells, a standard epifluorescence microscope or a confocal laser scanning microscope (CLSM) can be used.

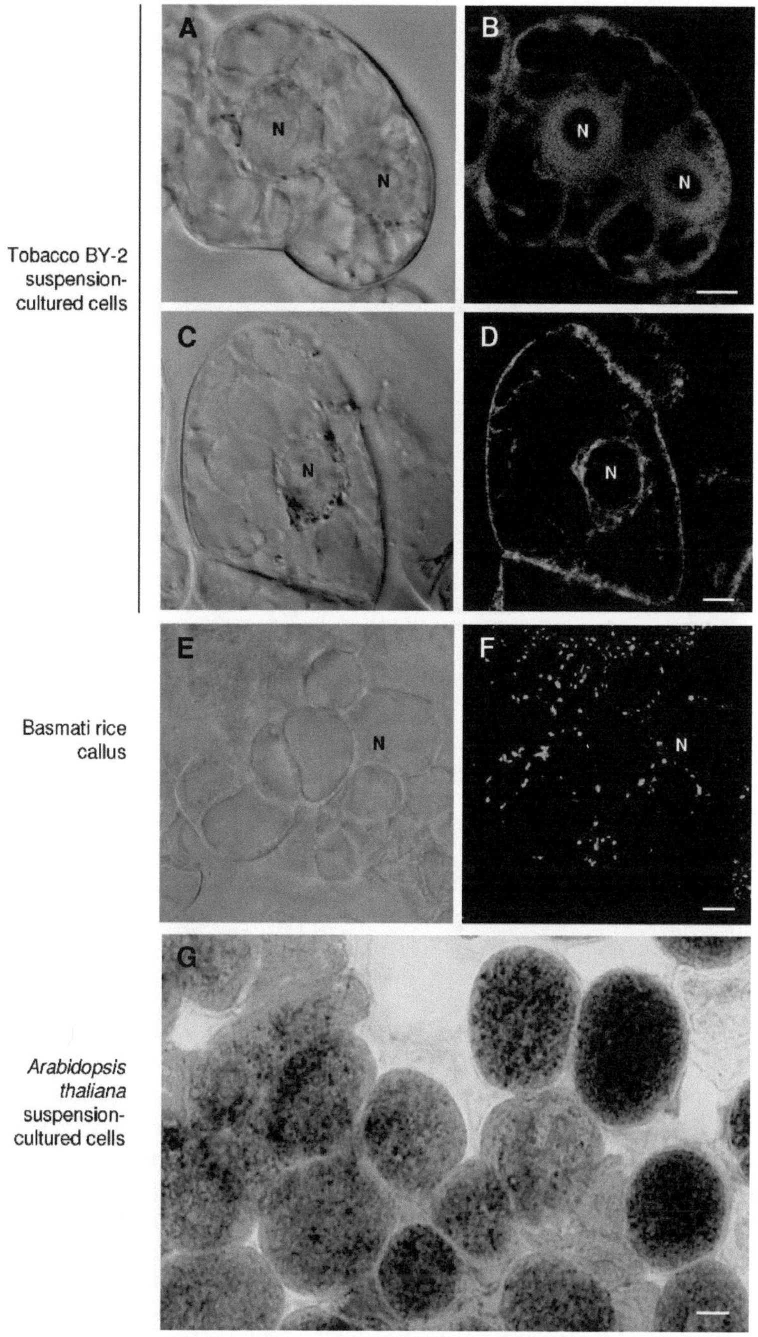

Fig. 9.1. Targeted expression of recombinant proteins in transgenic suspension-cultured cells (**A–F**) Suspension-cultured cells expressing GFP-fusions. Transmitted light (**A,B,E**). green fluorescent protein (*GFP*) expression (**B,D,F**). Bars = 8 μm, N, nucleus. When no signal peptide is added to a recombinant protein sequence, the protein is accumulated in the cytosol (BY-2 cells, **A,B**). When a signal peptide and an endoplasmic reticulum (*ER*) retention sequence are added, recombinant proteins are found mainly or exclusively in the *ER* (BY-2 cells, **C,D**). Proteins targeted to the *ER* bear high-mannose-type N-glycans. Recombinant proteins can also be retained in the Golgi apparatus (Rice cells, **E,F**). (**G**) *Arabidopsis* suspension cultured cells expressing a Gus-fusion. Bar = 20 μm. (*See Color Plates*)

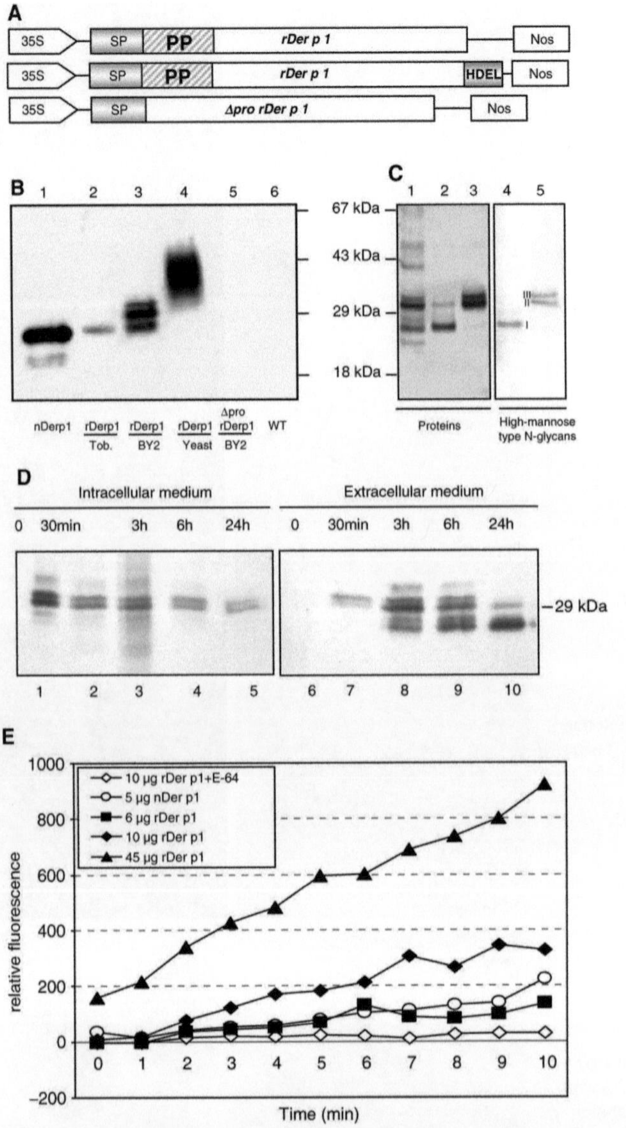

Fig. 9.2. Production and maturation of Der p 1 in tobacco BY-2 cells. (**A**) Der p 1 cDNA allergen constructs used for transformation of BY-2 tobacco cells and tobacco plants via *Agrobacterium tumefaciens*. *SP*, signal peptide; *PP*, propeptide. (**B**) Western blot analysis of Der p 1 from *Dermatophagoides pteronyssinus* (nDerp1lane 1), or recombinant Der p 1 (rDerp1) produced in transgenic tobacco plants (Tob., lane 2), transgenic tobacco BY-2 cells (BY-2, lanes 3 and 5) or transgenic *Pichia pastoris* (Yeast, lane 4). Lane 6 was loaded with a crude protein extract from wild-type tobacco cells. Each lane was loaded with 20 µg of crude protein extract, except for lane 1 loaded with 5 µg of purified nDer p 1. After sodium dodecyl sulfate-polyacrylamide gel electrophoresis (*SDS-PAGE*) and transfer on a nitrocellulose membrane, immunodetection was performed with a polyclonal serum directed against nDer p 1. (**C**) Mature (lanes 2 and 4) and intermediate (lanes 3 and 5) forms of rDer p 1 were separated by size exclusion chromatography from rDer p 1 immunopurified from the culture medium of tobacco BY-2 cells (lane 1). The different forms were separated by SDS-PAGE and either stained in the gel (proteins, lanes 1–3) or transferred on nitrocellulose and analyzed on blot for reactivity with a lectin specific for high-mannose-type N-glycans (lanes 4–5). (**D**) A pulse-chase experiment illustrates that rDer p 1 is matured after secretion in the culture medium of tobacco suspension-cultured cells. (**E**) Cysteine protease activity of nDer p 1 and rDer p 1 are compared for different amounts of recombinant proteins. In the experiments schematized with empty squares, cysteine proteinase inhibitor E-64 was incubated with recombinant allergen as a negative control. From Lienard et al. *(30)*.

For epifluorescence microscopy, excite GFP with blue light and look at the fluorescence. Epifluorescence microscopes have the advantage of being cheap and simple to use; however, the emitted light from all depth of the tissue is visible at the same time. This makes difficult to examine what is occurring in a single cell in a multilayered tissue or a callus.

For CLSM, excite GFP with a 488-nm argon ion laser and filter the emission spectrum with appropriate barrier filters to select against reflection and non-specific fluorescence of the chloroplasts and other elements of plant tissues. For example, if using a LEICA TCS SP2 AOBS microscope, set up photomultiplier for 493–538 nm (**Fig. 9.1**).

GFP can also be detected by immunoelectron microscopy with anti-GFP antibodies (for an example in BY-2 suspension-cultured cells, see **ref.** *42)*.

9.4. Notes

1. The plant cell wall has been described as an efficient barrier and trap for DNA molecules the size of a functional gene. We describe here methods to transform efficiently different growing cells using A. tumefaciens without the need to prepare protoplasts.
2. Large amounts of suspension-cultured cells can be produced from a single callus within 7 weeks after transformation. This large-scale production of transgenic tobacco BY-2 cells has facilitated our studies on recombinant protein targeting *(43)*, without the usual burden of cellular fractionation required with plant organs.
3. The high level of recombinant protein expression in BY-2 tobacco cells facilitates the biochemical analysis of protein maturation steps. For example, we have obtained sufficient amounts of recombinant allergens produced in transgenic suspension-cultured BY-2 cells to perform a detailed structural and immunological analysis of the protein post-translational maturation and (**Fig. 9.2**) *(30)*.
4. **Culture medium supplementation.** As observed for mammalian or insect cells expression systems, supplementation of the culture medium is an efficient way to influence glycosylation of pharmaceutical proteins produced in plant suspension cultures. For instance, we have recently observed that supplementation of BY-2 cell culture medium with glucosamine strongly reduces N-glycan heterogeneity (Gomord et al. unpublished results).

5. A very large panel of *Arabidopsis* mutants is now available and some of them offer interesting opportunities for improving either qualitatively or quantitatively the production of plant-made pharmaceuticals *(44)*. Currently, several groups are generating cell suspension cultures from complex N-glycan-deficient A. thaliana mutants for production of pharmaceutical glycoproteins deficient of plant-specific xylose and fucose residues.
6. Transgenic rice suspension cell cultures can provides large quantities of recombinant proteins in a matter of weeks or a few months. The use of rice cell culture as a protein expression system offers an additional advantage in addition to a relatively high growth rate, with minimal aggregate clumping, and easiness to cultivate. Significantly, rice is indeed a well-characterized food crop that does not produce any toxic secondary metabolites found in some plant cell cultures, such as those derived from tobacco *(45–47)*.
7. **Human proteins expressed and secreted by plant cells could be correctly folded into biologically active forms.** Heterologous proteins appear to be more prone to proteolysis, as they are regarded by cells as foreign *(48)*. Consequently, proteolytic stability is a significant factor influencing the final yields of recombinant proteins in all expression systems. Significantly, lower levels of protease have been detected in the culture medium of rice suspension cells compared to those in the culture medium of tobacco suspension-cultured cells *(48)*. The relatively low proteolytic activity present in rice cell culture medium is considered an important contributing factor to the 1000-fold increase in yield of secreted recombinant human granulocyte-macrophage colony stimulating factor (hGM-CSF) compared to tobacco cell culture medium *(49)*.
8. It appears that higher yields of secreted recombinant protein could be achieved by increasing in cell density in a bioreactor *(46)*.

Acknowledgements

Work on plant-made pharmaceuticals at the University of Rouen was supported by the Centre National de la Recherche Scientifique (CNRS) and by the "Ministère de la Recherche". This study was also partially supported by a grant from the "laboratoire P. Fabre", France. We thank present and former colleagues who contributed to the work described in this review and we also thank L. Faye and A.-C. Fitchette for critical reading of the manuscript.

References

1. Gomord, V., Sourrouille, C., Fitchette, A.C., Bardor, M., Pagny, S., Lerouge, P., and Faye, L. (2004) Production and glycosylation of plant made pharmaceuticals: the antibodies as a challenge. Plant Biotechnol. J. 2, 83–100.
2. Ma, J.K., Chikwamba, R., Sparrow, P., Fischer, R., Mahoney, R., and Twyman, R.M. (2005) Plant-derived pharmaceuticals—the road forward. Trends Plant Sci. 10, 580–585.
3. Ko, K., and Koprowski, H. (2005) Plant biopharming of monoclonal antibodies. Virus Res. 111, 93–100.
4. Schillberg, S., Twyman, R.M., and Fischer, R. (2005) Opportunities for recombinant antigen and antibody expression in transgenic plants—technology assessment. Vaccine 23, 1764–1769.
5. Gomord, V., Chamberlain, P., Jefferis, R., and Faye, L. (2005) Biopharmaceutical production in plants: problems, solutions and opportunities. Trends Biotechnol. 23, 559–565.
6. Faye, L., and Daniell, H. (2006) Pathway and freeways for protein import in the higher plant chloroplasts. Plant Biotechnol. J. 4, 275–279.
7. Sethuraman, N., and Stadheim, T.A. (2006) Challenges in therapeutic glycoprotein production. Curr. Opin. Biotechnol. 17, 341–346.
8. Gomord, V., and Faye, L. (2004) Post-translational modifications of therapeutic proteins in plants. Curr. Opin. Plant Biol. 7, 171–181.
9. Walsh, G., and Jefferis, R. (2006) Post-translational modifications in the context of therapeutic proteins. Nat. Biotechnol. 24, 1241–1252.
10. Merle, C., Perret, S., Lacour, T., Jonval, V., Hudaverdian, S., Garrone, R., Ruggiero, F., and Theisen, M. (2002) Hydroxylated human homotrimeric collagen I in *Agrobacterium tumefaciens*-mediated transient expression and in transgenic tobacco plant. FEBS Lett. 515, 114–118.
11. Theisen, M. (1999) Production of recombinant blood factors in transgenic plants. Adv. Exp. Med. Biol. 464, 211–220.
12. Ma, J.K.C., Hiatt, A., Hain, M., Vine, N.D., Wang, F., Stabila, P., van Dolleweerd, C., Mostov, K., and Lherner, T. (1995) Generation and assembly of secretory antibodies in plants. Science 268, 716–719.
13. Kieran, P.M., MacLoughlin, P.F., and Malone, D.M. (1997) Plant cell suspension cultures: some engineering considerations. J. Biotechnol. 59, 39–52.
14. Fischer, R., Emans, N., Schuster, F., Hellwig, S., and Drossard, J. (1999) Towards molecular farming in the future: using plant-cell-suspension cultures as bioreactors. Biotechnol. Appl. Biochem. 30, 109–12.
15. Doran, P.M. (2000) Foreign protein production in plant tissue cultures. Curr. Opin. Biotechnol. 11, 199–204.
16. Honda, G., Sakakibara, F., Yazaki, K., and Tabata, M. (1988) Isolation of deoxyshikonin, an antidermatophytic principle from *Lithospermum erythrorhizon* cell cultures. J. Nat. Prod. 51, 152–154.
17. Drapeau, D., Sauvaire, Y., Blanch, H.W., and Wilke, C.R. (1986) Improvement of diosgenin yield from *Dioscorea deltoidea* plant cell cultures by use of a non-traditional hydrolysis method. Planta Med. 52, 474–478.
18. Seki, M., Ohzora, C., Takeda, M., and Furusaki, S. (2000) Taxol (paclitaxel) production using free and immobilized cells of Taxus cuspidate. Biotechnol. Bioeng. 53, 214–219.
19. Sijmons, P.C., Dekker, B.M., Schrammeijer, B., Verwoerd, T.C., van den Elzen, P.J., and Hoekema, A. (1990) Production of correctly processed serum albumin in transgenic plants. Bio/Technology 8, 217–221.
20. Fischer, R., and Emans, N. (2000) Molecular farming of pharmaceutical proteins. Transgenic Res. 9, 279–99.
21. Kusnadi, A.R., Nikolov, Z.L., and Howard, J.A. (2000) Production of recombinant proteins in transgenic plants: practical considerations. Biotechnol. Bioeng. 56, 473–484.
22. Evangelista, R.L., Kusnadi, A.R., Howard, J.A., and Nikolov, Z.L. (1998) Process and economic evaluation of the extraction and purification of recombinant β-glucuronidase from transgenic corn. Biotechnol. Prog. 14, 607–614.
23. Firek, S., Draper, J., Owen, M.R.L., Gandecha, A., Cockburn, B., and Whitelam, G.C. (1993) Secretion of a functional single-chain Fv protein in transgenic tobacco plants and cell suspension cultures. Plant Mol. Biol. 23, 861–870.
24. Magnuson, N.S., Linzmaier, P.M., Reeves, R., An, G., HayGlass, K., and Lee, J.M. (1998) Secretion of biologically active human interleukin-2 and interleukin-4 from

genetically modified tobacco cells in suspension culture. Protein Expr. Purif. 13, 45–52.

25. James, E.A., Wang, C., Wang, Z., Reeves, R., Shin, J.H., Magnuson, N.S., and Lee, J.M. (2000) Production and characterization of biologically active human GM-CSF secreted by genetically modified plant cells. Protein Expr. Purif. 19, 131–138.
26. Kwon, T.H., Seo, J.E., Kim, J., Lee, J.H., Jang, Y.S., and Yang, M.S. (2003) Expression and secretion of the heterodimeric protein interleukin-12 in plant cell suspension culture. Biotechnol. Bioeng. 81, 870–875.
27. Sunil Kumar, G.B., Ganapathi, T.R., Revathi, C.J., Prasad, K.S., and Bapat, V.A. (2003) Expression of hepatitis B surface antigen in tobacco cell suspension cultures. Protein Expr. Purif. 32, 10–17.
28. Reggi, S., Marchetti, S., Patti, T., De Amicis, F., Cariati, R., Bembi, B., and Fogher, C. (2005) Recombinant human acid beta-glucosidase stored in tobacco seed is stable, active and taken up by human fibroblasts. Plant Mol. Biol. 57, 101–113.
29. Cramer, C.L., Weissenborn, D.L., Oishi, K.K., Grabau, E.A., Bennett, S., Ponce, E., Grabowski, G.A., and Radin, D.N. (1996) Bioproduction of human enzymes in transgenic tobacco. Ann. NY Acad. Sci. 792, 62–71.
30. Lienard, D., Tran Dinh, O., van Oort, E., Van Overtvelt, L., Bonneau, C., Wambre, E., Bardor, M., Cosette, P., Didier-Laurent, A., de Borne, F.D., Delon, R., van Ree, R., Moingeon, P., Faye, L., and Gomord, V. (2007) Suspension-cultured BY-2 tobacco cells produce and mature immunologically active house dust mite allergens. Plant Biotechnol. J. 5, 93–108.
31. Nebenfuhr, A., Gallagher, L.A., Dunahay, T.G., Frohlick, J.A., Mazurkiewicz, A.M., Meehl, J.B., and Staehelin, L.A. (1999) Stop-and-go movements of plant Golgi stacks are mediated by the acto-myosin system. Plant Physiol. 121, 1127–42.
32. Jefferson, R.A., Burgess, S.M., and Hirsch, D. (1986) P-Glucuronidase from *Escherichia coli* as a gene-fusion marker. Proc. Natl. Acad. Sci. USA 83, 8447–8451.
33. Jefferson, R.A., Kavanagh, T.A., and Bevan, M.W. (1987) GUS fusions: P-glucuronidase as a sensitive and versatile gene fusion marker in higher plants. EMBO J. 6, 3901–3907.
34. Hooykaas, P.J.J., and Schilperoort, R.A. (1992) Agrobacterium and plant genetic engineering. Plant Mol. Biol. 19, 15–38.
35. Pagny, S., Bouissonnié, F., Sarkar, M., Follet-Gueye, M.L., Driouich, A., Schachter, H., Faye, L., and Gomord, V. (2003) Structural requirements for Arabidopsis beta 1,2 xylosyltransferase activity and targeting to the Golgi. Plant J. 33, 189–203.
36. Laemmli, U. (1970) Cleavage of structural proteins during the assembly of the head of bacteriophage T4. Nature 227, 680–685.
37. Towbin, H., Staehelin, T., and Gordon, J. (1979) Electrophoretic transfer of proteins from polyacrylamide gels to nitrocellulose sheets: procedure and some applications. Proc. Natl. Acad. Sci. USA 76, 4350–4354.
38. Höfgen, R., and Willmitzer, L. (1988) Storage of competent cells for Agrobacterium transformation. Nucleic Acids Res. 16, 9877.
39. Hiei, Y., Komari, T., and Kubo, T. (1997) Transformation of rice mediated by *Agrobacterium tumefaciens.* Plant Mol. Biol. 35, 205–218.
40. Hoque, M.E., Mansfield, J.W., and Bennett, M.H. (2005) Agrobacterium-mediated transformation of Indica rice genotypes: an assessment of factors affecting the transformation efficiency. Plant Cell Tiss. Organ Cult. 82, 45–55.
41. Faye, L., Gomord, V., Lainé, A.-C., and Chrispeels, M.J. (1993) Affinity purification of antibodies specific for Asn-linked glycans containing αl,3 fucose or βl,2 xylose. Anal. Biochem. 209, 104–108.
42. Saint-Jore-Dupas, C., Nebenführ, A., Boulaflous, A., Follet-Gueye, M.L., Plasson, C., Hawes, C., Driouich, A., Faye, L., and Gomord, V. (2006) Plant N-glycan processing enzymes employ different targeting mechanisms for their spatial arrangement along the secretory pathway. Plant Cell 18, 3182–3200.
43. Gomord, V., Denmat, L.A., Fitchette-Lainé, A.-C., Satiat-Jeunemaitre, B., Hawes, C., and Faye, L. (1997) The C-terminal HDEL sequence is sufficient for retention of secretory proteins in the endoplasmic reticulum but promotes vacuolar targeting of proteins that escape the endoplasmic reticulum. Plant J. 11, 101–103.
44. Downing, W.L., Galpin, J.D., Clemens, S., Lauzon, S.M., Samuels, A.L., Pidkowich, M.S., Lorne Clarke, A., and Kermode, A.R. (2006) Synthesis of enzymatically active human α-l-iduronidase in Arabidopsis cgl (complex glycan-deficient) seeds. Plant Biotechnol. J. 4, 169–181.

45. Tyagi, A.K., and Mohanty, A.K. (2000) Rice transformation for crop improvement and functional genomics. Plant Sci. 158, 1–18.
46. Huang, L.F., Liu, Y.K., Lu, C.A., Hsieh, S.L., and Yu, S.M. (2005) Production of human serum albumin by sugar starvation induced promoter and rice cell culture. Transgenic Res. 14, 569–581.
47. Takaiwa, F., Takagi, H., Hirose, S., and Wakasa, Y. (2007) Endosperm tissue is good production platform for artificial recombinant proteins in transgenic rice. Plant Biotechnol. J. 5, 84–92
48. Enfors, S.O. (1992) Control of in vivo proteolysis in the production of recombinant proteins. Trends Biotechnol. 10, 310–315.
49. Shin, Y.J., Hong, S.Y., Kwon, T.H., Jang, Y.S., and Yang, M.S. (2003) High level of expression of recombinant human granulocyte-macrophage colony stimulating factor in transgenic rice cell suspension culture. Biotechnol. Bioeng. 82, 778–783.

Chapter 10

Chloroplast-Derived Vaccine Antigens and Biopharmaceuticals: Protocols for Expression, Purification, or Oral Delivery and Functional Evaluation

N. Dolendro Singh, Yi Ding, and Henry Daniell

Summary

Many vaccine antigens and biopharmaceutical proteins have been expressed at high levels via the chloroplast genome and their functionality has been evaluated using in vitro assays in cell cultures (i.e., macrophage lysis assay, inhibition of vesicular stomatitis virus-induced cytopathicity in baby hamster kidney cells, or inhibition of human HIV infection in TZM-BL cells) as well as protection after challenge with bacterial or viral pathogens or antitumor assays or delay the onset of insulitis in suitable animal models. Production of therapeutic proteins in chloroplasts eliminates the expensive fermentation technology. Moreover, oral delivery of chloroplast-derived therapeutic proteins eliminates expensive purification steps, cold storage, cold transportation, and delivery via sterile needles, thereby further decreasing their cost. In this chapter, we describe detailed protocols for chloroplast transformation including the construction of chloroplast transformation vectors, delivery of DNA into plant cells using particle bombardment, selection and regeneration of transformants by tissue culture, confirmation of transgene integration into the chloroplast genome and homoplasmy, evaluation of foreign gene expression, purification of foreign protein, or oral delivery via bioencapsulation, functional evaluation using in vitro and in vivo assays, and evaluation of immunity after challenge with pathogens in suitable animal models.

Key words: Chloroplast transformation, Transgene containment, Genetically modified crops, Oral delivery, Therapeutic proteins.

10.1. Introduction

Chloroplasts are prokaryotic compartments inside eukaryotic plant cells that sustain life on this planet. Besides photosynthesis, chloroplasts synthesize starch, certain amino acids, and lipids, and are involved in some key aspects of sulfur and nitrogen metabolism.

Loïc Faye and Véronique Gomord (eds.), *Methods in Molecular Biology, Recombinant Proteins From Plants, vol. 483*

DOI: 10.1007/978-1-59745-407-0_10

Each plant cell contains as many as 100 chloroplasts, each harboring ~100 identical copies of the plastid genome, resulting in up to 10,000 copies of the plastid genome in each cell. Also, most of the plastid genomes have two inverted repeats; therefore, integration of the transgene into the inverted repeats region of plastid genome can reach a copy number of ~20,000 per cell. This facilitates high level expression of foreign genes in plant cells. Transgenes have been integrated into chloroplast genome with accumulation of foreign proteins up to 46% of total leaf protein *(1)*. Therefore, chloroplast transformation is a powerful tool in the field of plant biotechnology.

In chloroplast transformation, foreign genes are stably integrated into the spacer region of two functional genes through homologous recombination. This allows site-specific integration and eliminates the concern of position effect, frequently observed in nuclear transgenic plants *(2)*. One of the major environmental concerns in genetically modified (GM) crops is the escape of foreign genes through pollen from transgenic crops to their weedy relatives or to other crops resulting in contamination of non-GM products *(3, 4)*. Chloroplast transformation system is highly desirable in situations where outcrossing between crops and weeds or among crops is a concern *(3, 4, 5)*. Chloroplast genomes are maternally inherited in most of the angiosperms *(6)*. Although pollen from plants shown to exhibit maternal plastid inheritance contains metabolically active plastids, the plastid DNA itself is lost during the process of pollen maturation and hence is not transmitted to the next generation *(3)*. In addition to this, cytoplasmic male sterility has added another possible method for transgene containment *(7)*. Another advantage of chloroplast transformation is the absence of gene silencing which is often a serious concern in nuclear transformation. There is no gene silencing in chloroplast transgenic lines at the transcriptional level despite accumulation of transcripts 150-fold higher than nuclear transgenic lines *(8, 9)* or at the translational level despite accumulation of foreign proteins up to 46% of total soluble protein (TSP) *(1)*. Another major advantage of engineering chloroplast genome is the expression of multiple transgenes as operons due to efficient translation of polycistronic messenger RNAs. This allows the engineering of multiple genes in a single step and several heterologous operons have been expressed in transgenic chloroplasts and polycistrons are translated without processing into monocistrons *(1, 10)*.

Plants have been extensively explored as renewable resources for production of vaccines and biopharmaceuticals. There are several advantages associated with plant-based vaccines including antigen protection through bioencapsulation, generation of systemic and mucosal immunity, elimination of expensive purification and fermenters, sterile injections upon oral delivery, heat stability, and improved safety and low-cost maintenance *(11, 12)*. The tobacco plant is often chosen because of its large biomass, yielding ~170 MT of biomass per hectare *(13, 14)*. In addition, it is easy to engineer tobacco chloroplast genome and regenerate transgenic lines within a few months. Each transgenic plant is capable of producing up to a million seeds and, therefore, it is possible to scale up from a single transgenic plant to 100 acres within 1 year. Moreover, tobacco is a non-food crop and is self pollinated, minimizing transgene escape. By using this technology, large quantities of vaccine antigen can be produced. Immunogenic *Bacillus anthracis* protective antigen (PA) has been expressed in transgenic tobacco chloroplast with high levels of expression *(15)* and chloroplast-derived PA was stable and fully functional *(16)*. Based on reported expression levels and yield of tobacco, it is possible to produce up to 300 million doses of the anthrax vaccine antigen in 1 acre of tobacco *(16)*. Moreover, expensive step for purifying the recombinant protein could be eliminated by using the oral delivery of therapeutic proteins. Transformation of the soybean, carrot, and lettuce plastid genomes has been accomplished recently using species-specific vectors *(17, 18, 19)*. Moreover, a human therapeutic protein has been already stably expressed in lettuce chloroplasts *(20)*. Oral delivery of cholera toxin B (CTB)-proinsulin (Pins) expressed in chloroplasts protected against development of insulitis in non-obese diabetic (NOD) mice *(20)*. This finding has opened the door for developing various other therapeutic proteins which can be orally administered. Several vaccine antigens have been already expressed in chloroplasts including the F1–V fusion antigen for plague *(21)*, the C-terminus of *Clostridium tetani (22)*, the CTB of *Vibrio cholerae (23)*, the *Lec*A from *Entamoeba histolytica (24)*, the 2L21 peptide from the canine parvovirus *(25)* and their immunity have been evaluated in suitable animal models. The chloroplast genome has been used for production of valuable therapeutic proteins, such as type I interferons *(26, 27)*, human somatotropin *(28)*, human serum albumin *(29)*, magainin, a broad-spectrum topical agent, systemic antibiotic, wound-healing stimulant, and a potential anticancer agent *(30)*.

Various steps are involved in chloroplast transformation and each step is equally important. Therefore, this chapter describes detailed procedures for chloroplast transformation including vector construction, delivery of DNA into chloroplast genome using particle delivery system, tissue culture techniques to recover transplastomic plants, confirmation of transgene integration and homoplasmy, extraction of total proteins, confirmation of transgene expression, expression of foreign proteins, quantitation of foreign proteins, purification of histidine (His)-tagged proteins by affinity chromatography, functional evaluation using in vitro and in vivo assays in suitable animal models.

10.2. Material

10.2.1. Construction of Chloroplast Transformation Vector

1. Fresh leaves.
2. Porcelain mortars and pestles.
3. Liquid nitrogen.
4. DNeasy Plant mini kit (Qiagen: MA, USA).
5. dNTPs (Invitrogen: CA, USA).
6. 10× *Pfu* buffer (Promega: CA, USA).
7. *Pfu* DNA polymerase (Promega: CA, USA).
8. Forward and reverse primers which anneal to left and right chloroplast flanking sequences of interest.
9. Thin wall 0.2 ml polymerase chain reaction (PCR) tubes (Midwest Scientific: MD, USA).
10. Autoclave distilled water.
11. Thermal cycler (PTC 100, Bio-Rad: CA, USA).
12. PCR cloning vector (Promega's pGEM®-T Easy Vector). Store all components at –20°C.
13. pUC19 plasmid (New England Biolabs: MS, USA).
14. Restriction endonucleases (avoid any contamination and store at –20°C).
15. T4 DNA polymerase (Invitrogen, store at –20°C).
16. Calf intestinal alkaline phosphatase (CIAP; Promega), store at –20°C.
17. T4 DNA ligase (Invitrogen, store at –20°C).
18. QIAquick gel purification kit (Qiagen, store at room temperature).
19. QIAquick PCR purification kit (Qiagen, store at room temperature).
20. Quantum Prep™ Plasmid Midiprep Kit (Bio-Rad).
21. Syringe-operated filter units, 0.2 μm pore size (Nalgene: NY, USA).
22. Antibiotic stocks: All stock solutions are prepared in distilled water, filter-sterilized under aseptic condition in the laminar airflow hood and stored at –20°C.

10.2.2. Particle Bombardment

1. Particle delivery system (PDS): PDS 1000 He (Bio-Rad), microcarriers (gold particle 0.6 μm), macrocarriers, stopping screen, rupture disks (650 psi, 900 psi, 1,100 psi, 1,350 psi and 1,550 psi).
2. 2.5 M $CaCl_2$ (sterilize by filtration with 0.2 μm filter unit and store at –20°C)

3. 0.1 M spermidine-free base (highly hygroscopic, make one time stock, filter-sterilize using 0.2 μm filter unit, aliquot into small amount and store at –20°C).
4. 50% (v/v) glycerol (sterilize by autoclaving).
5. 70% and 100% ethanol (molecular grade).
6. Circular Whatman filter paper 70 mm (sterilize by autoclaving).

10.2.3. Plant Tissue Culture Medium and Regeneration (for Tobacco)

1. MSO *(31)* medium for 1 l:1× Murashige and Skoog (MS, Caisson) salts *(32)*, 30 g sucrose, 6 g agar (Phytoblend, Caisson).
2. Regeneration medium of plants (RMOP) *(31)* medium for 1 l: 1× MS (Caisson) salts *(32)*, 100 mg myoinositol, 1 mg thiamine HCl, 1 mg benzyladenine purine (BAP, Sigma), 0.1 mg 1-naphthaleneacetic acid (NAA, Sigma), 30 g sucrose (Sigma: MO, USA), 6 g agar (Phytoblend, Caisson). Adjust pH to 5.8 before autoclaving medium. Add required specific antibiotics after autoclave and cooling down medium to 45–50°C. Use 500 μg/ml of spectinomycin for effective selection of transformants in tobacco (var. Petit Havana) and 300 μg/ml for low nicotine cultivar LAMD.
3. Scalpel or razor blade.
4. Sterile 100 × 25 mm Petri dishes and bottles.
5. Soil (Miracle Gro – Pot-mix).
6. Pots.
7. Greenhouse.

10.2.4. Molecular Characterization of Transplastomic Plants

10.2.4.1. Confirmation of Transgene Integration into the Chloroplast Genome by PCR

1. Genomic DNA from transplastomic and wild-type plants.
2. Sets of PCR primers (primer1 which anneals to the native chloroplast genome and primer 2 which anneals to the gene of interest) *(33)*.
3. Taq DNA polymerase and buffer (Bioline, supplied by vendors).

10.2.4.2. Southern Blot Analysis to Confirm Homoplasmy

1. Depurination solution (0.25 N HCl).
2. Transfer buffer (0.4 N NaOH, 1 M NaCl).
3. QuickHyb solution (Stratagene: TX, USA).
4. Nylon membrane (e.g., Duralon-UV™ Membrane, Stratagene).
5. Flanking probe (this can be isolated by restriction digestion or by PCR amplification of the flanking sequences).
6. ^{32}P-label (Amersham Pharmacia Biotech, Inc.). Extremely hazardous, special care is required while handling.

7. Ready-To-Go DNA labeling beads (Amersham Biosciences, Pittsburgh, PA).
8. GS GeneLinker UV Chamber (Bio-Rad).
9. Hybridization incubator (Fisher: DE, USA).
10. Salmon sperm DNA (Stratagene, store at –20°C).
11. 20× Sodium saline citrate (SSC): 3 M $NaCl_2$, 0.3 M sodium citrate trisodium salts.
12. Wash buffer-I: 2× SSC buffer and 0.1% (w/v) Sodium Dodecyl Sulfate (SDS); wash buffer-II: 0.1× SSC and 0.1% (w/v) SDS.
13. Autoradiography cassette (Fisher).
14. Film developer mini-medical series (AFP Imaging).

10.2.5. Western Blot Analysis to Confirm Expression of Foreign Proteins

1. Hand-operated homogenizer (Sigma) with sterile polypropylene pestle adapters (Sigma).
2. Plant extraction buffer (PEB) *(31)*: 100 mM NaCl, 10 mM EDTA (pH 8), 200 mM Tris-HCl, pH 8, 0.05% Tween-20, 0.1% SDS, 14 mM β-mercaptoethanol (BME), 200 mM sucrose, and 2 mM phenyl methyl sulfonyl fluoride (PMSF). Add PMSF just before use and can be stored at –20°C for up to a month.
3. 10% SDS. (Use glove when handling with SDS and PMSF. Avoid inhaling or contact with skin since they are harmful to health.)
4. Acryamide/Bis (ready-made mixture from Fischer, store at 4°C).
5. Laemmli sample buffer (Bio-Rad).
6. 10× Electrode buffer (EB): 30.3 g Tris base, 144 g glycine, 10 g SDS in 1 l dH_2O. Store at 4°C.
7. Tween-20 (Sigma).
8. Nonfat powdered milk (e.g., Carnation).
9. Bovine serum albumin (Sigma).
10. Bradford reagent (Bio-Rad).
11. Antibodies (store at 4°C).
12. 10× Phosphate-buffered saline (PBS) *(31)* for 1 l: 80 g NaCl, 2 g KCl, 26.8 g $Na_2HPO_{4,}$ $7H_2O$ or 14.4 g Na_2HPO_4, 2.4 g KH_2PO_4. Set the pH to 7.4 and keep in room temperature after autoclave.
13. 1× PBS (1:10 diluted from 10× PBS).
14. PBS-T (1× PBS containing 0.1% Tween-20).
15. Phosphate-buffered saline with Tween 20 and Milk (PTM) (PBS-T, 3% Milk).
16. Transfer buffer for 1.5 l: 300 ml 10× EB, 300 ml methanol (Fishers Scientific), 900 ml deionized water and 0.15 g SDS.

17. Transfer membrane (immune-blot polyvinylidene fluoride (PVDF) membrane or nitrocellulose membrane, Bio-Rad).
18. Electrochemiluminescent (ECL) substrates for Western blot (Pierce).
19. X-ray film.
20. Automated X-ray film processor.

10.2.5.1. Enzyme-Linked Immunosorbent Assay for Quantitation of Foreign Protein

1. Flat-bottom high-binding EIA/RIA plate (96 wells) (Costar).
2. Coating buffer: 15 mM Na_2CO_3, 35 mM $NaHCO_3$, 3 mM NaN_3, pH 9.6.
3. PBS containing 0.1% Tween 20 (PBS-T).
4. PBS containing 0.1% Tween 20 and 3% milk (PTM).
5. Antigen standard: purified *B. anthracis* PA (NIH); purified CTB subunit antigen (Sigma).
6. Antigen-specific primary antibody: mouse anti-PA primary antibody (Immunochemical labs); rabbit anti-CTB primary antibody (Sigma).
7. Secondary antibodies: horseradish peroxidase (HRP) – conjugated antimouse IgG antibody (Biomeda)/HRP-conjugated antimouse IgG antibody (Sigma).
8. 3, 3′, 5, 5′-Tetramethylbenzidine (TMB) substrate for enzyme-linked immunosorbent assay (ELISA) (American Qualex).
9. Sulfuric acid, dilute to 2 N (Sigma).
10. Microtiter plate reader (BioTek Instruments), equipped with 450-, 570- and 595-nm filters.

10.2.5.2. Purification of His-tagged PA by Affinity Chromatography

1. Nickel-Chelate-Charge column (Amersham Biosciences).
2. Binding buffers: 20 mM Na_2HPO_4, 0.5 M NaCl, 10 M imidazole, pH 7.4.
3. Elution buffer: 20 mM Na_2HPO_4, 0.5 M NaCl, 0.5 M imidazole, pH 7.4.
4. Ni-loading eluent: 100 mM $NiSO_4$ solution (Sigma).
5. Dialysis cassette (molecular weight, 10,000; Pierce).
6. Centricon 10,000-molecular-weight-cutoff ultrafiltration units (Millipore).

10.2.6. In Vitro Functional Assays *Anthrax PA: Macrophage Lysis Assays*

1. Purified *B. anthracis* PA (NIH).
2. Monoclonal anti-PA antibody (Immunochemical).
3. Cryopreserved cell lines: RAW264.7 cells (ATCC) derived from mice.
4. Culture medium for RAW264.7 cells: high-glucose DMEM (Dulbecco's Modified Eagle's medium) supplemented with

10% heat-inactivated fetal bovine serum (FBS; Mediatech Cellgro).

5. *B. anthracis* lethal factor (LF; NIH).
6. MTT [3-(4, 5-dimethylthiazo-2-yl)-2,5-diphenyltetrazolium bromide] (Sigma).

10.2.7. In Vivo Functional Assays

1. Outbred or inbred strains of pathogen-free mice (Charles River/Jackson Laboratory).
2. Commercial adjuvant: aluminum hydroxide (25% Alu-Gel-S; Sigma).
3. 1× PBS (pH 7.4).
4. Homogenizer.
5. 1 ml syringe.
6. Sterile 25-gauge needles.
7. Sterile 20-gauge feeding needles with ball end (1.5 in) (Popper & Sons).
8. Mobile Laboratory Animal Anesthesia System, VetEquip (VWR).
9. Induction Chambers, VetEquip (VWR).
10. Isoflurane (anesthetic agent, Sigma).
11. Microhematocrit tube.
12. BD microtainer serum separator collection tube.
13. Hand-operated homogenizer (e.g., Sigma) with sterile polypropylene pestle adapters (Sigma).
14. Purified mouse IgG (IgG1, IgG2a) or IgA isotype control (BD Pharmingen).
15. HRP conjugated IgG (IgG, IgG1, IgG2a, BD Pharmingen) or IgA antibodies (American Qualex).

10.3. Methods

10.3.1. Isolation and Cloning of Homologous Chloroplast Flanking Sequences

Chloroplast transformation uses two flanking sequences and through homologous recombination, inserts foreign DNA into the spacer region between the two functional genes of the chloroplast genome, thus targeting the foreign genes to a precise location. These flanking sequences can be obtained directly from chloroplast DNA or total plant DNA either by restriction digestion of the desired fragment or by amplifying the region of interest by PCR by using a polymerase with proofreading properties to minimize the error. PCR-based technology is more widely used due to limited number of chloroplast genome sequences available in databases. Design primers for 5′ left and 3′ right flanking complementary to the chloroplast sequence of interest based on the available genome sequence. Here, we designed a set of primers as an example to introduce foreign genes between highly conserved region of the 16S rRNA and 23S rRNA genes of the tobacco chloroplast genome.

1. Isolate genomic DNA from the fresh leaves using DNeasy Plant mini kit as described in the product protocol.
2. Prepare PCR reaction for 50 μl: 1 μl of 100 ng/μl genomic DNA, 5 μl of 10× PCR reaction buffer with 2 mM $MgSO_4$, 1 μl each of 10 μM primers, 200–250 μM of dNTPs, 1–2.5 units of *Pfu* DNA polymerase, adjust the final volume with sterile distilled water.
3. Set up the PCR reaction with initial denature at 94°C for 5 min followed by 30 cycles of denaturation at 94°C for 30 s, annealing at 55–60°C for 30 s and extension at 72°C for 2–4 min (1 min/kb) followed by final extension for 8–10 min at 72°C (*see* **Note 1**).
4. Examine the PCR product by running small aliquot in an agarose gel by electrophoresis and purify the remaining product using QIAquick PCR purification kit as described in vendor's procedure.
5. Incubate the purified DNA fragment with 1× Taq DNA polymerase buffer containing 2 mM $MgCl_2$, 200 μM dATP and 5U of Taq DNA polymerase at 70°C for 10–30 min for the addition of A-overhangs to facilitate the direct cloning of PCR product into pGEMT-easy cloning vector.
6. Set up ligation reaction of PCR amplified A-overhangs DNA fragment with pGEMT-easy cloning vector following manufacturer's protocol.
7. Take 2–5 μl of the ligation mixture and transform into competent *Escherichia coli* cells.
8. Screen the recombinant clone based on blue/white selection procedure following the manufacturer's protocol and isolate the recombinant plasmid DNA (*see* **Note 2**).
9. Digest the pGEMT-easy vector containing the flanking sequences with appropriate restriction enzyme to confirm anticipated DNA fragment. Sequencing of the recombinant clone flanking sequence is highly recommended.
10. Pull out the clone DNA fragment from the pGEMT-easy cloning vector by digesting with appropriate restriction enzyme(s).
11. Make blunt end the digested DNA fragment using T4 DNA polymerase following the manufacturer's procedure.

12. Digest pUC19 plasmid with *Pvu*II and dephosphorylate by calf intestinal alkaline phosphatase (Promega) following the venders's protocol. Purify the plasmid DNA using QIAquick gel purification kit (Qiagen) as described by the manufacturer.
13. Ligate the blunt-ended chloroplast flanking DNA into *Pvu*II digested or any other enzyme and dephosphorylated pUC19 plasmid using T4 DNA ligase.
14. Transform the ligated mixture into competent *E. coli* cells.
15. Confirm the recombinant clone by digesting with appropriate enzyme(s).

10.3.2. Insertion of Chloroplast Expression Cassette into Spacer Region

Chloroplast expression cassette consists of a strong chloroplast promoter, a selectable marker, 5′ and 3′ untranslated regions (UTR) to enhance the efficiency of transcription and translation of the introduced gene. Most commonly used 5′ regulatory sequences are derived from the *psb*A and *rbc*L genes while 3′ is from *psb*A, *rbc*L, and *rps*16 genes. The chloroplast promoter and regulatory sequences are directly amplified from the total cellular DNA or chloroplast DNA using specific primers designed on the basis of the specific sequence information available for the chloroplast genome. Approximately 1 kb flanking region on each side is adequate to facilitate homologous recombination. However, we use longer flanking sequences to enhance the efficiency of homologous recombination. Usually the chloroplast expression cassette is inserted at a unique site within the intergeneric spacer region of two functional genes to avoid disrupting any chloroplast genes. Here, this chloroplast expression cassette is inserted in the *Pvu*II unique site in between the spacer region of *trn*I – *trn*A (*see* **Note 3**).

1. Digest the pUC19 plasmid containing the chloroplast flanking sequences with *Pvu*II and dephosphorylate with calf intestinal alkaline phosphatase as described by manufacturer.
2. Pull out the chloroplast expression cassette from the vector by digesting with appropriate restriction endonuclease and make blunt ends using T4 DNA polymerase following the manufacturer's procedure.
3. Examine both the digested product on agarose gel and elute the specific required fragments using Qiagen's QIAquick Gel Extraction kit as instructed in the product protocol.
4. Prepare ligation reaction mixture which contains 1× T4 ligase buffer, chloroplast expression cassette (serve as insert), pUC19 plasmid containing chloroplast flank (vector), T4 DNA ligase and adjust the volume with sterile distill water accordingly. Continue the reaction at 14°C overnight.
5. Take 2–5 μl from the ligation mixture and transform into competent *E. coli* cells.

6. Confirm the recombinant clones containing the chloroplast expression cassette by digesting with appropriate enzyme(s). Make sure to confirm the right orientation. Western blot analysis to see the expression of foreign protein is highly recommended.
7. Select the confirmed recombinant clone, multiply the clone, and prepare the plasmid in large scale using BioRad midiprep kit as described in the product protocol for bombardment. Maintain glycerol stock of the same clone for future use.

10.3.3. Delivery of DNA by Particle Bombardment

10.3.3.1. Preparation of Gold Particle (Also Serve as Stock)

1. Suspend 50 mg of gold particles (0.6 μm) in 1 ml absolute alcohol into a 1.5 ml siliconized tube and vortex for 2 min.
2. Spin at 10,000×*g* for 3 min and discard the supernatant.
3. Resuspend the gold pellet in 1 ml of 70% ethanol by vortexing for 1 min.
4. Incubate the tube at room temperature for 15 min. Mix intermittently during the incubation.
5. Spin for 2 min at 5,000×*g*, and then discard the supernatant.
6. Resuspend the pellet in 1 ml of sterile distilled water by vortexing for 1 min.
7. Allow the particles to settle down for 1 min at room temperature and then centrifuge for 2 min.
8. Repeat the washing steps 6 and 7 for 3 additional times.
9. Resuspend the gold particle pellet in 1 ml of 50% (v/v) glycerol and store at –20°C until ready to use (*see* **Note 4**).

10.3.3.2. Coating of DNA to Gold Particles and Spreading onto Macrocarrier (Procedure for 5 Shots; see Note 5)

1. Take 50 μl of completely resuspended gold particle into a 1.5 ml siliconized tube.
2. While vortexing, add 5 μl of plasmid (1 μg/μl) followed by 50 μl of 2.5 M $CaCl_2$ and 20 μl of 0.1 M spermidine-free base. Keep on vortexing for 20 min at 4°C.
3. Collect the DNA-coated gold particles by centrifugation for 1 min at 3,000×*g*.
4. Wash the pellet with 200 μl of 70% ethanol followed by absolute ethanol at room temperature.
5. Finally, resuspend the pellet in 50 μl of 100% ethanol. Keep the DNA-coated particles on ice until ready to use.
6. Place the sterilized macrocarrier inside the macrocarrier holder by using insertion cap.
7. Vortex the DNA-coated gold particles to eliminate any clumps. Spread 10 μl of the DNA-coated gold particles onto center of each macrocarrier and let them dry in laminar airflow hood (*see* **Note 6**).

10.3.3.3. Bombardment of Tobacco Leaf Tissue (5 Shots)

1. Harvest five fully grown aseptically propagated tobacco leaves.
2. Place the abaxial side of the leaf up on a circular sterile Whatman filter paper (70 mm) soaked in agar-solidified RMOP thin plate without antibiotics.
3. Switch on power of the PDS and vacuum pump.
4. Open the knob on the top of the helium gas cylinder anticlockwise.
5. Adjust the screw bar by tightening slowly clockwise to bring the helium gas approximately to 200–250 psi above the desired pressure of the rupture disc.
6. Place the rupture disk in its holder by using a sterile forceps and screw tightly to the gas acceleration tube (*see* **Note 7**).
7. Place the stopping screen in the macrocarrier holder and place the DNA-coated macrocarrier facing downward toward the stopping screen. Screw the assembly with macrocarrier cover lid and place into its position in the gene gun chamber.
8. Place the uncovered Petri dish containing sample to be bombarded on the target plate holder and insert it into the third slot (9 cm) in the gene gun chamber from the top. Close the chamber door.
9. Create vacuum pressure inside the chamber by pressing the vacuum button toward up position (VAC) at least up to 28 in. Hg in the vacuum gauge display. Flip the switch swiftly to the third position (HOLD) to hold the vacuum and keep pressing the FIRE button up until the rupture disk bursts.
10. Free the FIRE button and flip the vacuum button to VENT position to release the vacuum. Remove the bombarded sample.
11. Repeat steps 6–10 for additional samples.
12. Shut down the system by rotating the knob of the gas cylinder clockwise. Create, vacuum as in step 9 to release the gas pressure remaining inside the helium tube. Repeatedly keep pressing and releasing the FIRE button until both the pressure falls to zero on the meter guage of the gas cylinder and loosen the screw bar turning counterclockwise. Release the vacuum and turn off the vacuum pump and PDS.
13. Wrap each Petri dish containing the bombarded leaves with parafilm and keep in the dark for 2 days inside the culture room.

10.3.3.4. Plant Regeneration and Selection of Transplastomic Shoots (see Note 8)

1. Cut each of the bombarded leaf into small (5 mm^2) pieces and place the abaxial side (bombarded side) touching RMOP *(31)* selection medium containing 500 μg/ml spectinomycin for tobacco (var. Petit Havana) and 300 μg/ml for LAMD. Wrap each Petri dish with parafilm and keep in the culture room

under appropriate growth conditions. Within 3–5 weeks of culture, putative transformants appears (**Fig. 10. 1a**).

2. Again cut the leaves of the primary regenerated PCR-positive putative transplastomic shoot into small pieces (2 mm^2) and subject to a second round of selection on fresh RMOP selection medium to achieve homoplasmy (**Fig. 10. 1b**).
3. Separate the regenerated shoots after second round of selection and transfer to MSO *(31)* selection medium containing 500 μg/ml (Petit Havana) and 300 μg/ml (LAMD) of spectinomycin for third round of selection (**Fig. 10. 1c**).

10.3.4. Molecular Characterization of Transplastomic Plants

10.3.4.1. Confirmation of Transgenes Integration into Chloroplast Genome by PCR

PCR technique is used to screen transgenic plants and distinguish transplastomic plants from mutants. The integration of transgene into site-specific chloroplast genome is determined by using sets of primers, one of which anneals to the native chloroplast genome beyond the flanking sequence and the other anneals to the transgene cassette *(33)*. No PCR amplification will be observed in mutants and nuclear transgenic plants with these primers. To confirm the site-specific integration of transgene cassette into chloroplast genome, perform the PCR using primer pairs 3P (5′–AAAACCCGTCCTCAGTTCGGATTGC–3′) and 3M (5′–CCGCGT-TGTTTCATCAAGCC TTACG–3′) *(33)*, which anneals to the native chloroplast genome and *aad*A genes, respectively (**Fig.10.2c**). Whereas to confirm the integration of gene of interest, perform PCR with primer pairs, 5P (5′–CTGTAGAAGTCACCATTGTTGTGC–3′) and 2M (5′–TGACTGCCCACCTGAGAGCGGACA–3′) *(33)*, which anneals to the *aad*A and *trn*A genes, respectively (**Fig. 10.2d**).

Fig. 10.1. Generation of transgenic plants. After 4–5 weeks of bombardment, putative transplastomic shoots appear on RMOP selection medium (a). Leaves from polymerase chain reaction (*PCR*)-positive putative transformants are transferred to RMOP selection medium for second round of selection. Within 2–3 weeks, several shoots appear (b). Regenerated shoots are transferred to MSO selection medium for third round of selection where roots appear in about 10 days (c). (*See Color Plates*)

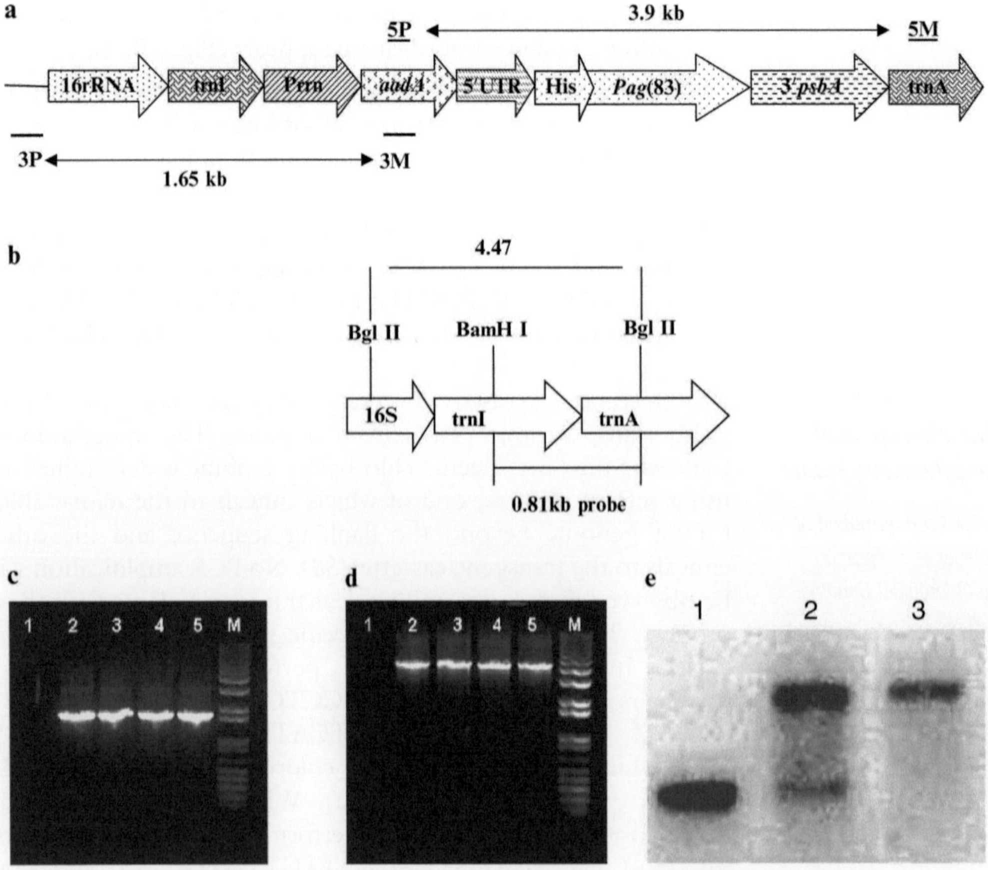

Fig. 10.2. Evaluation of transgene integration into the chloroplast genome by polymerase chain reaction (*PCR*) and homoplasmy by Southern blot. (a) Schematic representation of tobacco chloroplast transformation vector with the gene of interest (*pagA*), selectable marker gene (*aadA*), regulatory sequences, chloroplast flanking sequences for site-specific transgene integration, along with the primer annealing sites. (b) Flanking probe for Southern blot analysis showing expected product of untransformed and transplastomic plant. (c) *PCR* analysis of the transplastomic plants using 3P/3M for evaluation of site-specific integration of the *aadA* gene into chloroplast genome (Lane 1, untransformed; lanes 2–4 transplastomic lines; lane 5, positive control; lane 6, 1-kb DNA ladder. (d) Evaluation of transgene integration using 5P/2M primers (Lane 1, untransformed; lane 2–4, transplastomic lines; lane 5, positive control; lane 6, 1-kb DNA ladder. (e) Southern blot probed with the flanking sequence probe (lane 1, untransformed; lane 2, heteroplasmic plant; lane 3, homoplasmic plant).

1. Extract DNA from the leaf tissue of wild-type and putative transformants using the Qiagen's DNeasy Plant mini kit following manufacturer's protocol.
2. Prepare PCR reaction mixture for 50 μl: 1 μl of 100 ng/μl genomic DNA from the putative chloroplast transformants, 5 μl of 10×PCR reaction buffer; 2 μl of 50 mM $MgCl_2$, 1 μl each of 10 μM 3P and 3M primers or 1 μl each of 10 μM 5P and 2M primers, 1.5 μl of 10 mM dNTPs, 1–2.5 units of *Taq* DNA polymerase and adjust the final volume with sterile

distilled water. Include wild-type plant DNA in a separate PCR tube along with other PCR components to serve as control.

3. Perform PCR reaction with initial denaturation at 94°C for 5 min, followed by 30 cycles of denaturation at 94°C for 1 min, annealing at 58–60°C for 1 min and extension at 72°C for 1 min/kb followed by final elongation of 10 min at 72°C.
4. Analyze the PCR-amplified products in agarose gel by electrophoresis and visualized by staining with ethidium bromide. PCR analysis of putative transplastomic shoots with the 3P/3M and 5P/2M primer pairs yields PCR products of 1.6 kb which confirms site-specific integration of transgene cassette (**Fig. 10.2c**) and 3.9 kb (depending on the size of the inserted foreign gene) which confirms integration of gene of interest into chloroplast genome (**Fig. 10.2d**), respectively. The untransformed plant DNA shows no amplification.

10.3.4.2. Southern Blot Analysis to Confirm Homoplasmy

Initially, few copies of the chloroplast genomes receive transgene cassette and after several rounds of selection, untransformed genome copies are replaced by transgenic copies, leading to homoplasmy. This can be ascertained by Southern blot analysis of the total DNA isolated from the plants after third round of selection. Total plant DNA extracted from transplastomic leaves is digested with appropriate restriction endonuclease(s) and probed with radiolabeled chloroplast flanking sequences. Perform the following steps carefully.

Transfer of DNA to Membrane

1. Extract total cellular DNA from transplastomic as well as from wild-type plants using Qiagen's DNeasy Plant mini kit as described in product manual.
2. Digest 1–3 μg of total cellular DNA with suitable restriction enzyme and incubate at the prescribed temperature for 4–16 h depending on the enzyme. Include wild-type DNA as control.
3. Run the digested DNA on 0.8% agarose gel until completely separated.
4. Depurinate the DNA by soaking in 0.25 N HCl for 10 min and rinse twice with distilled water for 5 min each. This will facilitates efficient transfer of DNA fragments from the gel to membrane.
5. Denature the DNA by soaking the gel in transfer buffer for 20 min.
6. Turn the gel upside down so that the smooth surface faces the nylon membrane. Prewet the nitrocellular membrane in water then in transfer buffer for 5 min. Place the membrane

on top of the gel and apply gentle pressure by placing a stack of paper towels and a weight of 300 g on top to ensure even contact between gel and membrane. Allow the DNA to transfer overnight.

7. Remove the membrane with transferred DNA, rinse twice with 2× SSC buffer for 5 min each and air-dry on filter paper. The membrane is subject to cross-link by using GS GeneLinker UV Chamber (BioRad) at appropriate setting (C3).

Preparation of Flanking Probe

1. Pull out the flanking probe from the plasmid containing chloroplast flanking sequences by digesting with appropriate restriction enzyme(s) (*see* **Note 9**).
2. Denature 45 μl of the flanking DNA fragment (50–200 ng) at 95°C for 5 min and immediately keep on ice for 2–3 min.
3. Add the denatured probe to Ready-To-Go DNA labeling beads and mix gently by flicking the tube.
4. Add 5 μl of radioactive isotope ^{32}P dCTP to the probe mixture and incubate for 1 h at 37°C. Purify the radiolabeled probe by using ProbeQuant G50 Micro Columns (Amersham Pharmacia Biotech, Inc.) as described in the product protocol. Store at 4°C until ready to use.

Prehybridization and Hybridization

1. Place the blot in sufficient volume of QuickHyb solution (Stratagene) at 68°C for 1 h using a rotary hybridizing chamber.
2. Add 100 μl of salmon sperm DNA (10 mg/ml stock) to the labeled probe and heat at 95°C for 5 min.
3. Add labeled probe to QuickHyb solution for hybridization with the DNA on membrane at 68°C for 1 h in a rotary hybridizing chamber.
4. Discard the QuikHyb solution with the labeled probe and wash the membrane twice with wash buffer-I at room temperature for 15 min followed by wash buffer-II at 60°C for 15 min.
5. Expose X-ray film to the hybridized membrane in dark and keep at –80°C for 12–16 h depending upon the radioactive count.
6. Develop the exposed X-ray film in an automated X-ray film processor. Visualize the pattern of hybridization on the film. The transgenic plants show higher molecular weight ct-DNA fragment when compared to untransformed wild-type plants. Heteroplasmic transgenic plant will show the lower molecular weight fragment of the wild-type plant along with the higher molecular weight fragment. Absence of the wild-type fragment from transgenic plants confirms homoplasmy (**Fig. 10.2e**; *see* **Note 10**).

7. Transfer the confirmed homoplasmic plants to the pots containing sterilized soil and initially keep under high humidity for acclimatization and then move them to the greenhouse.

10.3.4.3. Extraction of Total Soluble Protein

1. Grind 100 mg leaf material in liquid nitrogen and add 200 μl of freshly prepared PEB on ice.
2. Homogenize the leaf tissue using the hand-held homogenizer for 5 min by keeping the samples on ice to prevent overheating.
3. Centrifuge at 10,000×*g* for 5 min at 4°C.
4. Aliquot the supernatant and proceed directly to protein quantitation.

10.3.4.4. Western Blot Analysis to Evaluate Expression of Foreign Proteins

Western blot analysis facilitates detection of expression of a specific protein within a tissue or mixture of protein. It can also be used to determine the quantity and the molecular size of the protein. This technique relies on specificity and quality of antibody used to probe the desired protein. Perform the following steps carefully to minimize error in the experiment (**Fig. 10.3**).

First day

1. Dilute various quantities of sample crude extracts (e.g., 100 μg, 10 μg, and 1 μg) in Laemmli sample buffer and boil for 4 min.
2. Immediately load 20 μl of samples (including unboiled control samples) into wells of a 12% sodium dodecyl sulfate-polyacrylamide gel. Include the protein marker in a separate well (*see* **Note 11**).

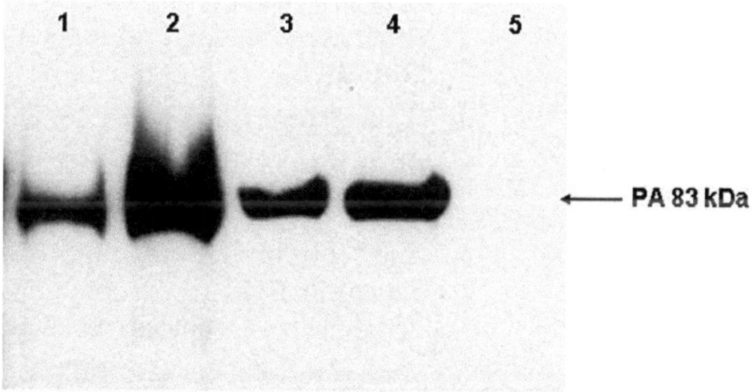

Fig. 10.3. Western blot analysis to confirm the expression of the anthrax protective antigen (*PA*) in transgenic plant crude extracts. Western blot with anti-PA monoclonal antibody shows full-length 83 kDa protein. Lane 1, 5 ng PA; lanes 2–4, transplastomic lines; lane 5, untransformed.

3. Run the gel with initial current at 85V for 30–40 min in 1× electrode buffer until proteins migrate into the resolving gel, then increase the current to 110 V for 45 min. Discontinue running when the dye has reached the bottom of the resolving gel.
4. Transfer the separated proteins to nitrocellulose or PVDF membrane in transfer buffer for 2–4 h at 80V, depending on the size of the protein or overnight at 15V (*see* **Note 12**).

Second day

1. Remove the membrane and rinse with water.
2. Incubate the membrane in PTM for 1 h at room temperature with gentle shaking.
3. Decant the PTM. Add primary antibody diluted (dilution ratio depends upon antibody titer) in PTM and incubate at room temperature for 2 h, with gentle rocking.
4. Wash the membrane one time with PBS-T for 5 min at room temperature, then add secondary antibody (conjugated to HRP) with appropriate dilution in PTM. Incubate for 1.5 h with gentle shaking.
5. Wash the membrane 3 times for 15 min each with PBS-T followed by one time washing with 1× PBS for 10 min. Add ECL substrate and incubate at room temperature for 5 min with gentle shaking.
6. Expose the membrane to X-ray film and develop the film using an automated X-ray film processor (*see* **Note 13**).

10.3.4.5. ELISA for Quantitation of Transgenic Proteins

ELISA is an enzyme-based immunoassay method which is widely used to detect the antigen or antibody concentration with high specificity and sensitivity (**Fig. 10. 4**).

1. Coat plates with 100 μl/well the sample crude extracts and antigen standards in coating buffer, and incubate plates either at room temperature for 4 h or at 4°C overnight (*see* **Note 14**).
2. Wash plates 3 times with PBS-T.
3. Block plates with 200 μl/well PTM for 2 h at 37°C.
4. Wash plates 3 times with PBS-T.
5. Add 100 μl/well of antigen-specific primary antibody diluted in PTM (appropriate dilution must be empirically determined). Incubate the plate 2 h at 37°C.
6. Wash plates 3 times with PBS-T.
7. Add 100 μl/well secondary antibody (conjugated to HRP) diluted in PTM (appropriate dilution must be empirically determined). Incubate the plate at 37°C for 1 h.
8. Wash plates 3 times with PBS-T.

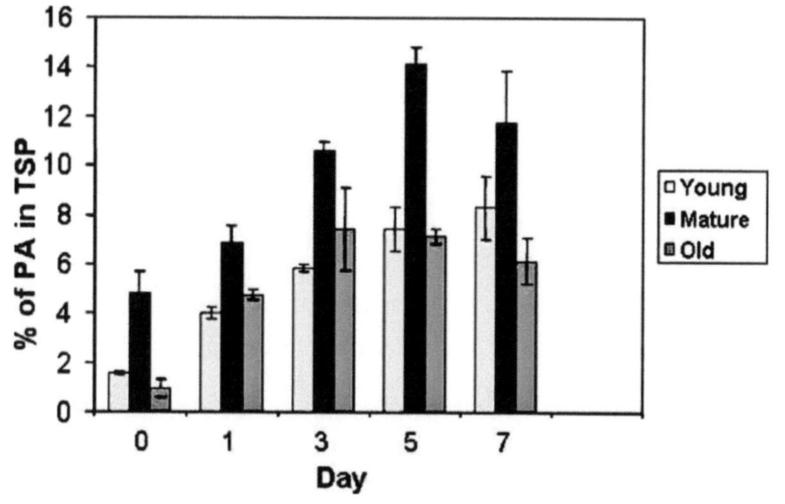

Fig. 10.4. Quantitation of chloroplast-derived anthrax protective antigen (*PA*) by enzyme-linked immunosorbent assay (*ELISA*) under normal and continuous illumination. The expression level increases under continuous illumination up to 5 days reaching a maximum of 14.2% of the total soluble protein (*TSP*) in mature leaves. Mature leaves have higher levels of *PA* than young or older leaves.

9. Develop with TMB substrate (100 μl/well). Stop the reaction by adding 50 μl 2 N sulfuric acid.
10. Read the plate immediately on a microtiter plate reader using a 450 nm filter.

10.3.4.6. Purification of His-Tagged PA by Affinity Chromatography

1. Purify the His-tagged PA by His affinity chromatography using nickel-chelate-charged column following the manufacturer's protocol.
2. Analyze the PA protein samples by ELISA.
3. Collect together the eluate fraction containing purified PA.
4. Set up dialysis reaction against 1× PBS (pH 7.4) using dialysis cassette.
5. Concentrate using Centricon 10,000-molecular-weight-cut-off ultrafiltration units as described in the product manual.

10.3.5. In Vitro Functional Assays

10.3.5.1. Anthrax PA: Macrophage Lysis Assay

Anthrax has been listed as category A biological agent due to its severity of impact on human health, high mortality rate, acuteness of disease, and potential for delivery as a biological weapon. Using chloroplast transformation technology, the PAs has been expressed in tobacco by inserting the *pagA* gene into tobacco chloroplast genome *(15)*. PA functionality was studied by macrophage lysis assays *(34)*. Macrophage lysis assays were performed on the crude leaf extracts partially purified chloroplast-derived PA and *B. anthracis*-derived PA. Cytotoxicity measurements in macrophage lysis assays showed that chloroplast-derived PA was equal in potency to PA produced in *B. anthracis* (**Fig. 10. 5**; *see* **Note 15**).

1. Plate macrophage cells RAW 264.7 in 96-well plates in 120 μl DMEM and grow the cells to 50% confluence.
2. Aspirate the medium and add 100 μl medium containing 250 ng/ml LF. A negative control well contains no LF in the culture medium and is used to assess the toxicity of the experimental samples.
3. In another 96 well plate, dilute serially experimental samples (e.g., crude leaf extracts, purified chloroplast-derived PA, and *B. anthracis*-derived PA) twofold and transfer each sample (in triplicate) onto the RAW264.7 cells in such a way that the first dilution of plant extract is 1:50 and PA is at 0.4 μg/ml.
4. Incubate the cells for 2.5–5 h and add MTT at a final concentration of 0.5 mg/ml and incubate for 40 min to asses cell viability.
5. Dissolve the blue pigment produced by the viable cells by aspirating the medium and adding 50 μl/well of a mixture of 0.5% (w/v) SDS and 25 mM HCl in 90% (v/v) isopropanol.
6. Shake the plates for 5 min before reading at 570 nm using a microplate reader.

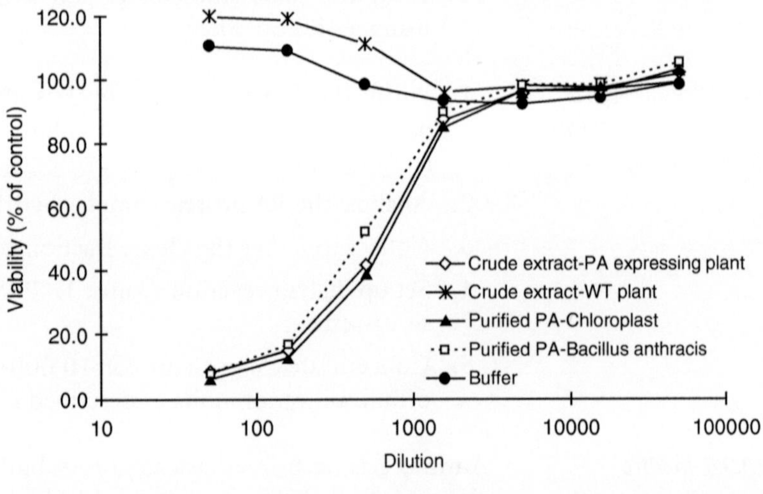

Fig.10.5. In vitro functional analysis of chloroplast-derived anthrax protective antigen (*PA*) by the macrophage cytotoxicity assay. Cytotoxicity of mouse macrophage RAW 264.7 cells treated with purified lethal factor (*LF*) and various dilutions *PA* samples: crude extracts of transgenic plants expressing *PA*; crude extracts of untransformed plants; 20 μg/ml of partially purified chloroplast-derived *PA*; or 20 μg/ml of fully purified *Bacillus anthracis*-derived PA; buffer alone (negative control). Crude extract of plant leaves expressing *PA* shows equal activity to that of fully purified *B. anthracis*-derived *PA*, indicating that *PA* expressed in chloroplasts is functional.

10.3.6. *In Vivo* Functional Assays

Animal models have been proven to be crucial to test the immunogenicity and effectiveness of vaccine antigens. While evaluating vaccines, a number of parameters can affect the results, including certain species of animals, doses of antigen, types of adjuvants, and routes of administration. We discuss below procedures for preparation of mice and vaccine, immunization, collection of mice serum and fecal samples, antibody detection, as well as lethal toxin (LT) challenge *(16, 20, 35, 36)*.

10.3.6.1. Preparation of Mice (see Note 16)

Selecting appropriate animals can reduce the variability which allows using of fewer animals and less work to achieve significant results. Generally, inbred animals are used because they are genetically identical *(37)*.

1. Pathogen-free female mice, 4–5 weeks of age, are purchased from Charles River/Jackson Laboratory.
2. Minimum of ten mice per group are set up for statistical comparisons.
3. House the mice under 12-h light/12-h dark cycle condition.
4. Start experiments after 3–5 days when the mice adapt the environment.

10.3.6.2. Preparation of Vaccine Antigen

An effective adjuvant should enhance the level of the immune response to a vaccine, compared to immunogen alone. Aluminum hydroxide is a commonly used adjuvant.

1. For subcutaneous (s.c.) immunization, adsorb adequate quantity of antigen to the adjuvant. For example, 5 μg of purified PA to 200 μl adjuvant including aluminum hydroxide (25% Alu-Gel-S; Sigma) 1:4 in PBS. Mix thoroughly in a shaker at 225 rpm at 4°C for 2–3 h to allow the PA to bind to the adjuvant *(38)*. Use a commercial standard (PA derived from *B. anthracis*) with adjuvant as positive control *(16, 39)*.
2. For oral gavage, grind the leaves in liquid nitrogen. Dilute the ground transgenic plant leaf which contains adequate doses of antigen or therapeutic protein (i.e., 14 μg of CTB-Pins protein in 200 μl of 1× PBS, pH 7.4). Negative control mice are gavaged with ground untransformed plant leaf material *(20, 40)*.

10.3.6.3. Immunization of Mice

Antigen delivery routes affect both the efficiency and the type of response obtained. Common routes of vaccine delivery are listed as follows: subcutaneous (s.c.), intradermal (i.d.), intramuscular (i.m.), intraperitoneal (i.p.), or intravenous (i.v.) injection; or by oral delivery; or by intranasal (i.n.) administration. Antigens administered s.c. generally elicit the strongest immune responses, but oral administration via plant cells are also very effective *(35, 36)*.

Subcutaneous Immunization

1. Place the prepared vaccine into 1 ml syringe and inject s.c. into lower flank of the mice, with a maximum dose of 100 μl per site.
2. Administer further booster doses at day 14, day 28, and day 140.
3. To measure PA-specific antibody responses, collect serum on days 43 and 155 (15 days after the third and fourth doses).

Oral Immunization

1. Prepare the vaccine in separate 1 ml sterile syringes, and attach a ball-end bent feeding needle.
2. Hold the mouse by the scruff of the neck and by the tail, and gently insert the feeding needle into the mouth. Inject plant cells once the ball-end of needle reaches proper position. Check animal periodically after administration of vaccine for any adverse effects (*see* **Note 17**).
3. Perform oral immunization of mice at 7-day intervals for a total 7 weeks.
4. Collect the serum samples at 12 weeks of age.

10.3.6.4. Collection of Mouse Serum and Fecal Samples

Serum and fecal samples are collected to measure antigen-specific antibody responses.

Serum Samples

Blood can be collected from retro-orbital sinus in anesthetized mice. Maximum 1% of the body weight may be collected every 2 weeks. Exceed blood collection will cause a physiologic compensatory response which will increase experimental variability *(35)*.

1. Set up the anesthetic circuit by connecting the anesthesia system with oxygen tank and the induction chamber.
2. Place the mice in the induction chamber.
3. Adjust the flowmeter which indicates the oxygen level to 0.4 L/min (according to manufacturer's recommendation).
4. Adjust the isoflurane level to 2% (according to manufacturer's recommendation; *see* **Note 18**).
5. Carefully monitor the respiration rate, body temperature, and the color of the skin.
6. Place the mouse on a flat surface and tight the skin of its head with the thumb and middle finger to slightly protrude the eye.
7. Put the microhematocrit tube at the inner corner of the eye beside the eyeball, and gently but firmly insert through the conjunctiva toward the back of the eye. Gently rotate the tube and blood should flow freely by capillary action (*see* **Note 19**).

8. After collection, withdraw the tube and apply pressure on the closed eyelids until bleeding stops completely.
9. Allow the blood to clot for 30 min in a BD microtainer serum separator collection tube at room temperature and separate serum by centrifugation at 13,000×*g* for 2 min.
10. Harvest the serum, aliquot, and store at –80°C.

Fecal Samples *(41)*

1. Collect the fecal samples for antibody detection from individual mice. Samples can be extracted immediately or stored at –80°C.
2. Place 0.1 g fecal material into 1.5 ml microcentrifuge tube with 1 ml (10 volumes, w/v) of PBS.
3. Crush the pellets using hand-operated homogenizer with mechanical pestle, one pestle per sample.
4. Vortex thoroughly and centrifuge at 10,000×*g* for 10 min at 4°C.
5. Save the supernatant, aliquote, and store at –80°C.

10.3.6.5. Antibody Detection and Quantitation in Mice (41, 42)

Antigen-specific antibody titers are determined by ELISA (**Fig. 10.6**).

1. Coat 100 μl/well standard antigen (2.0 μg/ml of PA and 5.0 μg/ml of insulin or CTB, respectively, in coating buffer, pH 9.6) in 96-well microtiter ELISA plates and incubate overnight at 4°C.
2. Wash with PBS-T, block the plates with PTM.
3. Add 100 μl/well serially diluted mouse serum or fecal samples and incubate for 1 h at 37°C, and then wash with PBS-T.
4. Determine the various immunoglobulin subclasses by incubating for 1 h at 37°C with corresponding HRP-conjugated antimouse IgG (IgG, IgG1, IgG2a), or IgA antibodies.
5. Add 100 μl/well of TMB as the substrate. Stop the reaction with 50 μl/well of 2 N sulfuric acid.
6. Read the plates on a plate reader at 450 nm.
7. Calculate the Titer values using a cutoff value equal to an absorbance difference of 0.5 between immunized and unimmunized mice.

10.3.6.6. Functional Evaluation: Lethal Toxin Challenge in Mice

Once the mice achieve the high immunity, LT or pathogen challenge is performed to test the protective efficacy of the vaccine (**Fig. 10. 7**).

1. Inject i.p. 150 μg LT (150 μg LF + 150 μg PA) in sterile 1× PBS (1 ml) to different groups of mice *(43)*.
2. Monitor the signs of malaise and mortality for every 8-h duration.

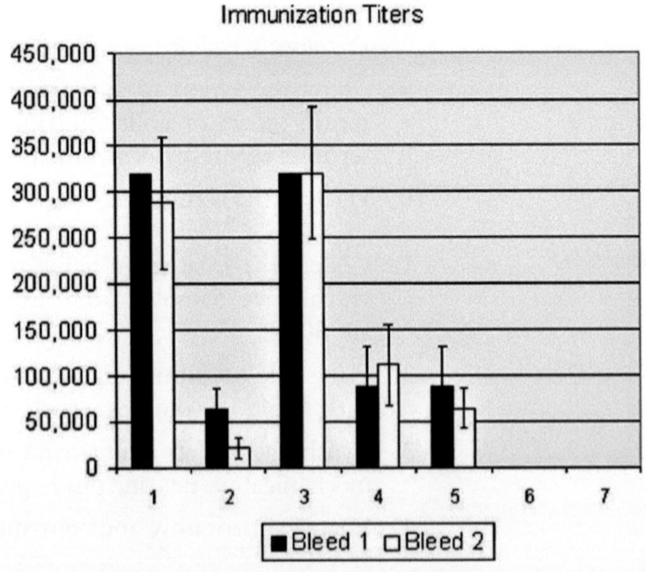

Fig. 10.6. Comparison of responses in serum samples of mice administered subcutaneously (s.c.) with chloroplast-derived anthrax protective antigen (*PA*) with adjuvant (column 1), chloroplast-derived *PA* alone (column 2), *Bacillus anthracis*-derived PA with adjuvant (column 3), *B. anthracis*-derived *PA* alone (column 4), *PA* expressing plant leaf crude extract with adjuvant (column 5), untransformed plant leaf crude extract with adjuvant (column 6), and nonimmunized mice (column 7). These data confirm that the plant-derived *PA* and *B. anthracis*-derived PA elicit significant immune responses.

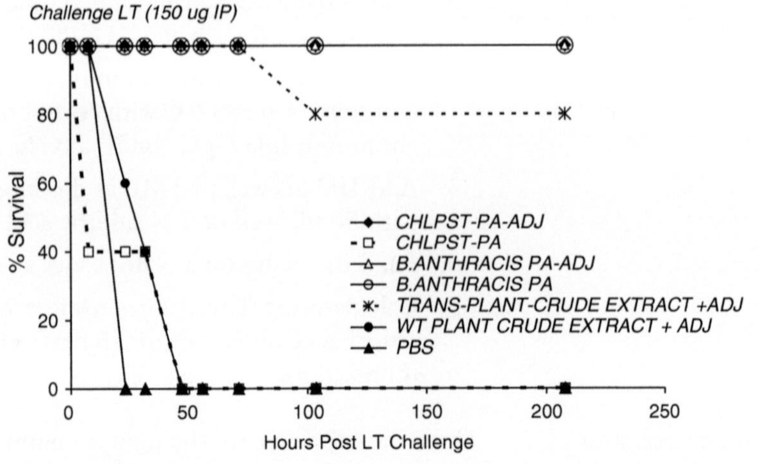

Fig. 10.7. Anthrax lethal toxin (*LT*) challenge in mice. Survival over time for different groups of mice following intraperitoneal (i.p.) challenge with 150 of the *LT*. *CHLPST*, chloroplast; *ADJ*, adjuvant; *WT*, untransformed. Mice immunized with the plant crude extracts expressing protective antigen (*PA*) showed survival rate similar to mice immunized with *Bacillus anthracis*-derived fully purified *PA*. These data demonstrate that the chloroplast-derived PA has the immunoprotective properties against anthrax *LT* similar to *B. anthracis*-derived *PA*.

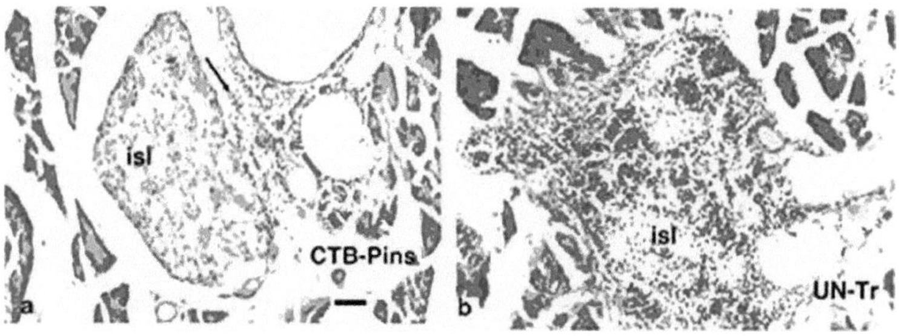

Fig. 10.8. Histochemical stained sections of nonobese diabetic (*NOD*) mouse pancreatic islets following oral delivery of cholera toxin B subunit fused to human proinsulin (*CTB–Pins*). Hematoxylin and eosin (H&E) stained sections of pancreas of mice orally delivered with (a) transgenic plant expressing CTB–Pins showing decreased lymphocytic infiltration in islet (isl) (b) Untransformed (UN-Tr) plant leaf material where severe lymphocytic infiltration in the islet is observed. (*See Color Plates*)

10.3.6.7. Cholera Toxin B Subunit: Human Proinsulin Fusion Protein

CTB–Pins therapy is assessed in 5-week-old nonobese diabetic mice. Mice are orally gavaged weekly for 7 weeks, using a tuberculin syringe fitted with a 20-G bulb-tipped gastric gavage needle. Negative controls are mice gavaged with untransformed leaf or those expressing CTB–green fluorescent protein (GFP) or interferon–GFP in chloroplasts.

1. At 12 weeks of age, sacrifice the mice.
2. Excise the pancreas and prepare it for histochemistry by fixing in 4% PFA and freezing in tissue freezing medium.
3. Prepare frozen sections of the pancreas, 10-μm thick, using a cryostat.
4. Stain the sections with hematoxylin and eosin (H&E) and dehydrate in serial graded alcohol solutions.
5. Set the slides with coverslips and analyze (**Fig. 10. 8**).

10.4. Notes

1. Proper designing of primer is very important. If the PCR product is different from the expected size, it could be due to nonspecific annealing of primers or mispriming. Try to set up the PCR reaction at different annealing temperature by increasing or decreasing or by increasing $MgCl_2$. Otherwise check the primer for repetitive sequences and design new set of primers.
2. Competency of *E. coli* cell is very important for getting efficient transformation. Thaw the *E. coli* cells on ice just

before use in transformation experiment and avoid refreezing. Check the competency of the *E. coli* cells with control ligated DNA and observe the colonies. If few or no white colony appears after transformation with the ligated product, it could be due to inefficient 3′ A overhang or the insert is not disturbing the *LacZ* gene. Check the blue colony for the presence of insert fragment. Otherwise, repeat the PCR and 3′ A overhang reaction.

3. High efficiency of chloroplast transformation of any species depends upon many factors including site of integration and regulatory sequences. In the present protocol, transgenes were integrated into the transcriptionally active spacer region between *trn*I/*trn*A genes within the *rrn* operon. High levels of foreign gene expression have been observed in this site of integration *(1)* and over hundred transgenes have been inserted at this site *(2, 12, 31, 33, 44, 45)*. Thus, it appears that this site of integration is important for chloroplast transformation.
4. Gold particles to be used for bombardment may be prepared in advance and stored at –20°C until ready to use.
5. While coating DNA to gold particle, strictly follow the order of addition such as DNA, $CaCl_2$ followed by spermidine. Spermidine is highly hygroscopic. Prepare the stock and aliquot into small volume to avoid refreezing and thawing. Take a small aliquot and discard the remaining after use.
6. Proper coating of DNA to gold particle and spreading of DNA-coated gold particle on macrocarrier are very important. Avoid any clumps while loading the DNA-coated gold particle because this will damage the leaf tissue as well as decrease the transformation frequency. Make sure to vortex the DNA-coated gold particles for at least 30 s before each spreading onto macrocarrier to avoid clumps. DNA-coated gold particle should be used within 2 h.
7. Make sure the disc is properly placed in its holder by tapping the holder. Different rupture disc is recommended depending on the tissue used for bombardment. Rupture disc of 1,100 psi is recommended for tobacco leaf transformation.
8. Bombarded side of the leaf should touch the RMOP selection medium. Avoid overcrowding of the leaf pieces because the leaf segments should expand. Always include the unbombarded leaf in the regeneration medium as well as in selection medium separately to serve as positive or negative control. This will tell us whether the regeneration medium or the selection system is working.
9. The chosen restriction enzyme should not be present in the introduced foreign DNA and should cut the chloroplast genome on either side of the integration cassette.

10. It is important to check the digestion efficiency of the genomic DNA before transferring to membrane. With successful digestion, all lanes will have identical smears and faint bands of similar intensity. Partial or incomplete digestion will hinder the transfer of DNA to membrane and give misleading results. Repeat the digestion until it shows uniform pattern. There will be no signal on X-ray if the probe is not labeled or probe was not denatured. To minimize the background, wash the gel tank before running the digested DNA and avoid overlapping of the blot during prehybridization and hybridization.
11. The exact percentage of the resolving gel will depend upon the size of the protein.
12. After transfer, only PVDF membranes wrapped in cling plastic wrap can be stored at –20°C for a few days.
13. If there is no signal in the X-ray film, there is possibility of poor transfer or low titers of primary/secondary antibodies or substrate are not working. If the positive standard control is working well, the expression level may be very low. Care should be taken during blocking and washing of the membrane to minimize background. The antibodies can bind to the protein of interest as well as the membrane. This is avoided by blocking all "empty" areas of the membrane with a blocking buffer (usually dry milk). Avoid air bubble in between the gel and membrane to reduce the unwanted spots on the film. Air bubbles can be removed by rolling a Pasteur pipette over the membrane before transfer. Keep the membrane wet.
14. Binding the antigen overnight is important to generate a good signal at the end.
15. If the cells are not lysed, it could be due to two reasons; (1) LF of the purified *B. anthracis* may not be functional or (2) quantity of the PA is incorrect. Recalculate the transgenic PA based on ELISA results, otherwise purchase new LF.
16. Animals with bite wounds, ulceration, necrosis, infected tumor, or display abnormal behavior should be excluded from the experiment.
17. For oral delivery, the mouse usually swallows when the feeding needle reaches the back of the throat, which facilitates its entry to the esophagus. If you encounter resistance, withdraw the feeding needle and try again. Do not force the needle to pass resistance. Improper gavage could result in injection to the lungs, rupture of the esophagus, or pharynx. Once it happens, the animal should be euthanized. Hold the animal firmly and carefully insert the feeding needle to avoid the adverse effects.

18. Anesthetic agent used must be approved from IACUC (Institutional Animal Care and Use Committee). Inhalant anesthetic agents provide a quicker induction, more stable maintenance, easier control of anesthetic depth, and smoother recovery than injectable anesthetic agent. Isoflurane as an inhalant anesthetic agent provides quicker induction and recovery than others.
19. While holding the animal, try not to block the animal's breathing. Over anesthesia will decrease the free flow of blood. Thus, it is very important to carefully monitor the depth of anesthesia. The depth of anesthesia should be controlled carefully as the mice cannot move during procedure, while maintaining adequate respiration. Prepare multiple microhematocrit tubes as needed to collect the total volume. Retro-orbital bleeding can cause blindness and eye injury *(46)*. If one eye is damaged, the second eye can be bled. For repeated blood collection, alternate eyes should be used. If the second eye is damaged, bleeding through this technique must cease.

Acknowledgment

The protocols reported in this article were supported in part by grants from USDA 3611-21000-021-02S and NIH R01 GM 63879 to Henry Daniell.

References

1. DeCosa, B., Moar, W., Lee, S. B., Miller, M. and Daniell, H. (2001) Overexpression of the Bt cry2Aa2 operon in chloroplasts leads to formation of insecticidal crystals. Nat. Biotechnol. 19, 71–74.
2. Grevich, J. J. and Daniell, H. (2005) Chloroplast genetic engineering: Recent advances and future perspectives. Crit. Rev. Plant Sci. 24, 83–107.
3. Daniell, H. (2007) Transgene containment by maternal inheritance: Effective or elusive? Proc. Natl. Acad. Sci. USA 104, 6879–6880.
4. Daniell, H. (2002) Molecular strategies for gene containment in transgenic crops. Nat. Biotechnol. 20, 581–586.
5. Daniell, H., Datta, R., Varma, S., Gray, S. and Lee, S. B. (1998) Containment of herbicide resistance through genetic engineering of the chloroplast genome. Nat. Biotechnol. 16, 345–348.
6. Hagemann, R. (2004) The sexual inheritance of plant organelles. InH Daniell, C Chase, eds., Molecular Biology and Biotechnology of Plant Organelles. Springer Publisher, Dordrecht, the Netherlands, pp. 93–113.
7. Ruiz, O. N. and Daniell, H. (2005) Engineering cytoplasmic male sterility via the chloroplast genome. Plant Physiol. 138, 1232–1246.
8. Lee, S. B., Kwon, H. B., Kwon, S. J., Park, S. C., Jeong, M. J., Han, S. E. and Daniell, H. (2003) Accumulation of trehalose within transgenic chloroplasts confers drought tolerance. Mol. Breed. 11, 1–13.

9. Dhingra, A., Portis, A. R. and Daniell, H. (2004) Enhanced translation of achloroplast expressed RbcS gene restores SSU levels and photosynthesis in nuclear antisense RbcS plants. Proc. Natl. Acad. Sci. USA 101, 6315–6320.
10. Quesada-Vargas, T., Ruiz, O. N. and Daniell, H. (2005) Characterization of heterologous multigene operons in transgenic chloroplasts: Transcription, processing, translation. Plant Physiol. 138, 1746–1762.
11. Kamarajugadda, S. and Daniell, H. (2006) Chloroplast-derived anthrax and other vaccine antigens: Their immunogenic and immunoprotective properties. Expert Rev. Vaccines 5, 839–849.
12. Daniell, H. (2006) Production of biopharmaceuticals and vaccines in plants via the chloroplast genome. Biotechnol. J. 1, 1071–1079.
13. Fischer, R. and Emans, N. (2000) Molecular farming of pharmaceutical proteins. Transgenic Res. 9, 279–299.
14. Cramer, C. L., Boothe, J. G. and Oishi, K. K. (1999) Transgenic plants for therapeutic proteins: Linking upstream and downstream strategies. Curr. Top. Microbiol. Immunol. 240, 95–118.
15. Watson, J., Koya, V., Stephen, H. L. and Daniell, H. (2004) Expression of Bacillus anthracis protective antigen in transgenic chloroplast of tobacco, a non-food/feed crop. Vaccine 22, 4374–4384.
16. Koya, V., Moayeri, M., Leppla, S. H. and Daniell, H. (2005) Plant-based vaccine: Mice immunized with chloroplast-derived anthrax protective antigen survive anthrax lethal toxin challenge. Infect. Immun. 73, 8266–8274.
17. Dufourmantel, N., Tissot, G., Goutorbe, F., Garcon, F., Muhr, C., Jansens, S., Pelissier, B., Peltier, G. and Dubald, M. (2004) Generation of fertile transplastomic soybean. Plant Mol. Biol. 55, 479–489.
18. Kumar, S., Dhingra, A. and Daniell, H. (2004) Plastid-expressed betaine aldehyde dehydrogenase gene in carrot cultured cells, roots, and leaves confer enhanced salt tolerance. Plant Physiol. 136, 2843–2854.
19. Kanamoto, H., Yamashita, A., Asao, H., Okumura, S., Takase, H., Hattori, M., Akiho Yokota, A. and Tomizawa, K. (2006) Efficient and stable transformation of Lactuca sativa L. cv. Cisco (lettuce) plastids. Transgenic Res. 15, 205–217.
20. Ruhlman, T., Ahangari, R., Devine, A., Samsam, M. and Daniell, H. (2007) Expression of cholera toxin B-proinsulin fusion protein in lettuce and tobacco chloroplasts—Oral administration protects against development of insulitis in non-obese diabetic mice. Plant Biotechnol. J. 5, 495–510.
21. Arlen, P. A., Singleton, M., Adamovicz, J. J. Ding, Y., Davoodi-Semiromi, A. and Daniell, H. (2008). Effective plague vaccination via oral delivery of plant cells expressing F1-V antigens in chloroplasts. Infect. Immun. 76, 3640–3650.
22. Tregoning, J. S., Nixon, P., Kuroda, H., Svab, Z., Clare, S., Bowe, F., Fairweather, N., Ytterberg, J., van Wijk, K. J., Dougan, G. and Maliga, P. (2003) Expression of tetanus toxin Fragment C in tobacco chloroplasts. Nucleic Acids Res. 31, 1174–1179.
23. Daniell, H., Lee, S. B., Panchal, T. and Wiebe, P. O. (2001) Expression of the native cholera toxin B subunit gene and assembly as functional oligomers in transgenic tobacco chloroplasts. J. Mol. Biol. 311, 1001–1009.
24. Chebolu, S. and Daniell, H. (2007) Stable expression of Gal/GalNAc lectin of Entamoeba histolytica in transgenic chloroplasts and immunogenicity in mice towards vaccine development for amoebiasis. Plant Biotechnol. J. 5, 230–239.
25. Molina, A., Hervas-Stubbs, S., Daniell, H., Mingo-Castel, A. M. and Veramendi, J. (2004) High-yield expression of a viral peptide animal vaccine in transgenic tobacco chloroplasts. Plant Biotechnol. J. 2, 141–153.
26. Arlen, P. A., Falconer, R., Cherukumilli, S., Cole, A., Cole, A. M., Oishi, K. K. and Daniell, H. (2007) Field production and functional evaluation of chloroplast-derived interferon alpha 2b. Plant Biotechnol. J. 5, 511–525.
27. Leelavathi, S. and Reddy, V. S. (2003) Chloroplast expression of His-tagged GUS-fusions: A general strategy to overproduce and purify foreign proteins using transplastomic plants as bioreactors. Mol. Breed. 11, 49–58.
28. Staub, J. M. Garcia, B.,, Graves, J.,, Hajdukiewicz,, P. T.,, Hunter,, P.,, Nehra,, N.,, Paradkar,, V.,, Schlittler,, M.,, Carroll,, J. A.,, Spatola,, L.,, Ward,, D.,, Ye,, G. , Russell,, D. A., (2000) High-yield production of a human therapeutic protein in tobacco chloroplasts. Nat. Biotechnol. 18, 333–338.
29. Fernandez-San Millan, A., Mingo-Castel, A., Miller, M. and Daniell, H. (2003) A chloroplast transgenic approach to hyper-express and purify human serum albumin, a protein highly susceptible to proteolytic degradation. Plant Biotechnol. J. 1, 71–79.

30. DeGray, G., Rajasekaran, K., Smith, F., Sanford, J. and Daniell, H. (2001) Expression of an antimicrobial peptide via the chloroplast genome to control phytopathogenic bacteria and fungi. Plant Physiol. 127, 852–862.
31. Daniell, H. (1997) Transformation and foreign gene expression in plants mediated by microprojectile bombardment. Methods Mol. Biol. 62, 453–488.
32. Murashige, T. and Skoog, F. (1962) A revised medium for rapid growth and bioassays with tobacco tissue cultures. Physiol. Plant. 15, 473–497.
33. Daniell, H., Dhingra, A. and Ruiz, O. N. (2004) Chloroplast genetic engineering to improve agronomic traits. Methods Mol. Biol. 286, 111–37.
34. Hanna, P. C., Acosta, D. and Collier, R. J. (1993) On the role of macrophages in anthrax. Proc. Natl. Acad. Sci. USA 90, 10198–10201.
35. Hau, J. and Van Hoosier, G. (2002) Handbook of Laboratory Animal Science. 2nd ed. Vol. I: Essential Principles and Practices, CRC Press LLC, Florida.
36. Janeway, C. A., Travers, P., Walport, M. and Shlomchik, M. J. (2001) Immunobiology: The Immune System in Health and Disease. 5th ed. Garland Publishing, New York.
37. Festing, M. F. (1976) Phenotypic variability of inbred and outbred mice. Nature 263(5574), 230–232.
38. McBride, B. W., Mogg, A., Telfer, J. L., Lever, M. S., Miller, J., Turnbull, P. C. and Baillie, L. (1998) Protective efficacy of a recombinant protective antigen against Bacillus anthracis challenge and assessment of immunological markers. Vaccine 16, 810–817.
39. Ramirez, D. M., Leppla, S. H., Schneerson, R. and Shiloach, J. (2002) Production, recovery and immunogenicity of the protective antigen from a recombinant strain of Bacillus anthracis. J. Ind. Microbiol. Biotechnol. 28, 232–238.
40. Arakawa, T., Yu, J., Chong, D., Hough, J., Engen, P. C. and Langridge, W. (1998) A plant-based cholera toxin B subunit-insulin fusion protein protects against the development of autoimmune diabetes. Nat. Biotechnol. 16, 934–938.
41. de Vos, T. and Dick, T. A. (1991) A rapid method to determine the isotype and specificity of coproantibodies in mice infected with Trichinella or fed cholera toxin. J. Immunol. Methods 141, 285–288.
42. Quinn, C. P., Semenova, V. A., Elie, C. M., Romero-Steiner, S., Greene, C. et-al. (2002) Specific, sensitive, and quantitative enzyme-linked immunosorbent assay for human immunoglobulin G antibodies to anthrax toxin protective antigen. Emerg. Infect. Dis. 8, 1103–1110.
43. Moayeri, M., Martinez, N. W., Wiggins, J., Young, H. A. and Leppla, S. H. (2004) Mouse susceptibility to anthrax lethal toxin is influenced by genetic factors in addition to those controlling macrophage sensitivity. Infect. Immun. 72, 4439–4447.
44. Dhingra, A. and Daniell, H. (2006) Chloroplast genetic engineering via organogenesis or somatic embryogenesis. Methods Mol. Biol. 323, 245–262.
45. Kumar, S. and Daniell, H. (2004) Engineering the chloroplast genome for hyperexpression of human therapeutic proteins and vaccine antigens. Methods Mol. Biol. 267, 365–383.
46. Van Herck, H., Baumans, V., Brandt, C. J., Boere, H. A., Hesp, A. P., van Lith, H. A., Schurink, M. and Beynen, A. C. (2001) Blood sampling from the retroorbital plexus, the saphenous vein and the tail vein in rats: Comparative effects on selected behavioural and blood variables. Lab. Anim. 35(2), 131–139.

Chapter 11

Protein Body Induction: A New Tool to Produce and Recover Recombinant Proteins in Plants

Margarita Torrent, Imma Llop-Tous, and M. Dolors Ludevid

Summary

Stable accumulation of storage proteins, lipids and carbohydrates is a hallmark of the plant seed, and is a characteristic that is typically deficient in existing platforms for recombinant protein manufacture. One of the biological sequestration mechanisms that facilitate the folding, assembly and stabilization of plant seed storage proteins involve the de novo formation of unique intracellular organelles, the endoplasmic reticulum (ER)-derived protein bodies (PBs). In cereals, such as maize, PBs are formed directly in the lumen of the ER of endosperm cells and contain zeins, a group of polypeptides, which account for more than half of the total seed protein mass. The 27 kD γ zein protein localizes to the periphery of the PBs surrounding aggregates of other zeins (including α zein and δ zein). Heterologous expression of γ zein has been shown to result in the formation of PB-like structures, and the N-terminal proline-rich domain of γ zein (Zera®), containing eight PPPVHL repeats and a Pro-X sequence is by itself capable of directing ER retention and PB formation in non-seed tissues. We present a novel approach to produce recombinant proteins in plants based on the ability of γ zein-Zera domain to store recombinant proteins inside PBs. Zera domain fused to several proteins, including a enhanced cyan fluorescent protein (ECFP), calcitonin (Ct) and epidermal growth factor (EGF), were cloned into vectors for transient or stable transformation of tobacco plants. In tobacco leaves, we observed the formation of dense, ER-localized structures containing high concentrations of the respective target proteins. The intact synthetic organelles containing Zera fusions were readily isolated from cellular material using density-based separation methods.

Key words: Protein bodies, Transgenic plants, Protein storage organelle, Fluorescent protein, Proline-rich domains, γ zein domain, Fusion proteins.

11.1. Introduction

Transgenic plants have come to be recognized as viable and efficient bioreactors for the large-scale production of recombinant peptides and proteins, including pharmaceuticals and industrial

Loïc Faye and Véronique Gomord (eds.), *Methods in Molecular Biology, Recombinant Proteins From Plants, vol. 483*

DOI: 10.1007/978-1-59745-407-0_11

enzymes *(1, 2)*. Plants offer an alternative to microbial fermentation and animal cultures for the production of recombinant proteins. For plant pharmaceuticals, plant systems are inherently safer than recombinant animal sources and even native sources *(1, 3)*. In addition, post-translational modifications, such as glycosylation and disulfide bonds, which cannot be achieved with bacterial fermentation, can be successfully accomplished using plants *(4)*. The main advantages foreseen for plant systems are reduced production costs and the easy control of production scale. This is particularly well suited to the production of edible oral vaccines and industrial enzymes for biomass treatment *(5)*. Several approaches have been used for the improvement of yields and stability of recombinant proteins. Among others, genetic approaches, a variety of plant transformation vehicles, tissue expression and subcellular targeting of recombinant proteins *(6, 7, 8, 9, 10)*. The stability and sorting of the target molecule in specific subcellular organelles seems to be essential for protein accumulation *(11, 12, 13)*.

The endoplasmic reticulum (ER) is the crucial site where newly synthesized proteins enter the secretory pathway and it provides the machinery responsible for various protein processing steps including folding, disulfide bond formation, glycosylation and oligomerization *(14)*. In addition, the ER of plant cells plays an essential role in quality control of proteins which enter the secretory pathway *(15, 16)*. If an endogenous or recombinant protein cannot fold in the ER, even after several rounds of chaperone binding, it is targeted for ER-associated degradation (ERAD) *(17)* and finally degraded by proteasome in the cytosol. This control mechanism is conserved in all eukaryotic cells.

To date, a number of common retention mechanisms have been characterized for both transmembrane and luminal ER-resident proteins. Most soluble ER proteins contain a tetrapeptide at the C-terminus H/KDEL motif *(18)* that prevents their secretion, returning escapees from the pre-Golgi compartments to the ER by means of the ERD2p receptor *(19, 20)*. Sorting of recombinant proteins to the ER by using the H/KDEL ER localization signal have been widely used for recombinant protein accumulation in plants *(21)*. Usually, this ER retention driven by the C-terminal H/KDEL extension increases the recombinant protein levels *(22)*. However, it is also well known that ER retention via KDEL is a saturable mechanism since it is mediated by a KDEL receptor that ensures that the protein is retrieved from the Golgi and transported to the ER *(20)*. When receptor fails by competition with other KDEL ligands, the KDEL-tagged recombinant protein either secretes or is transported to the lytic vacuole in plants *(23, 24)*.

To overcome this limitation, we developed a fusion protein-based system to accumulate recombinant peptides or proteins in ER-derived protein bodies (PBs) in plants. Plant seeds store large amounts of proteins to feed the embryo during germination and seedling growth. These proteins accumulate in membrane-bound organelles called PBs, which are derived either from vacuoles or from the ER *(25, 26)*. The maize storage protein γ zein, accumulates with α, β and δ zeins in large spherical PBs (1–2 μm) formed directly in the ER lumen in endosperm cells *(27, 28)*. In seeds, γ zein is an ER-resident protein that does not have the retention signal H/KDEL *(29)*. When expressed in a non-seed tissue, such as transgenic *Arabidopsis* leaves, it is able by itself to form PB-like structures within the ER *(30)*. The information for ER retention and the self-assembly ability of γ zein that leads to PB formation, seems to be contained in its N-terminal proline-rich domain *(30, 31)* . Although the mechanism of PB formation is not fully understood, it appears that γ zein lends stability to other storage proteins inside the ER. For example, α zeins are degraded in transgenic plants but when are co-expressed with γ zein, both proteins, accumulate properly in ER-derived PBs *(32)*.

Therefore, PBs formation is an excellent example in nature by which maize seeds prevent the entry of zeins to the ERAD pathway and their export from the ER to other compartments of the secretory pathway. Hence, storage proteins remain encapsulated surrounded by membranes and protected from proteases during all seed development.

Here, we describe a protocol of a new protein fusion-based approach for heterologous protein production in newly formed PBs. We decided to exploit the natural mechanism of protein accumulation, which occurs in seed, to produce recombinant proteins in non-seed plant tissues. Our approach relies on fusion proteins that contain the γ zein signal peptide, the γ zein proline-rich domain (Zera®) and a protein of interest fused to the C-terminus of the proline-rich region. We generated chimeric genes for their expression in tobacco plants and we tested the capacity of recombinant proteins to induce PBs. This approach presents important advantages over other protein production methods; (i) proteins remain stable within membrane-bound organelles, (ii) the newly formed PB-like vesicles are dense organelles which can be easily recovered by density and lastly, (iii) the presence of PBs seems not to disrupt the normal growth or development of plants.

The company ERA Biotech (www.erabiotech.com) has developed a proprietary downstream process based on the high density of PBs. This method allows, with very few and cheap steps, to obtain a sample highly pure and concentrated in the protein of interest, resulting in a very cost-effective procedure.

The methods presented or mentioned in this chapter are covered by families of patents, and assigned to ERA Biotech. Enquire at ERA Biotech for the modalities by which Zera assembler peptides, StorPro™ organelles and methods can be used for R&D or for commercial uses.

11.2. Materials

11.2.1. Plant Material

Wild-type *Nicotiana benthamiana* plants are used for transient transformation. Plants are grown for 4–6 weeks in a greenhouse at 18–28°C, 55–65% relative humidity and a photoperiod of 16 h.

For stable plant transformation *Nicotiana* tabacum var. Wisconsin 38 plants are grown in an in vitro growth chamber at 24–26°C with a 16-h photoperiod or in a greenhouse under the conditions described above.

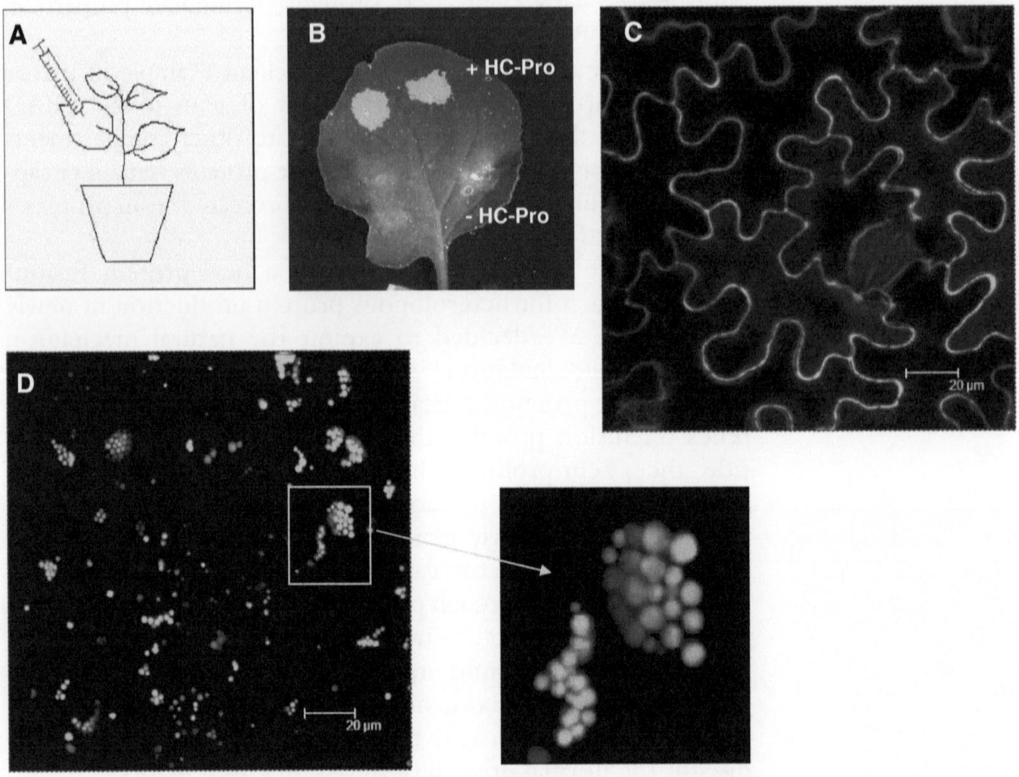

Fig. 11.1. Protein body induction in *Nicotiana benthamiana* leaves. (**A**) Schematic drawing showing young, fully expanded leaves infiltrated with *Agrobacterium* cultures carrying the desired constructs. (**B**) Effect of the silencing suppressor HC-Pro on the levels of Zera-*ECFP* (enhanced cyan fluorescent protein) expression. *Agrobacterium* cultures harbouring Zera-ECFP construct were infiltrated into leaves, either alone (*lower half of the leaf*) or mixed with an *Agrobacterium* culture transformed with the suppressor of silencing HC-Pro (*upper half*). The fluorescence was assessed under UV illumination on the fourth day after agroinfiltration. (**C**) Confocal fluorescent image of epidermal cells agroinfiltrated with SPg-ECFP + HC-Pro. The absence of the Zera partner results in the secretion of the recombinant protein. (**D**) Subcellular fluorescence localization in epidermal cells agroinfiltrated with Zera-ECFP + HC-Pro. Confocal microscope observation reveals that the recombinant protein accumulates in numerous protein bodies (*PBs*) in each cell. A higher magnification of the enclosed area is shown on the right. (*See Color Plates*).

11.2.2. Transient Nicotiana Benthamiana Transformation

11.2.2.1. Equipment

1. Laminar flow hood
2. Autoclave
3. 1 ml sterile syringes (without needle)
4. UV (ultraviolet)-Spectrophotometer
5. 28°C shaking incubator

11.2.2.2. Consumables and Reagents

1. Agrobacterium tumefaciens strain EHA 105 transformed with the desired constructs (see **Notes 1** and **2**)
2. Filter-sterilized agroinfiltration buffer: 10 mM 2-(N-morpholino) ethanesulfonic acid (MES) buffer (pH 5.7), 10 mM $MgCl_2$, 200 μM 5′dimethoxy-4′-hydroxyacetophenone (acetosyringone)
3. DMSO (to prepare 0.2 M acetosyringone stock)
4. 1 M KOH (to adjust pH of MES buffer)
5. LB (Luria-Bertani) medium: 1% tryptone, 0.5% yeast extract and 1% NaCl, pH 7.5
6. Antibiotics (i.e. rifampicin and kanamycin) (see **Note 3**)

11.2.3. Stable Tobacco Plant Transformation

11.2.3.1. Equipment

1. Plant culture room
2. Laminar flow hood
3. Dissecting microscope
4. Autoclave

11.2.3.2. Consumables and Reagents

1. Agrobacterium tumefaciens strain LBA4404 transformed with the desired constructs (see **Notes 1 and 2**)
2. Murashige and Skoogs (MS) basal salts and vitamins (this medium can be purchased as ready mixed power)
3. MS1 medium: 4.4 g/L MS salt and vitamins, 1% sucrose and MES at 0.5 g/L, pH 5.8. To make solid MS1 medium, add 0.2% Gelrite (or 0.8% agar) (see **Note 4**)
4. MS2 medium: 4.4 g/L MS salt and vitamins, 3% sucrose, MES 0.5 g/L, pH 5.8., 0.2% Gelrite, 1 mg/L 6-benzylaminopurine (BAP) and 0.1 mg/L alpha-naphthalene acetic acid (NAA)
5. MS3 medium: MS2 plus the appropriate antibiotics (i.e. timentine and kanamycin) (see **Note 3**)
6. MS4 medium: 2.2 g/L MS salts and vitamins, 3% sucrose, 0.5 g/L MES, pH 5.8, 0.2% Gelrite and the appropriate antibiotics (see **Note 3**)
7. 1.0 M KOH (to adjust MS-derived media pH)
8. LB medium (see **Subheading 11.2.2.2**)

11.2.4. Subcellular Fractionation and Protein Analysis

11.2.4.1. Equipment

1. Ultracentrifuge equipped with a swinging rotor (i.e. an ultracentrifuge Beckman provided with SW40Ti or SW28 rotors)
2. Low-speed centrifuge
3. Polytron homogenizer or mortar with pestle

11.2.4.2. Consumables and Reagents

1. HB (homogenization buffer): 100 mM Tris-HCl, pH 8, 50 mM KCl, 6 mM $MgCl_2$, 1 mM EDTA, 0.4 M NaCl
2. Protease inhibitors (aprotinina, leupeptina, pepstatina, PMSF)
3. Sucrose 20, 30, 42 and 56% (w/w) in HB

11.2.5. Immunocyto-chemistry and Fluorescence Imaging

11.2.5.1. Equipment

1. Hand-held UV lamp (see **Note 5**)
2. Confocal laser microscope (see **Note 6**)
3. Transmission electron microscope
4. A -20°C camera fitted with UV lamp
5. Ultramicrotome
6. Appropriate forceps and scalpel
7. Vacuum bomb

11.2.5.2. Consumables and Reagents

1. Fixative: 1% glutaraldehyde, 2.5% paraformaldehyde, 20 mM phosphate buffer, pH 7.4
2. Phosphate buffer (20 mM): 3.2 mM NaH_2PO_4, 16.8 mM Na_2HPO_4, pH 7.5
3. Phosphate-buffered saline (PBS): 1.6 mM NaH_2PO_4, 8.4 mM Na_2HPO_4, 150 mM NaCl, pH 7.5
4. Blocking solution BS: PBS, 0.1% Tween-20 and 1% ovoalbumin
5. PBST1: PBS, 0.1% Tween-20
6. PBST2: PBS, 0.5% Tween-20
7. Protein A-colloidal gold (5–15 nm particles diameter)

11.3. Methods

11.3.1. Plants Transformation

11.3.1.1. Constructs

Constructs contain chimeric genes obtained by fusing in frame the N-terminal Pro-rich γ zein DNA sequence (coding for Zera peptide) with the desired protein sequence (note that the Zera sequence is able to induce PBs in vegetative tissues). These constructs are introduced in binary vectors (such as pBin19 or pCAMBIA) under the control of a constitutive promoter (such as the enhanced CaMV 35S promoter) (see **Note 1**).

The binary plant vectors carrying the constructs of interest were used to transform A. tumefaciens competent cells (strains EHA 105 or LBA4401) (see **Note 2**) by standard procedures *(33)*.

11.3.1.2. Transient Transformation of Nicotiana Benthamiana

Individual transformed Agrobacterium cultures (EHA 105 strain) were used either alone or mixed with Agrobacterium cultures carrying the HC-Pro suppressor of silencing (see **Note 7**). Agrobacterium transformants were delivered to plant leaves through the stomata by the Agrobacterium-mediated transient gene expression system, referred to as agroinfiltration. This system represents a fast, simple and highly reproducible approach to obtain transiently transformed plant tissues (Fig. 11.1).

Preparation of Recombinant Agrobacterium Cultures for Agroinfiltration

1. Grow EHA 105 transformed Agrobacterium harbouring the constructs of interest for 2 days at 28°C on LB plates containing the selection antibiotics (see **Note 3**).
2. Use isolated colonies to inoculate 5 ml LB medium supplemented with suitable antibiotics (see **Note 3**). In addition, it is recommended to start a culture with transformed Agrobacterium containing a silencing suppressor (such as HC-Pro) (see **Note 7**).
3. Incubate the cultures 20–24 h at 28°C with shaking (150 rpm).
4. Use 100 µl of the previous primary culture to inoculate 20 ml LB medium supplemented with suitable antibiotics.
5. Incubate overnight at 28°C with shaking (150 rpm).
6. Measure the optical density of the culture at 600 nm (it should be around 1.0).
7. Spin down bacteria at 3,000 rpm for 15 min at room temperature.
8. Pour off the supernatant. Resuspend the pellet in agroinfiltration buffer by gently pipetting up and down. The volume of agroinfiltration buffer added should be such as to have a final OD_{600} of 1.0.
9. Incubate Agrobacterium cultures at room temperature for 3–5 h without shaking.
10. It is recommended to mix each individual Agrobacterium culture containing the transgene of interest with an equal volume of the Agrobacterium culture containing the suppressor of silencing (HC-Pro) (see **Note 8**).

Agroinfiltration Procedure

1. Label 2 or 3 fully expanded young leaves of 4- to 6-week-old *N. benthamiana* plants.

2. Place the Agrobacterium culture in a 1 ml sterile syringe (without needle). Press the tip of the syringe into the abaxial side of the selected leaves while applying a counterpressure on the other side of the leaf (see **Note 9**).
3. Cover the *Nicotiana* plants with wet tissue paper for ~1 h to prevent them from drying.
4. Transfer agroinfiltrated plants to the greenhouse.
5. At 4–7 days post-infiltration (see **Note 10**), observe fluorescence of the agroinfiltrated leaves under a UV lamp (see **Note 5**) and collect leaf samples for confocal observation.

11.3.1.3. Stable Tobacco Transformation

We use the transformation method via A. tumefaciens as described by Draper et al. *(34)* with some modifications. Seeds, callus and shoots must be manipulated in a laminar flow hood using sterilized tools and equipment.

Recombinant Agrobacterium Preparation

1. Grow transformed LBA4404 Agrobacterium for 3 days at 28°C on LB plates containing the selection antibiotics (see **Note 3**).
2. Use isolated colonies to inoculate 100 mL of liquid LB medium containing the selection antibiotics.
3. Incubate at 28°C and 150 rpm until the culture has an absorbance of 0.4–0.8 at 600 nm (usually takes a night).
4. Centrifuge at 3000 × g for 15 min and resuspend the pellet with 25 mL of liquid MS1.
5. Incubate 20–30 min at 28°C and 150 rpm.
6. Use this solution to transform tobacco leaf sections (Subheading tobacco leaves infection and callus induction).

Plant Material Preparation

1. Sterilize tobacco seeds by washing as follows:
   70% ethanol for 30 s
   Sterile water for 1 min
   20% bleach (or 2% NaOCl) plus 0.1% Tween-20 for 10 min
   Sterile water 5 times, 30 s each wash
2. Sterilized seeds on MS1 plates.
3. Incubate at 26°C, 16-h photoperiod with a 1700 Lux light intensity.
4. Transfer 10- to 12-day-old plantlets into jars with MS1 solid medium. Plantlets must show the first two true leaves expanded.
5. Incubate as in point 3 until plants are 6–8 weeks old.

Tobacco Leaves Infection and Callus Induction

1. Select young developed leaves (third and fourth from the apex) of 6- to 8-week-old plants.

2. Cut leaves into around 1 cm squares and immerse into 7 mL MS1 liquid medium in Petri dishes. Usually around 50 sections are prepared per transformation event.
3. Add 3 mL of transformed Agrobacterium culture prepared as described (**Subheading** Recombinant Agrobacterium Preparation) and incubate for 5 min gently swirling the plate at regular intervals.
4. Remove leaf sections using flamed forceps and remove the excess of liquid with a sterile filter paper.
5. Place the sections upside down on Petri dishes containing MS2 solid medium to induce callus formation.
6. Incubate in the dark for 2 days at 26°C (alternative: 3 days at 23°C).
7. Transfer the sections to MS3 medium for transformants selection.
8. Incubate the samples at 26°C, 16-h photoperiod with a 1700 Lux light intensity for 2–3 weeks.
9. Excise calli (with size around 2–3 mm) that appear in the section borders and transfer them into fresh MS3.
10. Incubate as described above (step 8) for 2–3 additional weeks until transgenic shoots develop from calli.

Regeneration and Selection of Transgenic Tobacco Plants

1. Excise shoots showing 2–3 leaves developed from transgenic calli and place them in MS4 medium to regenerate roots.
2. Incubate at 26°C, 16-h photoperiod with a 1700 Lux light intensity until roots appear.
3. Transfer rooted plantlets to jars with MS1 solid medium and grow them at 26°C, 16-h photoperiod with a 1700 Lux light intensity.
4. For screening of transformants, take leaf sections of the growing plants using sterile forceps and scalpel.
5. Extract proteins from the dissected leaf sections and analyze the extracts by SDS-polyacrylamide gels and immunoblot using specific antibodies against the fusion-expressed proteins (see **Notes 11** and **12**).
6. Select four to six plants showing the highest recombinant protein levels and transfer them to the greenhouse to obtain T1 seed generations.
7. If desired, grow T1 seeds on selective medium to obtain T2 generation (and use the same procedure with successive selections to obtain homozygous generations).

11.3.2. Subcellular Fractionation

1. Grind transgenic tobacco leaves at 4°C in the HB containing 10% (w/v) sucrose and protease inhibitors, using a polytron (alternative: grind in a mortar at 4°C).
2. Filter the homogenate through four layers of Miracloth (22–24 μm) to remove tissue debris.
3. Centrifuge the filtrate at 50×g for 5 min at 4°C.
4. Load the resulting supernatant onto a multistep 20/30/42/56% (w/w) sucrose gradient buffered with HB.
5. Centrifuge the gradient at 4°C for 2 h at 80,000×g by using a swinging bucket rotor.
6. Collect the supernatant, interphases and pellet fractions.
7. Separate the proteins obtained from equivalent amounts of the fractions by SDS-polyacrylamide gels.
8. Analyze the pattern of proteins by Coomassie or silver stain. Additionally, label the recombinant protein in immunoblot by using specific antibodies (see **Note 11**). Antibodies against subcellular protein markers can also be used (see **Note 13**) (**Fig. 11.2**).

11.3.3. Immunocytochemistry and Fluorescence Imaging

We describe here the subcellular localization of recombinant fusion proteins by fluorescence using confocal microscopy or by immunocytochemistry using electron microscopy.

11.3.3.1. Confocal Microscopy

Sections (4–8 mm^2) of leaf tobacco tissues transiently transformed with fluorescent-tagged protein sequences are simply mounted in water to observe epidermal tissue cells.

Micrographs from the transfected tissues are obtained using a confocal laser scanning microscope fitted with spectrophotometers for emission band wavelength selection (see **Note 14**).

11.3.3.2. Electron Microscopy

Fixation, Embedding and Sectioning

1. Cut around 1 mm^2 leaf sections from wild-type and stably transformed tobacco plants in a drop of fixative.
2. Fix the samples immersed in the fixative by vacuum infiltration for 1 h at room temperature.
3. Wash 4 times for 15 min with 20 mM phosphate buffer, pH 7.5.
4. Dehydrate the samples through ethanol series:

   30% ethanol, 30 min at 4°C
   50% ethanol, 1 h at –20°C
   70% ethanol, 1 h at –20°C (alternative: overnight at 4°C)
   100% ethanol, 1 h to overnight at –20°C
   100% ethanol, 1 h at –20°C
5. Embed in Lowicryl K4M resin. All the embedding steps are performed at –20°C. Incubate the samples as follows:

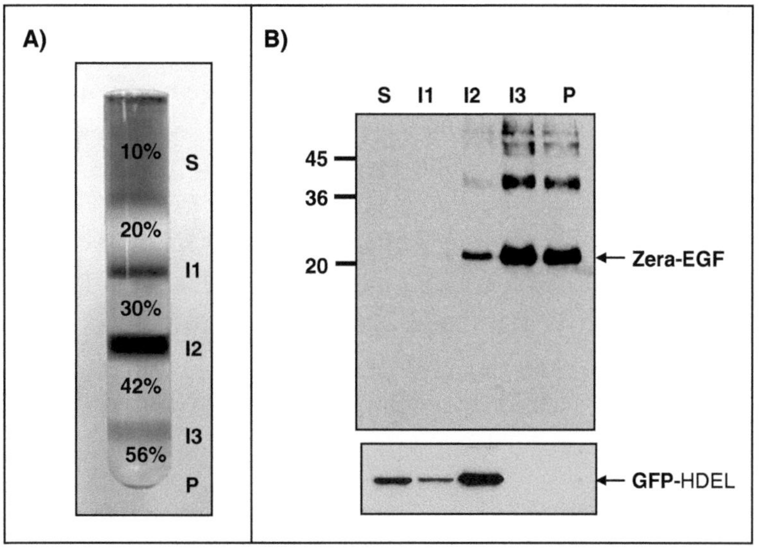

Fig. 11.2. Protein body isolation from transgenic tobacco leaves by using density gradients. **(A)** Photograph of the gradient after ultracentrifugation. The sucrose (w/w) steps in the gradient are indicated. Supernatant (S), interfaces (I1, I2 and I3) and pellet (P) fractions are also indicated. **(B)** Immunoblots of gradient fractions by using the anti-R8 (*upper panel*) and anti-*GFP* (green fluorescent protein) antibodies (*lower panel*). The upper panel shows fractions obtained from Zera-fusion expressing plants. Note the preferential accumulation of the fusion Zera-*EGF* (epidermal growth factor) in the dense fraction I3 indicating the presence of dense organelles. The lower panel corresponds to an independent gradient from GFP-HDEL expressing plants. As expected, in this case no protein bodies (*PBs*) are induced and the endoplasmic reticulum (*ER*)-marker GFP-HDEL was preferentially detected in the I2 fraction where normally the *ER* accumulates.

Resin/absolute ethanol 1/3 (v/v) for 1 h
Resin/absolute ethanol 1/1 (v/v) for 1 h
Resin/absolute ethanol 3/1 (v/v) for 1 h
Pure resin for overnight

6. Place the samples into fresh resin for polymerization and leave them under UV light at –20°C for 3 days.
7. Obtain ultrathin sections using an ultramicrotome and collect them on gold grids.

Immunodetection

1. Incubate sections collected on gold grids (see **Subheading** Fixation, Embedding and Sectioning) with distilled water for 5 min at room temperature.
2. Incubate with 0.1 N HCl for 15 min at room temperature.
3. Wash 2 times, 10 min each, with PBS.
4. Block for 1 h at room temperature with blocking solution BS.
5. Incubate sections with the specific antibody (see **Note 11**) diluted in PBST1 for 2 h at room temperature (alternative:

overnight at 4°C). The usual range of antibody dilution is 1/100–1/500.

6. Wash 4 times with PBST1, 15 min each wash.
7. Incubate the samples for 1 h at room temperature with protein A coupled to colloidal gold particles diluted in PBST1. The usual gold particle diameter range is 5–15 nm.
8. Wash grids 5 times, 10 min per wash, in PBST1.
9. Wash grids 5 times, 10 min per wash, in PBS.
10. Wash twice, 5 min each in distilled water.
11. After drying to air, stain the sections with 2% aqueous uranyl acetate for 15 min and wash 4 times in distilled water (Fig. 11.3).

11.4. Notes

1. The constructs used in this work were as follows. The synthetic gene corresponding to the mature calcitonin (Ct) sequence was obtained as described (patent WO2004003207). The synthetic gene encoding the 53 amino acids of mature epidermal growth factor (EGF) was obtained by primer overlap extension polymerase chain reaction (PCR) method, using 4 oligonucleotides of about 60-mer, with 20 overlapping bases. The cDNA corresponding to the enhanced cyan fluorescent protein (ECFP) was obtained by PCR from plasmid pECFP-N1 (Clontech). The three cDNAs were fused in frame to the Zera (patent WO2004003207), obtaining Zera-Ct, Zera-EGF and Zera-ECFP fusion sequences (**Fig. 11.3**).

   Plasmid pECFP-N1 was also used as the template to obtain SPg-ECFP where SPg is the γ zein signal peptide coding sequence. This chimeric gene was achieved by PCR using two 5′ overlapping primers to introduce SPg sequence. All four constructs, Zera-Ct, Zera-EGF, Zera-ECFP and SPg-ECFP, were introduced into effective plant transformation vectors containing the enhanced cauliflower mosaic virus (CaMV) 35S promoter, the tobacco etch virus (TEV) translation enhancer and the 3′ polyadenylation sequences from the CaMV. Finally, plasmid pBIN mgfp5-ER containing the sequence for an optimized green fluorescent protein (GFP) containing the HDEL C-terminal ER-retention sequence *(35)* was used to express a luminal protein marker.
2. Several Agrobacterium strains are suitable for plant transformation. However, the highly virulent strain EHA105 has been shown to be optimal for *N. benthamiana* agroinfiltration *(36)*

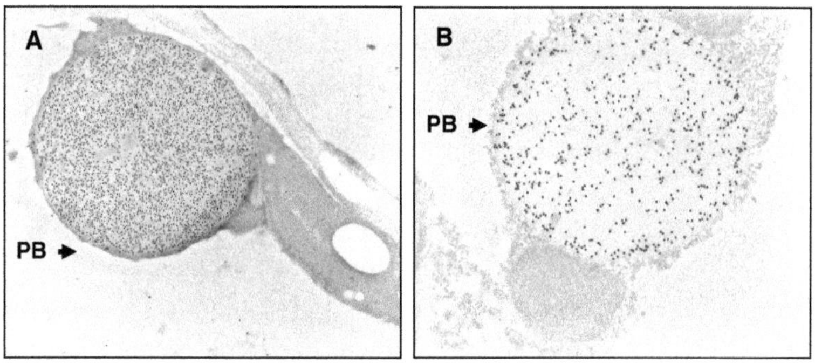

Fig. 11.3. Immunolocalization of Zera-*Ct* (calcitonin) in mesophyll cells of transgenic tobacco plants. (**A**) Electron micrograph of a protein body immunolabelled with the anti-R8 antibody and protein A-colloidal gold showing a high accumulation of the fusion Zera-Ct. (**B**) Electron micrograph of a protein body immunolabelled with the anti-BiP antibody and protein A-colloidal gold. The presence of BiP inside the protein bodies (*PBs*) sustains the endoplasmic reticulum (*ER*) origin of this kind of induced *PBs*. It should be noted that no any similar structure can be detected in wild-type plants.

whereas less virulent strain LB4404 is frequently the choice for stable transformation.

3. Antibiotics used:
   Rifampicin at 100 μg/mL. Most Agrobacterium strains (including EHA105 and LBA4404) carry a rifampicin resistance gene for selection. The use of this antibiotic prevents bacterial contaminants in Agrobacterium cell cultures. Timentine at 100 μg/mL to eliminate Agrobacterium during induction of transformed calli. pCAMBIA 2300 and pBin19 plant expression vectors contain nptII selector gene. We use kanamycin 50 μg/mL for selection of transformed Agrobacterium cells and 100 μg/mL of kanamycin for transformed plant selection media.
4. The culture media must be autoclaved or filter-sterilized. The heat-sensitive components (hormones and antibiotics) must be filtered and added after cooling the media around 50°C.
5. The fluorescence in *N. benthamiana* agroinfiltrated leaves can be visualized by epi-illumination with a hand-held long-wave UV lamp.
6. Fluorescent images can also be taken by using a fluorescence microscope fitted with appropriated filter sets (i.e. Zeiss AxioPhot epifluorescence microscope).
7. Plants frequently respond to Agrobacterium-mediated transformation by post-transcriptionally silencing the transgene expression, limiting the efficiency of recombinant protein production. This general plant response

can be counteracted by co-expressing viral suppressors of gene silencing. The co-expression of Zera-fusion constructs with the suppressor of silencing HC-Pro from tobacco etch virus *(37)*, highly enhances the expression of the recombinant protein.

8. When various independent Agrobacterium cultures are to be co-infiltrated, the optical density at OD_{600} of each individual Agrobacterium culture must be around 1.0.
9. Usually round-shaped agroinfiltrated areas of ~1–2 cm diameter are obtained. Each single leaf can be used for up to four independent agroinfiltrations. If different independent Agrobacterium cultures are used to infiltrate one particular leaf, it is essential to set up a criterion to unequivocally identify each agroinfiltrated area.
10. In the absence of a suppressor of silencing, the level of transgene expression usually peaks at 60–72 h post-infiltration and declines thereafter. However, co-expression of a suppressor of silencing results in a more persistent transgene expression *(38)*. In our experiments, optimal results were obtained at 4–7 days after agroinfiltration.
11. Antibodies used in this work:
    Anti-R8 antiserum raised in rabbits injected with the synthetic γ zein repeat domain (PPPVHL×8) coupled to the keyhole limpet haemocyanin (KLH) protein used as a carrier. This antibody recognizes all the Zera-fusion proteins analyzed.
    Anti-Ct antiserum raised in rabbits injected with a synthetic Ct peptide coupled to the Keyhole limpet haemocyanin protein used as a carrier.
    Anti-GFP IgGs from Molecular Probes.
    Anti-BiP antiserum raised in rabbits injected with the recombinant C-terminal half of tobacco BiP sequence produced in Escherichia coli.
12. When fluorescent-tagged proteins are used, the selection of the most promising stably transformed plants can also be done by measuring the fluorescence present in the leaf protein extracts by using a fluorimeter fitted with the appropriated filters.
13. Here, we use anti-BiP antibodies to detect the ER protein marker BiP. The protein BiP is an abundant luminal resident ER chaperone.
14. Here, we use ECFP as a marker. Images are collected at 458-nm excitation using an emission 470–530 nm window. Optical sections are 0.5 microns thick. Digital images and projections are recorded by using the confocal microscope software.

References

1. Fisher, R., Stoger, E., Schillberg, S., Christou, P. and Twyman, R.M. (2004) Plant-based production of biopharmaceuticals. Curr. Opin. Plant Biol. 7, 1–7.
2. Giddings, G. (2001) Transgenic plants as protein factories. Curr. Opin. Biotechnol. 12, 450–454.
3. Streatfield, S.J. (2007) Approaches to achieve high-level heterologous protein production in plants. Plant Biotechnol. J. 5, 2–15.
4. Gomord, V., Chamberlain, P., Jefferis, R. and Faye, L. (2005) Biopharmaceutical production in plants: problems, solutions and opportunities. Trends Biotechnol. 23, 559–565.
5. Xue, G.P., Patel, M., Johnson, J.S., Smyth, D.J. and Vickers, C.E. (2003) Selectable marker-free transgenic barley producing a high level of cellulase (1,4-β-glucanase) in developing grains. Plant Cell Rep. 21, 1088–1094.
6. Staub, J.M. and Maliga, P. (1995) Expression of a chimeric uidA gene indicates that polycistronic mRNAs are efficiently translated in tobacco plastids. Plant J. 7, 845–848.
7. Gleba, Y., Klimyuk, V. and Marinollet, S. (2007) Viral vectors for the expression of proteins in plants. Curr. Opin. Biotechnol. 18, 134–141.
8. Twyman, R.M., Stoger, E., Schillberg, S., Christou, P. and Fischer, R. (2003) Molecular farming in plants: host systems and expression technology. Trends Biotechnol. 21, 570–578.
9. Takaiwa, F., Takagi, H., Hirose, S. and Wakasa, Y. (2007) Endosperm tissue is good production platform for artificial recombinant proteins in transgenic rice. Plant Biotechnol. J. 5, 84–92.
10. Stoger, E., Ma, J.K., Fisher, R. and Christou, P. (2005) Sowing the seeds of success: pharmaceutical proteins from plants. Curr. Opin. Biotechnol. 16, 167–173.
11. Fernandez-San Millan, A., Mingo-Castel, A., Miller, M. and Daniell, H. (2003) A chloroplast transgenic approach to hyper-express and purify human serum albumin, a protein highly susceptible to proteolytic degradation. Plant Biotechnol. 1, 77–79.
12. Tregoning, J.S., Nixon, P., Kuroda, H., Svab, Z., Clare, S., Bowe, F., Fairweather, N., Ytterberg, J., Van Wijk, K.J., Dougan, G. and Maliga, P. (2003) Expression of tetanus toxin fragment C in tobacco chloroplasts. Nucleic Acids Res. 31, 1174–1179.
13. Shillberg, S., Zimmermann, S., Voss, A. and Fischer, R. (1999) Apoplastic and cytosolic expression of full-size antibodies and antibody fragments in Nicotiana tabacum. Tansgenic Res. 8, 255–2633.
14. Hadlington, J.L. and Denecke, J. (2000) Sorting of soluble proteins in the secretory pathway of plants. Curr. Opin. Plant Biol. 3, 461–468.
15. Kirst, M.E., Meyer, D.J., Gibbon, B., Jung, R. and Boston, R. (2005) Identification and characterization of endoplasmic reticulum-associated degradation proteins differentially affected by endoplasmic reticulum stress. Plant Physiol. 138, 218–231.
16. Ellgaard, L. and Helenius, A. (2003) Quality control in the endoplasmic reticulum. Nat. Mol. Cell Biol. 4, 181–191.
17. Tsai, B., Ye, Y. and Rapoport, T.A. (2002) Retro-translocation of proteins from the endoplasmic reticulum into the cytosol. Nat. Rev. Mol. Cell Biol. 3, 246–255.
18. Munro, S. and Pelham, H.R. (1987) A C-terminal signal prevents secretion of luminal ER proteins. Cell 48, 899–907.
19. Lewis, M.J., Sweet, D.J. and Pelham, H.R. (1990) The ERD2 gene determines the specificity of the luminal ER protein retention system. Cell 61, 1359–1363.
20. Semenza, J.C., Hardwick, K.G., Dean, N. and Pelham, H.R. (1990) ERD2, a yeast gene required for the receptor-mediated retrieval of luminal ER proteins from the secretory pathway. Cell 61, 1349–57.
21. Ko, K., Tekoah, Y., Rudd, P.M., Harvey, D.J., Dwek, R.A., Spitsin, S., Hanlon, C.A., Rupprecht, C., Dietzschold, B., Golovkin, M. and Koprowski, H. (2003) Function and glycosilation of plant derived antiviral monoclonal antibody. Proc. Natl. Acad. Sci. USA 100, 8013–8018.
22. Schouten, A., Roosien, J., Van Engelen, F., De Jong, G., Borst-Vrenssen, A., Zilverentant, J., Bosch, D., Stiekema, W., Gommers, F., Schots, A. and Bakker, J. (1996) The C-terminal KDEL sequence increases the expression level of a single-chain antibody designed to be targeted to both the cytosol and the secretory pathway in transgenic tobacco. Plant Mol. Biol. 30, 781–793.
23. Frigerio, L., Pastres, A., Prada, A. and Vitale, A. (2001) Influence of KDEL on the fate

of trimeric or assembly-defective phaseolin: selective use of an alternative route to vacuoles. Plant Cell 13, 1109–1126.

24. Pimpl, P., Taylor, J.P., Snowden, C., Hillmer, S., Robinson, D.G. and Denecke, J. (2006) Golgi-mediated vacuolar sorting of the endoplasmic reticulum chaperone BiP may play an active role in quality control within the secretory pathway. Plant Cell 18, 198–211.
25. Herman, E.M. and Larkins, B. (1999) Protein storage bodies and vacuoles. Plant Cell 11, 601–613.
26. Galili, G., Altsschuler, Y. and Levanony, H. (1993) Assembly and transport of seed storage proteins. Trends Cell Biol. 3, 437–442.
27. Lending, C.R. and Larkins, B.A. (1989) Changes in the zein composition of protein bodies during maize endosperm development. Plant Cell 1, 1011–1023.
28. Ludevid, M.D., Torrent, M., Martinez-Izquierdo, J.A., Puigdomenech, P. and Palau, P. (1984) Subcellular localization of glutelin-2 in maize (Zea mays L.) endosperm. Plant Mol. Biol. 3, 227–234.
29. Prat, S., Cortadas, J., Puigdomenech, P. and Palau, J. (1985) Nucleic acid and amino acid sequences of the maize endosperm protein glutelin-2. Nucleic Acids Res. 13, 1493–1504.
30. Geli, M.I., Torrent, M. and Ludevid, D. (1994) Two structural domains mediate two sequential events in γ zein targeting: protein endoplasmic reticulum retention and protein body formation. Plant Cell 6, 1911–1922.
31. Mainieri, D., Rossi, M., Archinti, M., Belluci, M., De Marchis, F., Vavassori, S., Pompa, A., Arcioni, S. and Vitale, A. (2004) Zeolin. A new recombinant storage protein constructed using maize γ-zein and bean phaseolin. Plant Physiol. 136, 3447–3456.
32. Coleman, C.E., Herman, E.M., Takasaki, K. and Larkins, B.A. (1996) The maize γ zein sequesters α zein and stabilizes its accumulation in protein bodies of transgenic tobacco endosperm. Plant Cell 8, 2335–2345.
33. Annamalai, P. and Rao, A.L.N. (2006) Delivery and expression of functional viral RNA genomes in plants by agroinfiltraion. In: Current protocols in Microbiology. Downey (ed.) Vol 1. John Wiley and Sons Inc., Hoboken, N.J.
34. Draper, J., Scott, R. and Hamil, J. (1988) Transformation of dicotyledonous plant cells using the Ti plasmid of Agrobacterium tumefaciens and the Ri plasmid of Agrobacterium rhizogenes. In: Plant Genetic Transformation and Gene Expression. A Laboratory Manual (Draper, J., Scott, R., Armitage, P. and Walden, R. Eds.) Oxford: Blackwell Scientific Publications pp. 69–160.
35. Haseloff, J., Siemering, K.R., Prasher, D.C. and Hodge, S. (1997) Removal of a cryptic intron and subcellular localization of green fluorescent protein are required to mark transgenic Arabidopsis plants brightly. Proc. Nat. Acad. Sci. USA 94, 2122–2127.
36. Wydro, M., Kozubek, E. and Lehmann, P. (2006) Optimization of transient Agrobacterium-mediated gene expression system in leaves of Nicotiana benthamiana. Acta Biochim. Pol. 53, 289–298.
37. Goytia, E., Fernández-Clavino, L., Martínez-Garcia, B., López-Abella, D. and López-Moya, J.J. (2006) Production of plum pox virus HC Pro functionally active for aphid transmission in a transient expression system. J. Gen. Virol. 87, 3413–3423.
38. Voinnet, O., Rivas, S., Mestre, P. and Baulcombe, D. (2003) An enhanced transient expression system in plants based on suppression of gene silencing by the p19 protein of tomato bushy stunt virus. Plant J. 33, 949–956.

Chapter 12

A Case Study for Plant-Made Pharmaceuticals Comparing Different Plant Expression and Production Systems

Guy Vancanneyt, Manuel Dubald, Werner Schröder, Jörg Peters, and Johan Botterman

Summary

Over the last decade, plant-based production of pharmaceuticals has made remarkable progress as the expression of a diverse set of proteins has been demonstrated in a range of plant crops. Although the commercial exploitation is still pending, today various plant-based expression technologies have reached significant milestones through clinical testing in humans. Each of the protein manufacturing platforms in plants has specific benefits and drawbacks. We have engaged in comparing some of these production systems with respect to their performance: protein yield and quality. Using a specific tester protein (aprotinin), it was shown that functional aprotinin can be manufactured in plants in substantial amounts, as illustrated in this chapter.

Key words: Aprotinin, Transient expression, Nuclear expression, Chloroplast expression, *Nicotiana benthamiana*.

12.1. Introduction

During the last decades, several expression systems have been explored for their feasibility to produce recombinant proteins, including plants (**reviewed in refs.** *1, 2)*. Although the commercialization of recombinant proteins produced in plants is still very limited, a whole range of different plant species have been or are being explored today as a potential production system. These plant production systems can be classified in 'seed crops', such as maize, rice, safflower, legumes and others (**reviewed in ref.** *3)* and 'leaf biomass' plants such as alfalfa, duckweed (*Lemna, Spirodela*), tobacco (*Nicotiana*), thalecress (*Arabidopsis*) and others. **Table 12.1** provides an overview of current production systems

Loïc Faye and Véronique Gomord (eds.), *Methods in Molecular Biology, Recombinant Proteins From Plants, vol. 483*

DOI: 10.1007/978-1-59745-407-0_12

Table 12.1
Overview of major crops and companies involved in plant-made pharmaceutical (PMP) production

Organisations	Production Systems											Expression		
	Nicotiana	cell culture	alfalfa	*lemna*	peas	moss	corn	safflower	barley	rice	*Arabidopsis*	nuclear	viral	chloroplast
Chlorogen	x													x
Bayer	x												x	
Planet Biotech	x											x		
Dow AgroScience		x										x		
Protalix Biotherapeutics		x										x		
Medicago			x									x		
Biolex				x								x		
Novoplant					x							x		
Greenovation						x						x		
Meristem Therapeutics							x					x		
SemBioSys								x				x		
ORF Genetics									x			x		
Ventria									x	x		x		
Cobento											x	x		

being explored by the major companies involved in plant-made pharmaceuticals (PMP).

Several groups have communicated successes with a protein of interest in their plant production system of interest. However, very few studies have compared the established plant-based protein expression system in terms of their production potential and protein quality. Model proteins such as the green fluorescent protein (GFP) have been expressed in a wide range of plant species. But qualitative and quantitative comparisons across different species have not been reported so far.

We have engaged in setting up a comparison of different plant expression systems using aprotinin, a commercial pharmaceutical protein, as a test system. The main argument to select this protein is the technical simplicity of aprotinin: a small and well-characterized polypeptide (lacking N-glycosylation).

Aprotinin, also known as bovine pancreatic trypsin inhibitor (BPTI), is a small protein belonging to the Kunitz-type family of proteins (**reviewed in refs.** *4*). It is a potent inhibitor of some serine proteases such as trypsin, plasmin and kallikrein. Although of bovine origin, aprotinin inhibits these enzymes in a variety of species including humans. Aprotinin has been on the market for more than 40 years and has been in clinical use for a variety of indications. Today, it is, for example, used for surgical blood loss reduction during coronary artery by-pass graft surgery (CABG), an open heart surgery procedure. Aprotinin consists of a single

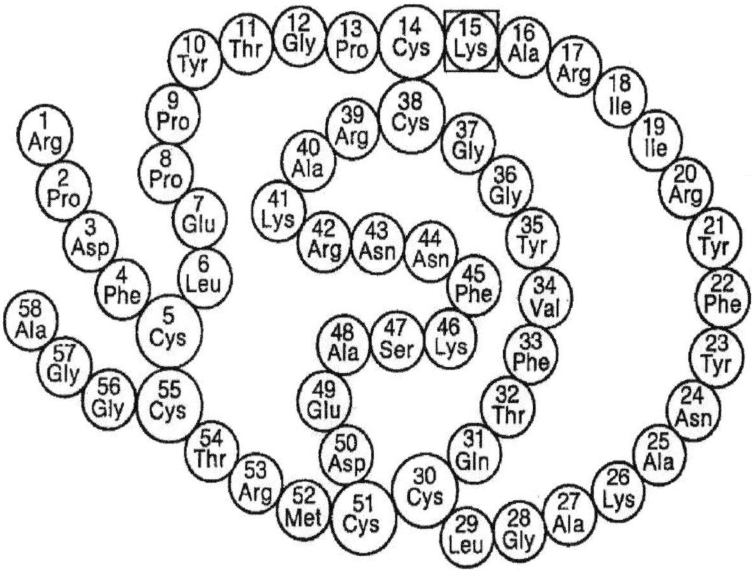

Fig. 12.1. Structure of aprotinin. Amino acid sequence of mature aprotinin indicating the disulfide bonds and the active center (Lys15), according to Kassel and Laskowski *(17)*.

chain polypeptide covering a chain of 58 amino acids that are cross-linked by three disulfide bridges (**Fig. 12.1**)

Although natural aprotinin can be extracted from bovine lungs, a number of recombinant production processes have been reported. For example, in yeast, it was shown that variants of aprotinin with improved properties in terms of enzyme inhibition can be efficiently produced in *Saccharomyces cerevisiae (5)*. The expression of aprotinin in plants has also been reported and will be discussed below.

In previous studies, 'seed' crops such as maize have been used for the expression of aprotinin *(6)*. In this study, we have focused on 'leafy' species such as tobacco (*Nicotiana tabacum*), *Nicotiana benthamiana*, alfalfa (*Medicago sativa*) and duckweed (*Spirodela oligorrhiza*). These species will allow earlier harvesting of recombinant biomass compared to the typical seed crops, and hence will results in shorter production times for industrial applications. Aprotinin has been expressed in these different plant systems by having the gene construct located on the nuclear-, on the chloroplast- or on a viral genome.

12.2. Expression in Tobacco from a Nuclear Transgene

Based on the bovine amino acid sequence, the nucleotide sequence for aprotinin has been designed according to the preferred tobacco nuclear codon usage, chemically synthesized and cloned

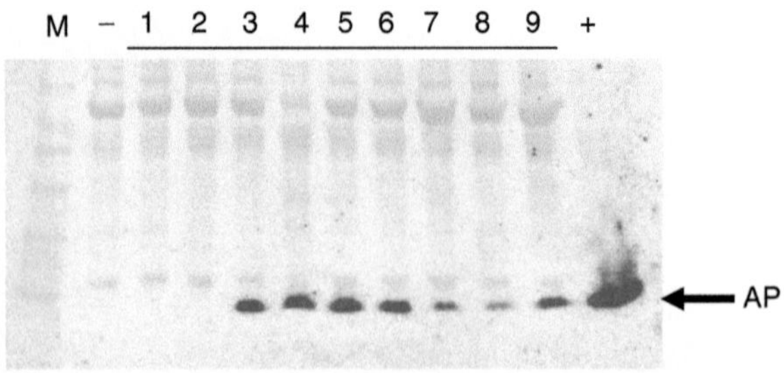

Fig. 12.2. Nuclear aprotinin expression in tobacco. Western blot analysis of total leaf extracts of tobacco. Lanes 1–9 represent protein extracts derived from independent transgenic lines. Controls: ' + ' commercial aprotinin (10 ng), '-'non-transgenic line, M: protein molecular weight marker. Membrane was stained with Ponceau (red) as a control for protein loading. Aprotinin (*AP*) indicates the molecular weight for aprotinin.

in appropriate T-DNA vectors. The constructs have been designed to target the protein to the different cellular compartments using the appropriate targeting sequences: cytoplasm (no targeting signal), chloroplast (transit peptide), apoplast (signal peptide from the pathogenesis-related protein PR1a) and endoplasmic reticulum [signal peptide PR1a plus a C-terminal retention signal (KDEL)]. All gene constructs were placed under control of the 35S promoter from the Cauliflower Mosaic Virus (CaMV). Using Agrobacterium-Ti-plasmid mediated-plant transformation, at least 30 independent transgenic lines were generated per construct. Expression levels of the individual lines have been estimated on the basis of Western blotting. No relevant protein expression could be detected when aprotinin was expressed in the cytoplasm or targeted to the chloroplast compartment. A higher expression level was observed when aprotinin was retained in the endoplasmic reticulum and estimated about 0.03% of the total soluble plant protein (**Fig. 12.2**). On average 1 kg of tobacco green biomass will contain about 10 g of soluble protein. Hence, the aprotinin expression based on a transgene in the nucleus is in the order of about 3 mg of aprotinin per kilogram of plant biomass.

12.3. Expression in Tobacco from a Chloroplast Transgene

The amino acid sequence from aprotinin has been reverse-translated into a nucleotide sequence according to the preferred tobacco chloroplast codon usage. Different protein fusion strategies were tested to be able, after cleavage or processing, to produce genuine aprotinin. Five different N-terminal fusion partners have been selected (**Table 12.2**), and the chimeric genes placed under the control of the tobacco psbA promoter. Constructs were introduced into the chloroplast genome between the rbcl and accD genes using particle bombardment and independent transgenic lines were selected on spectinomycin, essentially as described by *(7)*.

Low or no recombinant protein expression was observed on Western blots in tobacco when aprotinin was fused either to a small histidine containing tag or to an intein. For the three other constructs, the fusion protein was easily detected in leaf extracts with expression levels in the 1–2% range. One unexpected observation in those lines was a strong variation in expression during plant development, between independent transformants generated with the same construct, and the quite frequent appearance of variegated sectors in leaves. Variation in expression could be correlated to unexpected patterns in Southern blots, and the maintenance of wild-type plastid genome. This suggests that aprotinin expression in the stroma might have some phytotoxic effect.

Thioredoxin and maltose-binding protein fusions could be purified from tobacco leaves by affinity, but the release of aprotinin by genenase and particularly enterokinase proved to be difficult and incomplete. The most promising strategy for aprotinin expression in chloroplasts was observed with the ubiquitin fusion, which is processed very efficiently after organelle disruption by a cytosolic protease, as described by Staub et al. *(8)* for human somatotropin. Unfortunately, in this case also, the cleavage gave heterogenous N-terminal variants and occurred predominantly not at the expected site.

Table 12.2
Overview of molecular strategies for the expression of aprotinin in tobacco chloroplasts

N-terminal partner	Expression systems	Purification	Cleavage	Expression level
Thioredoxin	Novagen (pET-32a)	S-tag	Enterokinase	0.5–1% TSP
Maltose-binding protein	NEB (pMALc2g)	Amylose resin	Genenase	1% TSP
Intein	NEB (pTWIN1)	Chitin beads	pH and T	<0.01% TSP
MK-His *(6)*	Quiagen	Nickel beads	Exodipeptidase	Not detected
Ubiquitin			During extraction	1–2% TSP

TSP total soluble protein

12.4. Expression in Alfalfa from a Nuclear Transgene

For expression in alfalfa, the aprotinin coding sequence from bovine has been used and placed under the control of the plastocyanin promoter from alfalfa. The protein was targeted to the apoplast with either the original bovine targeting signal or by the signal sequence from the protein disulfide isomerase (PDI) gene from alfalfa. With both signal sequences similar levels of aprotinin expression have been obtained, expression levels being in the range of 0.1% total soluble protein (TSP) range (or 5 mg/kg fresh biomass) (**Fig. 12.3**). Aprotinin was purified from plant extracts using its binding feature to trypsin. N-terminal sequencing has confirmed the identity of the plant expressed aprotinin. Aprotinin expression in alfalfa was performed at Medicago, Quebec (Canada).

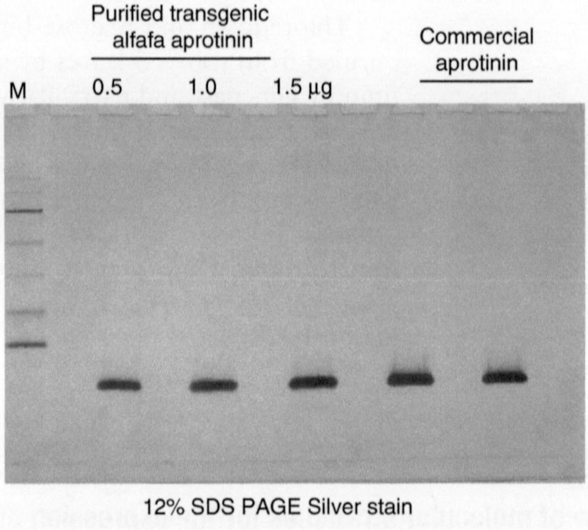

Fig. 12.3. Transgenic expression of aprotinin in alfalfa. Sodium dodecyl sulphate-polyacrylamide gel electrphoresis (SDS-PAGE) analysis of aprotinin purified from a transgenic alfalfa line. Aprotinin was purified according to scheme represented in Fig. 12.5. Different amounts of purified protein were loaded and compared to commercial aprotinin. Total proteins were visualized by silver staining.

12.5. Expression in Duckweed from a Nuclear Transgene

For expression in Lemnaceae, the amino acid sequence was used to design a nucleotide sequence according to the preferred *Spirodela* codon usage, chemically synthesized and cloned in

T-DNA vectors. The protein was targeted to the apoplast (using the signal peptide of PR1a) and to the chloroplast (using a transit peptide). The transgenes were driven either by the 35S promoter from CaMV or by the actin promoter from rice. Using Agrobacterium-mediated transformation, several transgenic lines have been generated and screened for aprotinin expression by Western analysis *(9)*. Production of aprotinin was only detected when the protein was targeted to the apoplast in combination with the CaMV 35S promoter. No significant production has been observed in the other combinations. The highest expression levels were estimated by Western analysis between 1.9% and 3.7% of the TSP. Interestingly, aprotinin was secreted in the fermentation medium (up to 0.67 mg/L). Furthermore, the secreted protein was processed properly as judged from N-terminal sequence analysis of the aprotinin purified from the fermentation medium *(9)*. Aprotinin expression in duckweed was performed at *Lemna*-Gene, Lyon (France).

12.6. Nuclear Expression in Maize

The company ProdiGene has reported the production of aprotinin in maize (*Zea Mays*) *(10)*. They have used a synthetic aprotinin gene, optimized according to preferred maize codon usage and under the transcriptional control of a seed-specific ubiquitin promoter (containing an intron) from maize. The protein was targeted to the apoplast using the a-amylase signal peptide and was reported accumulate aprotinin in the seeds up to 8.9% of the TSP *(11)*. The identity of the maize-purified aprotinin was confirmed by N-terminal sequence analysis.

12.7. Transient Expression in *Nicotiana Benthamiana*

N. benthamiana, a model plant for plant virology, has been used for the application of two different transient expression systems, both based on the use of components of the TMV-related viruses. The company Large Scale Biology Corporation (LSBC) reported the use of viral vector system as a transient expression system for the production of aprotinin *(12)*. The gene encoding aprotinin has been incorporated into the viral genome of tobacco mild green mosaic virus (TMGMV) under the control of a viral subgenomic promoter. The protein was targeted to the apoplast. Optimal aprotinin expression was obtained between 12 and 21 days after plant inoculation with the virus. Quantitative expression levels have not been reported, except that aprotinin was visible as a band on a sodium dodecyl sulphate-polyacrylamide gel electro phoresis (SDS-PAGE) analysis. Aprotinin production was shown to be effective both in the field (N. excelsiana) and indoor (*N. benthamiana*). Detailed protein analysis was performed on plant-purified aprotinin and compared to the commercial product. The amino acid sequence was shown to be identical.

The company Icon Genetics has used a transient expression based on Agrobacterium-mediated delivery of viral vectors to express aprotinin in *N. benthamiana (13)*. The original bovine gene encoding aprotinin was introduced in the T-DNA vectors. The protein was targeted either to the chloroplast using a transit peptide or to the apoplast using different plant-derived signal sequences, such as calreticulin (tobacco), a-amylase (rice) and a-amylase (bean). No protein expression was observed when aprotinin was targeted to the chloroplast, whereas significant expression was observed when targeted to the apoplast. The a-amylase signal peptide turned out to be the most efficient leader sequence in terms of protein quantity. Aprotinin expression levels range between 2% and 4.2% of TSP (or up to 400 mg/kg fresh biomass) (**Fig. 12.4**). An interesting added feature of this expression system is the speed for protein production. For optimal aprotinin production, only 7 days are needed between plant infection and biomass harvest. This transient protein expression system seems to be very demanding for the cellular machinery of the plant cells. This was evidenced by the appearance of symptoms (necrotic lesions) on the leaves when expressing aprotinin. Symptom development during protein expression has already been reported for the other proteins using the expression system *(14)*. The level of these is probably related to the nature of the protein of interest.

12.8. Protein Quality of the Plant-Produced Aprotinin

In order to evaluate the quality of the plant expressed protein, aprotinin has been extracted and purified from biomass harvested from different species used (*N. benthamiana*, alfalfa, maize). Commonly used principles for aprotinin purification are based on the aprotinin-binding capacity to trypsin or chymotrypsin; alternatively, it has been purified by classical chromatographic techniques. Reported purification efficiencies (recovery) are in the range of 40–50%. Purified aprotinin from plants has been compared with the commercially available protein in terms of impurities and biological activity as judged by the ability to inhibit trypsin (TIU, trypsin inhibition unit).

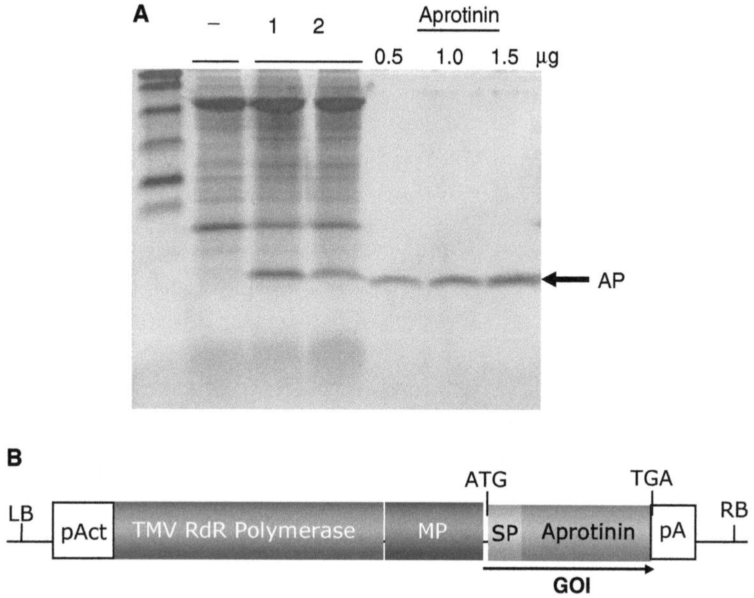

Fig. 12.4. Transient expression of aprotinin using magnification in *Nicotiana benthamiana*. 4A: Sodium dodecyl sulphate-polyacrylamide gel electrphoresis (SDS-PAGE) (12%) analysis of protein extracts derived from *N. benthamiana* plants infected with Agrobacterium tumefaciens (7 days post-infection). Total proteins have been visualized by Coomassie staining. Aprotinin (*AP*) indicates the molecular weight for aprotinin. Lanes 1 and 2 indicate independent plants, '-'indicates a non-transformed *N. benthamiana* plant, 'aprotinin' stands for different amounts of commercial aprotinin. 4B: Schematic representation of the molecular components contained within the T-DNA expressing apoplastic aprotinin. *pAct* actin promoter from rice, *MP* movement protein from TMV, *SP* apoplastic signal peptide, *LB/RB* left and right T-DNA border, *GOI* gene of interest.

LSBC reported that aprotinin expressed by the viral transient system has identical biological activity (TIU) compared to the commercial-grade protein *(12)*. In terms of impurities, it was shown that a higher degree of single oxidized aprotinin was observed in the plants samples. This oxidized form was much less pronounced in samples derived from plants grown indoor compared to plants grown in the field.

ProdiGene has reported complications of the aprotinin purification due to the co-purification of a seed-specific trypsin inhibitor from maize (CTI) *(6)*. An ion-exchange step was essential to separate both protein species *(15)*.

We have extracted and purified aprotinin from two different plant sources: transgenic alfalfa samples and tissue from infected *N. benthamiana* leaves. Frozen leaf material was disintegrated in a blender filtered and further purified in two steps: cation-exchange chromatography followed by reversed-phase chromatography. Aprotinin purified from *N. benthamiana* with a purity greater than 98%, as judged by reversed-phase high performance liquid chromatography (HPLC) analysis, was lyophilized in glass vials. The purified samples were analyzed by N-terminal sequence analysis (17 cycles), SDS-PAGE and Western blot analysis, trypsin-inhibition assays and methionine sulfoxidation determination. The biological activity and immunological behaviour of the plant purified aprotinin was identical to the commercial-grade protein, both for the protein purified from *Nicotiana* and alfalfa (**Fig. 12.5.**). The only recognized difference between the plant and the animal-derived aprotinin relates to the degree of methionine sulfoxide which was higher both in alfalfa and *N. benthamiana* in comparison to aprotinin derived from bovine lungs.

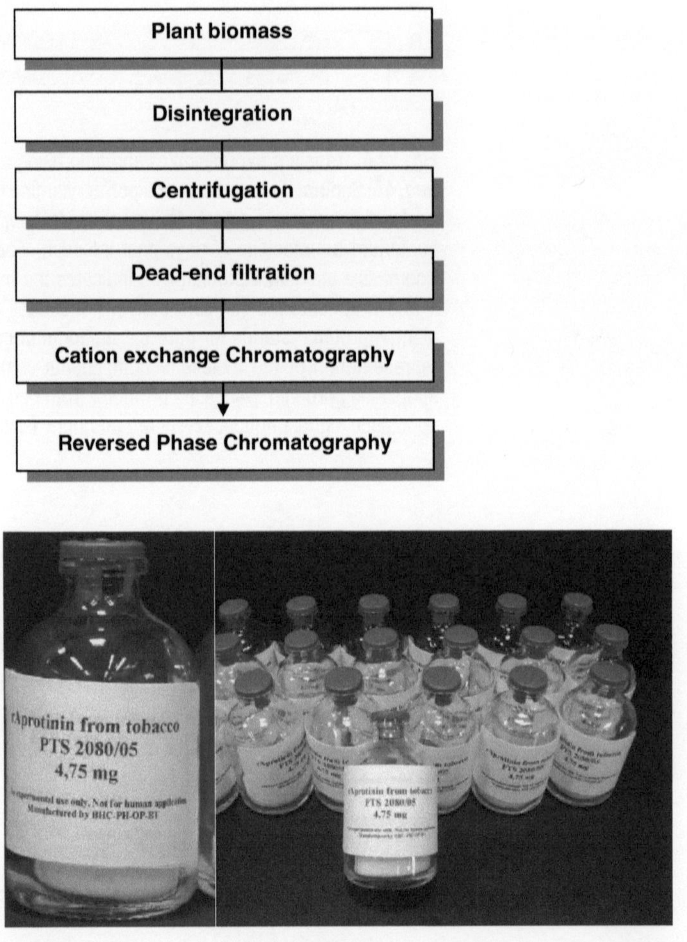

Fig. 12.5. Purification scheme of aprotinin from plants. Top: Scheme representing the major steps in the purification process optimized for plants. Bottom: Bayer's first batch of purified plant-made pharmaceutical (PMP) protein: Aprotinin was purified from *Nicotiana benthamiana* (Magnicon expression) biochemically qualified and aliquoted in protein batches of 4.75 mg purified aprotinin.

12.9. Discussion and Conclusions

Data in the literature and our data have shown that aprotinin can be produced in a wide variety of plants, including monocotyledonous (duckweed, maize) and dicotyledonous species (alfalfa, *N. benthamiana*). The highest level of protein expression in green biomass has been obtained with the transient system based on Agrobacterium delivery of viral vectors in *N. benthamiana*: up to 4% of the TSPs. This system also provides a high level of expression within a very short time frame from infection of the plants up to harvest of the biomass.

Traditionally, people in the field have reported the expression levels relative compared to the total amount of soluble proteins after extraction (% TSP). Today, this does not seem to be a very objective measure for comparison, especially when comparing different experimental setups. Obviously, the solubility of proteins depends on the particular extraction conditions (ionic strength, pH, detergents, temperature, etc.). We therefore favour using a less subjective measure: amount of expressed protein per fresh biomass (milligram protein/kilogram fresh biomass). In the case of aprotinin synthesis, the maximal obtained expression was in the range of 400 mg/kg fresh biomass.

Originally, aprotinin has been chosen for this case study because it has previously been characterized very well. We had anticipated that expression levels would not be limiting in most plant systems to be tested. However, when compared to other tester proteins the overall expression of aprotinin was rather modest, up to 4% TSP. Marillonnet et al. *(13)* reported protein expression of the green fluorescent protein up to 40% TSP. There seems to be an upper expression limit for this particular protein which is probably being directed by plant host cell interactions. The functionality of aprotinin as a protease inhibitor in plants cells has actually been shown in potato *(16)*. These authors have shown that apoplastically expressed aprotinin can stabilize other proteins such as neomycin phosphotransferase (NPTII). Taken together, it seems more appropriate to choose a molecule with no biological activity as a model protein. In this respect, aprotinin might not have been the better choice for this case study.

The choice of the appropriate codon usage for protein codification has been discussed as one relevant parameter for good protein expression. In the case of aprotinin, both plant optimized and non-optimized original gene sequences have been tested in several plant species. For this particular protein, codon usage seems not to be that critical, as evidenced by the fact that the highest expression levels have been obtained with gene constructs carrying the original bovine coding sequence.

\

The plant-produced aprotinin has similar biochemical and biological characteristics as compared to animal-derived manufacturing processes. A single difference in the aprotinin produced in animals is the degree of methionine oxidation, being higher in plant-derived compared to animal-derived aprotinin (bovine). For aprotinin, the higher level of methionine oxidation has not affected the biological activity of the molecule since the methionine residues are not directly involved in enzyme activity. This could be different with other proteins of interest, and hence it might be relevant to find out what is the basis for this higher degree of oxidation. The LSBC data have indicated that controlling of the growth conditions of the plants might remediate this problem.

Today, only few comparative PMP studies across species are reported. Aprotinin has been setting the stage. More 'real' case studies with other molecules will be needed to estimate the advantages and disadvantages of any particular plant expression system.

References

1. Fischer, R., Stoger, E., Schillberg, S., Christou, P., and Twyman, R.M. (2004). Plant-based production of biopharmaceuticals. Curr. Opin. Plant Biol. 7, 152–158.
2. Kermode, A.R. (2006). Plants as factories for production of biopharmaceutical and bioindustrial proteins: lessons from cell biology. Can. J. Bot. 84, 679–694.
3. Stoger, E., Ma, J.K.-C., Fischer, R., and Christou, P. (2005). Sowing the seeds of success: pharmaceutical proteins. Curr. Opin. Biotechnol. 16, 167–173.
4. Hewlett, G. (1990). Apropos aprotinin: a review. Biotechnology 8, 565–568.
5. Apeler, H., Peters, J., Schröder, W., Schneider, K.-H., Lemm, G., Hinz, V., Rossouw, G.J., and Dembowsky, K. (2004). Expression, purification, biochemical and pharmacological characterization of a recombinant aprotinin variant. Arzneim.-Forsch./Drug Res. 54(8), 483–497.
6. Azzoni, A.R., Kusnadi, A.R., Miranda, E.A., and Nikolov, Z.L. (2002). Recombinant aprotinin produced in transgenic corn seed: extraction and purification studies. Biotechnol. Bioeng. 80(3), 268–276.
7. Svab, Z., and Maliga, P. (1993). High-frequency plastid transformation in tobacco by selection for a chimeric aadA gene. Proc. Natl. Acad. Sci. USA 90(3), 913–917.
8. Staub, J.M., Garcia, B., Graves, J., Hajdukiewicz, P.T., Hunter, P., Nehra, N., Paradkar, V., Schlittler, M., Carroll, J.A., Spatola, L., Ward, D., Ye, G., and Russel, D.A. (2000). High-yield production of a human therapeutic protein in tobacco chloroplasts. Nat. Biotechnol. 18(3), 333–338.
9. Rival, S., Wisniewski, J.-P., Langlais, A., Kaplan, H., Freyssinet, G., Vancanneyt, G., Vunsh, R., Perl, A., and Edelman, M. (2007). Spirodela (duckweed) as alternative production system for pharmaceuticals: a case study, aprotinin. Transgenic Res. 17(4), 503–513.
10. Zhong, G.-Y., Peterson, D., Delaney, D.E., Bailey, M., Witcher, D.R., Register, J.C., III, Bond, D., Li, C.-P., Marshall, L., Kulisek, E., Ritland, D., Meyer, T., Hood, E.E., and Howard, J.A. (1999). Commercial production of aprotinin in transgenic seeds. Mol. Breed. 5(4), 345–356.
11. Delaney, D., Jilka, J., Barker, D., Irwin, P., Poage, M., Woodard, S., Horn, M., Vinas, A., Beifuss, K., Barker, M., Wiggins, B., Drees, C., Harkey, R., Nikolov, Z., Hood, E., and Howard, J. (2002). Production of aprotinin in transgenic maize seeds for the pharmaceutical and cell culture markets. In: Vasil, I.K. (ed.) Plant biotechnology 2002 and beyond. Kluwer, Dordrecht, pp. 393–394.
12. Vojdani, F.S., Palmer, K.E., Garger, S.J., and Pogue, G.P. (2006). Plant produced recombinant aprotinin and aprotinin variants. Patent application US2006/0218667 A1.
13. Marillonnet, S., Thoeringer, C., Kandzia, R., Klimyuk, V., and Gleba, Y. (2005).

Systemic Agrobacterium tumefaciens-mediated transfection of viral replicons for efficient transient expression in plants. Nat. Biotechnol. 23(6), 718–723.

14. Giritch, A., Marillonnet, S., Engler, C., van Eldik, G., Botterman, J., Klimyuk, V., and Gleba, Y. (2006). Rapid high-yield expression of full-size IgG antibodies in plants coinfected with noncompeting viral vectors. Proc. Natl. Acad. Sci. USA 103(40), 14701–6.
15. Azzoni, A.R., Takahashi, K., Woodard, S.L., Miranda, E.A., and Nikolov, Z.L. (2005). Purification of recombinant aprotinin produced in transgenic corn seed: separation from CTI utilizing ion-exchange chromatography. Braz. J. Chem. Eng. 22 (3), 323–330.
16. Rivard, D., Anguenot, R., Brunelle, F., Quy Le, V., Vézina, L.-P., Trépanier, S., and Michaud, D. (2006). An in-built proteinase inhibitor system for the protection of recombinant proteins recovered from transgenic plants. Plant Biotechnol. 4, 359–368.
17. Kassel, B., and Laskowski, M. (1965). The basic trypsin inhibitor of bovine pancreas. V. The disulfide linkages. Biochem. Biophys. Res. Commun. 20, 463–468.

Chapter 13

Glycosylation of Antibody Therapeutics: Optimisation for Purpose

Roy Jefferis

Summary

Recombinant antibody therapeutics represent a significant success story in terms of clinical benefit delivered and revenue (profit) generated within the biopharmaceutical industry. Additionally, it is estimated that ~30% of new drugs likely to be licensed during the next decade will be based on antibody products. High volume production with the maintenance of structural and functional fidelity of these large biological molecules results in high "cost of goods" that can limit their availability to patients, due to the strain it puts on national and private health budgets. The challenge in reducing cost of goods is that each antibody is unique, both in structure and function. Optimal clinical efficacy will require engineering of antibody genes to deliver products with enhanced activities produced by cell lines engineered to deliver antibody homogeneous for pre-selected post-translational modifications, that is, protein structures and glycoforms. A "universal" production vehicle cannot meet these demands and several production mammalian cells are now available, alternatives to mammalian cell lines are also reaching maturity. Advances in downstream processing also need to be realised whilst chemical changes during processing and storage must be minimised.

Key words: Therapeutic antibodies, Human IgG subclasses (1,2,3,4), Effector functions, IgG-Fc glycosylation, IgG-Fab glycosylation, IgG-Fc receptors, ADCC, CDC, Pharmacokinetics, Fucosylation, Bisecting *N*-acetylglucosamine, Galactosylation, Aglycosylated antibody.

13.1. Introduction

Recombinant antibody therapeutics represent a significant success story in terms of clinical benefit delivered and revenue (profit) generated within the biopharmaceutical industry. Additionally, it is estimated that ~30% of new drugs likely to be licensed during

Loïc Faye and Véronique Gomord (eds.), *Methods in Molecular Biology, Recombinant Proteins From Plants, vol. 483*

DOI: 10.1007/978-1-59745-407-0_13

the next decade will be based on antibody products. However, the establishment of protocols and facilities for the production, downstream processing, formulation, etc. of metric tonnes of product has resulted in very high "cost of goods" (CODs). The high CODs can limit the availability of these drugs to patients due to the strain they put on national and private health budgets. This has resulted in the establishment of somewhat controversial criteria for determining which products are made available for which disease indications and to which patient groups. The result can be rather rigid patient selection and/or international, national and regional differences in drug availability. The perceived benefits that could flow from lower CODs has acted as an incentive for innovation in the development of improved and novel production vehicles. An added incentive to reduce CODs is provided by the increasing capability of low labour cost economies to compete with and undercut production costs of high labour cost economies.

In this chapter, I explore the potential to achieve lower CODs by developing improved antibody products and production vehicles, employing protein and cell engineering.

13.1.1. Post-Translational Modifications

An understanding of human complexity is being sought not in the total number of genes but in the protein products of those genes. The proteome exceeds the genome and transcriptome due to differential RNA splicing, the introduction of post-translational modifications (PTMs), etc. Possibly the most frequent and diverse PTM is glycosylation; oligosaccharides may be N-linked, through asparagine residues, or O-linked through serine or threonine. The importance of glycosylation is evident from the fact that congenital disorders of glycosylation can result in serious medical consequences *(1, 2)*. Additionally, changes in the glycosylation profiles of specific proteins may serve as disease markers *(3–5)*. Glycosylation and other PTMs have been shown to be species, tissue and gender specific *(6–8)*.

It will be evident that a recombinant protein should, ideally, exhibit the same PTMs as the endogenous protein product; however, it is important to recognise that the structure determined for an endogenous protein is that of molecules that have had a residence time in a body compartment/fluid before being subject to multiple isolation and purification protocols. The structure of this purified product could differ from that of the nascent molecule secreted from its tissue of origin. Similarly, recombinant proteins are synthesised in an "alien" tissue, are exposed to the culture medium, products of the host cell line, and subjected to rigorous downstream and formulation processes. This can result in lack of structural fidelity, which can impact on function, stability and immunogenicity. An immune response may impact therapeutic efficacy and/or result in harmful side-effects *(9, 10)*.

Considerable success has been reported for increased productivity of antibody in Chinese hamster ovary (CHO) cell lines, with levels of 5 g/l being achieved and 10 g/l being set as a goal *(11)*. However, high production levels may overwhelm the PTM machinery resulting in poor product quality; it is essential, therefore, to include product characterisation as a parameter influencing clone selection, to optimise both productivity and quality. The nature of the glycoprotein product may dictate the producer cell line selected; thus, CHO, NS0 and Sp2/0 cell lines produce an endogenous carboxypeptidase-b and may be particularly inappropriate for the production of the complement proteins C3a, C4a, etc. that bear functionally significant C-terminal arginine residues; the desArg forms of these proteins having a unique activity profile.

The development of alternative production vehicles to mammalian cells has been hindered by their inability to effect human-type PTMs. Thus, plants differentially glycosylate proteins but consistently add α-(1-3) fucose and β-(1-3) xylose sugars that are reported to be immunogenic/allergenic in humans *(8)*. Yeast systems add high mannose and insect cells add pauci-mannose oligosaccharides, whilst most prokaryotic systems (e.g., *Escherichia coli*) are not able to effect glycosylation.

13.1.2. Recombinant Antibody Therapeutics

The biopharmaceutical industry has met the challenge to produce rMAbs; however, productivity, cost and potency remain to be optimised. All antibody therapeutics currently licensed are produced by mammalian cell culture, utilising CHO, mouse NSO or mouse Sp2/0 cell lines. All rMAbs have evidenced a potential for immunogenicity whether presented as mouse, chimeric, humanised or fully human sequences. These responses are referred to as human anti-mouse antibody (HAMA), human anti-chimeric-antibody (HACA) or human anti-human antibody (HAHA) *(12)*; the promise that fully human antibodies may not be immunogenic has not been realized for Humira (adalimumab), generated by phage display from a human heavy and light chain library, since a 12% of patients have been shown to produce anti-Humira antibodies. Such antibody responses will prejudice treatment if they are neutralizing, lead to clearance of the therapeutic or sensitize the patient for severe reactions on re-exposure *(10)*.

The effectiveness of rMAb in oncology depends on sensitizing target cells for subsequent killing by the mechanisms of antibody-dependent cellular cytotoxicity (ADCC) and/or complement-dependent cytotoxicity (CDC). It is unequivocally established that these effector functions are dependent on appropriate glycosylation of the rMAb *(7)*. This presents a continuing challenge for the biopharmaceutical industry to further develop cell lines and culture conditions for increased productivity. Studies of producer cell line proteomes are being initiated in attempts to identify parameters influencing production levels and product quality *(13)*. To date attention has focused on rMAb glycosylation at a single site (Asn 297); however, ~20% of normal human antibodies have an additional glycosylation site, as does one licensed therapeutic antibody (Erbitux) *(7)*.

13.1.3. Structural and Functional Characteristics of Human Antibodies

The four human IgG subclasses are enumerated according to their relative concentrations in normal serum; thus, IgG1, IgG2, IgG3 and IgG4 account for ~60%, 25%, 10% and 5%, respectively. IgG antibody predominates in human blood, equilibrates with extravascular space and activates a wide range of effector functions resulting in the killing, removal and destruction of pathogens. Each IgG subclass exhibits a unique profile of effector functions when evaluated by in vitro assays. Thus, the choice of IgG subclass when developing a rMAb is predicated on the presumption that one can extrapolate from *in vitro* activities to function *in vivo*, in man *(7, 14, 15)*. The broad generalisation can be made that IgG1 and IgG3 responses predominate for protein antigens, IgG2 for carbohydrate antigens and IgG4 may predominate as a result of chronic antigen stimulation; http://www.researchd.com/rdikits/rdisubbk.htm.

The predominance and accessibility of IgG1 subclass proteins and antibodies allowed for comprehensive study of their structure and structure/function relationships. Consequently, the IgG1 subclass appeared to be the natural choice in development of rMAbs. This assumption is now under review as it is appreciated that the optimal mode of action, *in vivo*, will vary according to the disease indication. Devising assays to determine the effector functions activated *in vivo* remains a considerable challenge; multiple parameters impact on the outcome when a given rMAb therapeutic is delivered to an individual patient for a given disease indication.

The IgG1 and IgG3 subclasses are essentially equivalent in their abilities to activate each of the leucocyte Fc receptors and the classical complement cascade; however, they have different serum half-life of ~21 and ~7 days, respectively. It is recognised that activation of inflammatory cascades may be detrimental, for some antibody based therapies. This has resulted in contemplation of the respective merits of the IgG2 and IgG4 subclasses. Two IgG4 rMAbs have been developed, presumably selected for reduced effector activities, and received regulatory approval. This presumption is open for re-evaluation since it has been demonstrated that IgG4 can activate FcγRI and FcγRIIIa, depending on the glycoform of the antibody and the allotype of the receptor *(16)*. In contrast, IgG2 has been shown only to activate one allotypic variant of the FcγRIIa receptor. Neither IgG2 nor IgG4 have been convincingly shown to activate the classical complement pathway when only human complement proteins are employed *(14)*; http://www.researchd.com/rdikits/rdisubbk.htm.

13.2. The Structure of Human IgG Antibodies

In its simplest form an individual IgG molecule is composed of two identical light chains and two identical heavy chains composed of repeating structural motifs (homology regions) of ~110 amino acid residues. The tertiary structure of each homology region defines the immunoglobulin fold or domain – a ubiquitous structure that further defines the immunoglobulin superfamily *(14, 15, 17)*. Domains of the light and heavy chains pair in covalent and non-covalent association to form three independent protein moieties connected thorough a flexible linker (the hinge region) **Fig. 13.1**. Two of these moieties, referred to as Fab regions, are of identical structure and each expresses a specific antigen-binding site; the third, the Fc, expresses interaction sites for ligands that activate clearance mechanisms. These effector ligands include three structurally homologous cellular Fc receptor types (FcγRI, FcγRII, FcγRIII) *(14, 18, 19)*, the C1q component of complement and the neonatal Fc receptor (FcRn) *(9, 14, 15)*. Activation of Fc receptors and the C1q component of complement

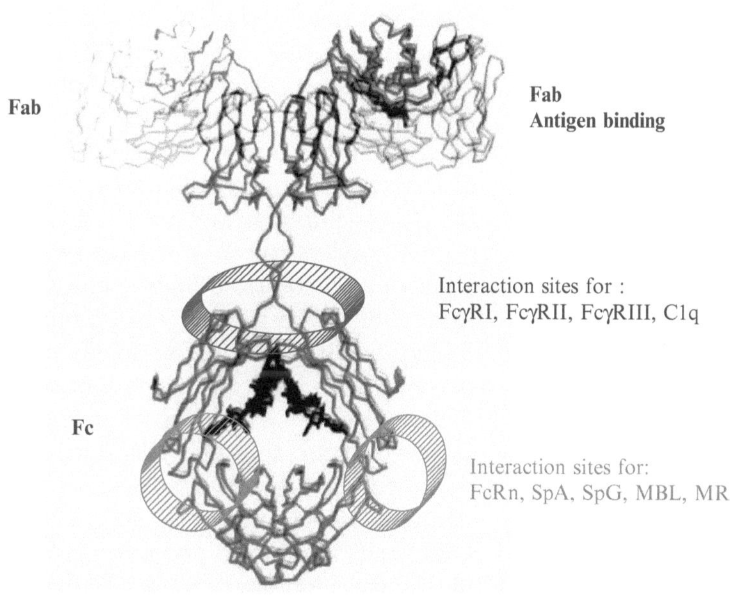

Fig. 13.1. Alpha carbon backbone structure of an IgG1 molecule. Digestion within the hinge region, by papain, releases the Fab (Fragment antigen binding) and Fc (Fragment crystallisable) fragments. (*See Color Plates*)

initiate inflammatory cascades that combat and resolve episodes of infection by pathogens. These activities are critically dependent on IgG-Fc N-linked glycosylation and, in part, on individual antibody glycoforms *(6, 7, 16, 20)*. In contrast, activation of FcRn, which determines antibody half-life and placental transport, is independent of glycosylation status.

The IgG-Fc region is a homodimer composed of covalent inter-heavy chain disulphide bonded hinge regions and non-covalently paired CH3 domains; the CH2 domains are glycosylated through covalent attachment of oligosaccharide at asparagine 297 (Asn-297). X-ray crystallographic analysis reveals discreet structure for the oligosaccharide that is integral to the IgG-Fc structure and forms multiple non-covalent interactions with the protein surface of the CH2 domain; thus, the protein and oligosaccharide exert reciprocal influences on the conformation of each other *(7, 14, 15, 17)*. There is cumulative evidence that interaction sites on IgG-Fc for FcγRI, FcγRII, FcγRIII and C1q effector ligands are composed of the protein moiety only; however, generation of the essential IgG-Fc protein conformation is dependent on the presence of the oligosaccharide. Thus, effector mechanisms mediated through FcγRI, FcγRII, FcγRIII and C1q are severely compromised or ablated for aglycosylated or deglycosylated forms of IgG *(7, 21, 22)*.

13.2.1. IgG-Fc Glycosylation

The oligosaccharide of normal polyclonal human IgG-Fc is of the diantennary complex type and shows considerable heterogeneity; a "core" heptasaccharide can be defined with variable addition of outer arm sugar residues, **Fig. 13.2**. A total of 32 different oligosaccharides may be attached generating, potentially, more than 400 glycoforms, given random pairing of heavy chain glycoforms. Analysis of the oligosaccharide released from normal polyclonal IgG-Fc shows a paucity of sialylation (<10%) and 12 of the possible 16 neutral oligosaccharides to predominate *(23)*, providing the potential to generate a total of 72 glycoforms, **Fig. 13.3**.

Glycosylation is a co-translational PTM initiated by attachment of a glucosylated high mannose oligosaccharide (GlcNAc-$_2$Man$_9$Glu$_3$). The glycosylated protein binds chaperones that aid folding fidelity and exercise quality control. The oligosaccharide is trimmed to a GlcNAc$_2$Man$_9$Glu structure that transits to the Golgi apparatus where the oligosaccharide is initially trimmed, by glycosidases, to a GlcNAc$_2$Man$_5$ structure, before the action of successive glycosyltransferases, to generate the complex diantennary structures *(8, 24)*. Early studies established that CHO cells are able to produce rMAb IgG with the majority of glycoforms identical to those of polyclonal human IgG; however, minor populations of abnormally glycosylated products may be present that are potentially immunogenic and unacceptable as

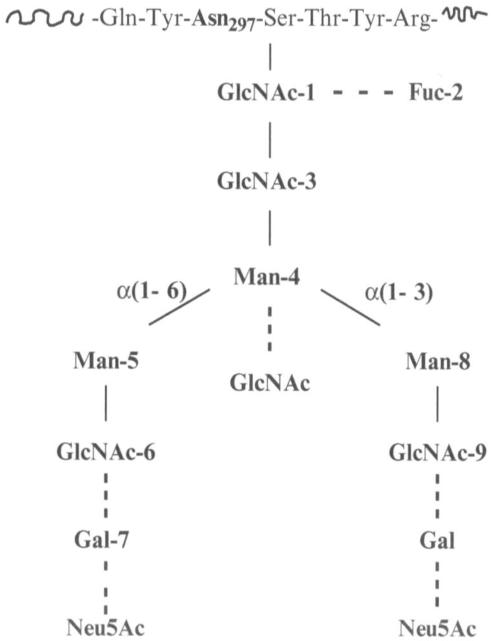

Fig. 13.2. Potential heterogeneity of diantennary oligosaccharide structures. The core G0 heptasaccharide is shown in blue, and additional sugar residues that may be present shown in red. *GlcNAc* *N*-acetylglucosamine, *Fuc* fucose, *Man* mannose, *Gal* galactose, *Neu5Ac* *N*-acetyl neuraminic acid. Shorthand nomenclature: G0 + fucose = G0F; G0 + galactose = G1; G0 + galactose + fucose = G1F; G0 + bisecting GlcNac = G0b; G0 + 2 galactose residues = G2; G2 + fucose = G2F; etc. N.B. structures bearing bisecting *N*-acetylglucosamine in the absence of fucose are not present.

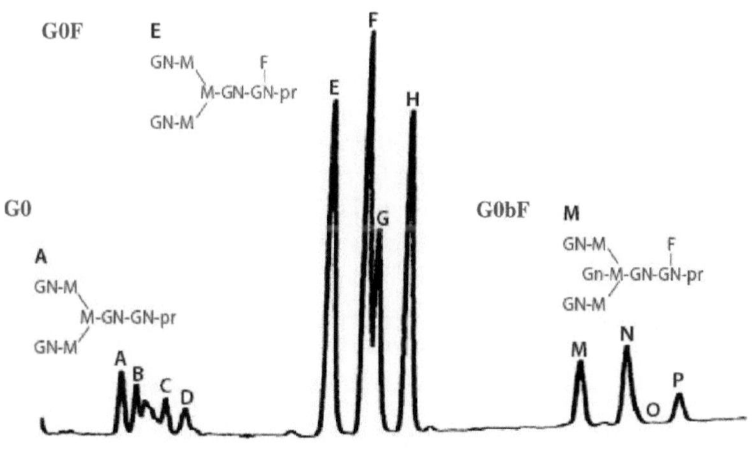

Fig. 13.3. HPLC profile of neutral oligosaccharides released from polyclonal human IgG *(23)*.

therapeutics, for example, galactose α(1-3) galactose and *N*-glycolylneuraminic acid structures *(7, 8)*. A major achievement of the biopharmaceutical industry has been the development of cell lines that can be expanded in serum- and protein-free media to produce rMAb whilst maintaining glycosylation fidelity and minimising the content of abnormal glycoforms. Regulatory authorities require that the glycoform profile of the marketed antibody product be maintained within strict limits, set by the glycoform profile of the material that gained licensing approval.

13.2.2. IgG-Fab Glycosylation

It is established that 20–30% of polyclonal human IgG molecules bear N-linked oligosaccharides within the IgG-Fab region, in addition to the conserved glycosylation site at Asn 297 in the IgG-Fc *(4, 5, 7)*. When present they are attached in the variable regions of the kappa (V_κ), lambda (V_λ) or heavy (V_H) chains, sometimes both. In the immunoglobulin sequence, database ~20% of IgG V regions have N-linked glycosylation consensus sequences (Asn-X-Thr/Ser, where X can be any amino acid except proline). Interestingly, these consensus sequences are mostly not germline encoded but result from somatic mutation – suggesting positive selection for improved antigen binding. The functional significance for IgG-Fab glycosylation of polyclonal IgG has not been fully evaluated, but data emerging for monoclonal antibodies suggest that V_κ, V_λ or V_H glycosylation can have a neutral, positive or negative influence on antigen binding *(7)*.

Analysis of polyclonal human IgG-Fab reveals the presence of diantennary oligosaccharides that are extensively galactosylated and substantially sialylated, in contrast to the oligosaccharides released from IgG-Fc *(5, 7)*. This demonstrates that the glycosylation processing machinery is fully competent and that the relative lack of galactosylation and sialylation of IgG-Fc must be due to a lack of accessibility to specific transferases. The licensed antibody therapeutic cetuximab (Erbitux) bears an N-linked oligosaccharide at Asn 88 of the VH region; interestingly, there is also a glycosylation motif at Asn 41 of the VL but it is not occupied *(25)*. Different glycoforms predominate at each site; the IgG-Fc bearing fucosylated G0 and G1 oligosaccharides and the IgG-Fab fucosylated G2 and sialylated oligosaccharides. Whilst the IgG-Fc oligosaccharides could be released on exposure to PNGase F, the IgG-Fab oligosaccharide linkage was refractory. Thus, the rMAb and the Sp2/0 cell line appear to replicate the situation observed for polyclonal human IgG produced *in vivo*.

Other experiences of IgG-Fab glycosylation have been reported. A detailed analysis of the glycoforms of a humanised IgG rMAb, expressed in Sp2/0 cells, bearing oligosaccharides at Asn 56 of the VH and Asn 297 reveals the expected IgG-Fc oligosaccharides profile of predominantly fucosylated G0 and G1; however, 11 oligosaccharides were released from the IgG-Fab, including triantennary, and other oligosaccharides not observed for normal human IgG *(26)*. All IgG-Fab oligosaccharides were extensively sialylated but with *N*-glycolylneuraminic acid, rather than *N*-acetylneuraminic acid. The consistent observation of higher levels of galactosylation and sialylation for IgG-Fab N-linked oligosaccharides, in comparison to IgG-Fc, is thought to reflect increased exposure and/or accessibility.

In conclusion, IgG-Fab glycosylation can impact differentially on the structural and functional characteristics of IgG. It may be exploited to increase the solubility and stability of antibodies, limiting aggregation and hence immunogenicity. However, controlling glycoform fidelity at two sites offers a further challenge to the biopharmaceutical industry.

13.2.3. Cell Engineering to Influence Glycoform Profiles

Production CHO, NS0 and Sp2/0 cell lines yield rMAb having a restricted glycoform profile, relative to that observed for normal polyclonal human IgG, with G0F and G1F glycoforms predominating. Concern has been expressed at the relatively low levels of galactosylation and its possible impact on activation of the classical complement pathway; it has been reported that galactosylation impacts positively on the ability of rituximab to lyse CD20 expressing cells, however, to date, this appears to be a solitary experience [http://www.fda.gov/cder/biologics/review/ritugen112697-r2.pdf].

Development of a CHO cell line expressing the GNTIII transferase and able to add bisecting *N*-acetylglucosamine residues resulted in the production of an anti-neuroblastoma rMAb that exhibited a 15–20-fold improvement in ADCC *(20, 27)*. A similar improvement in ADCC was reported by for the rMAb rituximab when bisecting *N*-acetylglucosamine residues were incorporated *(28)*. Alternatively, a profound improvement in ADCC efficacy was reported for herceptin produced in a mutant CHO cell line (LEC 13) that is deficient in the addition of fucose; the LEC 13 cell line was not considered to be suitable for development as a production vehicle *(29)*. A similar improvement in ADCC was reported for the non-fucosylated fraction of a recombinant anti-human IL-5 receptor (rhIL-5-R) antibody *(6)* produced in the rat derived YB2/0 cell line. Physical separation of the non-fucosylated and bisecting *N*-acetylglucosamine glycoforms suggested that it was the absence of fucose rather than the presence of bisecting *N*-acetylglucosamine that resulted in enhanced ADCC. An α(1-6) fucosyltransferase "knock-out" cell line was subsequently generated that allows the production of non-fucosylated IgG antibodies *(30)*. Importantly, it has been shown that FcγRIIIa-mediated ADCC is amplified for the non-fucosylated glycoform of each of the IgG subclasses *(16)*. The increase in FcγRIIIa-mediated ADCC has been correlated with an increase in the affinity of the antibody for the receptor and the non-fucosylated antibody overcomes the deficit evident for the homozygous FcγRIIIa Phe158 allotype.

13.3. Interaction Sites for Effector Ligands

Whilst it is established that N-linked glycosylation of the IgG-Fc is essential for optimal expression of biological activities mediated through FcγRI, FcγRII, FcγRIII and the C1q component of complement *(7, 14)*, present evidence suggests that it does not influence interactions with FcRn and consequently, presumably, the catabolic half-life or transport across the placenta. The IgG-Fc binding properties of bacterial proteins, for example, SpA, SpG, are also unaffected *(7, 14, 17)*. Protein engineering, employing alanine scanning, has been used to "map" amino acid residues deemed to be critical for FcγR and C1q binding. These studies "map" the binding site for all four of these ligands to the hinge proximal or lower hinge region of the CH2 domain *(7, 14, 15)*.

X-ray crystallographic analysis of the IgG-Fc fragment, residues 216–446, reveals electron density for residues 238–443 only *(17)*; thus, the lower hinge region would appear to be mobile and without defined structure. This might appear to be incompatible with the suggestion that the lower hinge region is directly involved in the generation of structurally distinct interaction sites for the FcγR and C1q ligands; however, we have proposed that this region of the molecule is not without structure but is composed of an equilibrium of multiple conformers, resulting from reciprocal interactions between the oligosaccharide and the protein moiety such that individual conformers are compatible with specific ligand recognition *(7)*. X-ray crystallographic analysis of IgG-Fc in complex with a soluble recombinant form of the receptor, rFcγRIIIa, provided proof of the direct involvement of the lower hinge regions and hinge proximal CH2 domain residues *(31, 32)*. The interaction site on the IgG-Fc is seen to include asymmetric binding to discrete conformations of the lower hinge residues of each heavy chain. A critical requirement is explained by this structure – that the IgG-Fc should be univalent for the FcγR. This is essential since if monomeric IgG were divalent it could cross-link cellular receptors and hence constantly activate inflammatory reactions.

13.3.1. Pharmacokinetics and Placental Transport

The catabolic half-life of human IgG is very extended, relative to other serum proteins; IgG1, IgG2 and IgG4 proteins having a half-life of ~21 days and IgG3 proteins ~7 days. Catabolism is mediated through the unique neonatal Fc receptor, FcRn, pathway. This membrane bound receptor is widely expressed throughout the body. An X-ray crystal structure of IgG-Fc in complex with recombinant FcRn has shown the binding site to be at the CH2/CH3 interface with residues of both domains contributing to binding *(15)*. Extensive mutational studies have identified residues crucial to FcRn binding and mutants with higher binding affinities that exhibit extended half-lives *(15, 33)*. These studies may provide a route to optimising the *in vivo* half-life of an antibody for a given therapeutic application. Importantly, interactions with FcRn appear to be independent of IgG-Fc glycoform.

The long half-life of IgG antibodies is being exploited through the generation of fusion proteins, for example, single chain Fv-Fc $(scFv\text{-}Fc)_2$, cytokine-IgG-Fc therapeutics. The presence of the IgG-Fc region contributes to improved stability, pharmacokinetics and pharmacodynamics. A further development opens a new route for administration. It has been shown that FcRn is expressed in the central and upper airways and that drug-IgG-Fc fusion proteins delivered to these sites can be transferred, by transcytosis, to the systemic circulation. This is an exciting development with considerable promise and significance [*(34)*; http://www.syntnx.com/home.php].

13.4. Glycosylation Engineering and Non-mammalian Production Vehicles

Subtle structural parameters also influence the glycosylation profile. Thus, alanine scanning studies showed that single amino acid replacements in IgG-Fc could result in gross changes in the glycoform profile of product, resulting in increased galactosylation and sialylation *(35)*. It will be evident that an ability to produce selected homogenous glycoforms of recombinant antibody molecules would be advantageous. Manipulation of culture conditions can have a limited, but significant, influence on the glycoform profile of product *(36)* and may allow for manipulation of the glycoform profile over the time of a production run. Cell engineering is being undertaken to "knock-out" and/or "knock-in" genes encoding for selected glycosyltransferases, as illustrated above.

13.4.1. Plants

It has been comprehensively demonstrated that human proteins can be produced in plants *(8)*. However, PTMs present a problem, particularly for glycosylation. Thus, whilst plants can add human-type "core" diantennary oligosaccharides they also add α(1-3) fucose and β(1-2) xylose sugar residues that are not expressed in humans and are reported to be immunogenic and, possibly, allergenic. Several strategies have been employed to overcome this disadvantage. Thus, the addition of endoplasmic reticulum retaining KDEL sequences to the C-terminus of heavy and light chain genes results in the synthesis of a high mannose form of antibody. This may be harvested and converted to homogeneous glycoform products by in vitro enzymatic manipulation. This technique has been applied to the production of glycoproteins in tobacco plants. A further problem with the natural tobacco plant is that the glycoprotein glycoform profile varies with the maturity of the leaves *(8, 37)*.

A more economic route would appear to be available by employing RNA inhibition (RNAi) of the appropriate fucose and xylose transferases. The production of homogenous rMAb glycoforms in the plant Lemmna (duckweed) has been reported *(38)*. Successful double "knock-out" of the α(1-3) fucose and β(1-2) xylose transferases has been reported for the production of human vascular endothelial growth factor in *Physcomitrella patens* (moss) *(39)* and endogenous glycoproteins in *Arabidopsis thaliana (40)*. Each of these systems may yield rMAb with enhanced ADCC due to the non-addition of α(1-6) fucose residues.

13.4.2. Insect Cells

The utility of the Sf2 insect cell line has been extended with the generation of the SfSWT-1 line that is transgenic for genes encoding five mammalian glycosyltransferases, including sialyltransferases. In a further development a transgenic line was generated that can also synthesise cytidine monophosphate (CMP) sialic acid and hence sialylate glycoproteins in the absence of an exogenous source of sialic acid, for example, bovine serum *(41)*.

13.4.3. Yeast

Yeasts synthesise glycoproteins bearing high mannose oligosaccharides, the number of mannose residues varying between individual strains. *Pichia pastoris* has been successfully engineered such that antibodies with selected homogeneous glycoforms of human IgG-Fc may be produced *(42)*.

13.4.4. Prokaryotes

Several small protein therapeutics are produced in *E. coli*, for example, insulin. The production of full-length IgG antibodies has also been reported; however, they are not glycosylated and lack the potential to activate effector functions *(43)*. It has recently been reported that *Campylobacter jejuni* has the capacity to add oligosaccharides to proteins and that this capacity can be transferred to *E. coli (44)*. The resulting glycosylation patterns are radically different to those of mammalian cells; however, it suggests that it may be possible to engineer *E. coli* to express mammalian-type oligosaccharides.

13.5. "Bio-similar" and "Follow-On" Biologics

Proteins and vaccines are classed as "biologics" and the Food and Drug Administration (FDA) has different rules for their production as generics, as apposed to small molecule "drugs". Thus, any change in the production process for a biologic is considered to have a possible impact on product potency. Therefore, the concept of "comparability" has been introduced as a pragmatic evaluation. This concept is particularly relevant to the production of protein biologics in non-mammalian production vehicles. The debate over "follow-on" biologics centres on the complexity of a protein/glycoprotein and the manufacturing process; it is argued that the product of a given process in one facility can never be exactly reproduced in another. If sustained there would be a requirement for a "follow-on" biologic to be subject to new clinical trials before it could be licensed, which would compromise any potential cost advantage *(45)*; http://www.fda.gov/cder/guidance/compare.htm.

13.6. Concluding Remarks

It is evident that a universal cell line for the production of recombinant protein therapeutics is not available and that engineering a cell line to optimise one product, for example, antibodies, may reduce applicability to the production of other products. This chapter has focused on glycosylation but >350 PTMs have been characterised and individual molecules may bear the stamp of multiple PTM events. In addition, whilst we focus on the known biologic activity of a given protein therapeutic, each may exercise several roles *in vivo*, that is, are functionally pleomorphic, and each may be influenced by particular PTMs *(46, 47)*. The present and future challenge is to develop means of monitoring mechanisms activated *in vivo* and individual variations determined by the interaction of multiple polymorphisms – it is called "systems biology".

References

1. Freeze, H.H. (2002) Human disorders in *N*-glycosylation and animal models. Biochim Biophys Acta 1573:388–393.
2. Butler, M., Quelhas, D., Critchley, A.J., Carchon, H., Hebestreit, H.F., Hibbet, R.G., Vilarinho, L., Teles, E., Matthijs, G., Schollen, E., Argibay, P., Harvey, D.J., Dwek, R.A. and Rudd, P.M. (2003) Detailed glycan analysis of serum glycoproteins of patients with congenital disorders of glycosylation indicates the specific defective glycan processing step and provides an insight into pathogenesis. Glycobiology 13:601–622.
3. Axford, J.S., Cunnane, G., Fitzgerald, O., Bresnihan, B. and Frears, E.R. (2003) Rheumatic disease differentiation using immunoglobulin G sugar printing by high-density electrophoresis. J Rheumatol 12:2540–2546.
4. Youings, A., Chang, S.C., Dwek, R.A. and Scragg, I.G. (1996) Site-specific glycosylation of human immunoglobulin G is altered in four rheumatoid arthritis patients. Biochem J 314:621–630.
5. Holland, M., Yagi, H., Takahashi, N., Kato, K., Savage, C.O.S., Goodall, D.M. and Jefferis, R. (2006) Differential glycosylation of

polyclonal IgG, IgG-Fc and IgG-Fab isolated from the sera of patients with ANCA associated systemic vasculitis. Biochim Biophys Acta 1760:669–677.

6. Shinkawa, T., Nakamura, K., Yamane, N., Shoji-Hosaka, E., Kanda, Y., Sakurada, M., Uchida, K., Anazawa, H., Satoh, M., Yamasaki, M., Hanai, N. and Shitara, K. (2003) The absence of fucose but not the presence of galactose or bisecting N-acetylglucosamine of human IgG1 complex-type oligosaccharides shows the critical role of enhancing antibody-dependent cellular cytotoxicity. J Biol Chem 278:3466–3473.
7. Jefferis, R. (2005) Glycosylation of recombinant antibody therapeutics. Biotechnol Prog 21:11–16.
8. Gomord, V., Chamberlain, P., Jefferis, R. and Foye, L. (2005) Biopharmaceutical production in plants: problems, solutions and opportunities. Trends Biotechnol 23:559–565.
9. Sinclair, A.M. and Elliott, S. (2005) Glycoengineering: The effect of glycosylation on the properties of therapeutic proteins. J Pharm Sci 94:1626–1635.
10. Smalling, R., Foot, M., Molineux, G., Swanson, S.J. and Elliott, S. (2004) Drug-induced and antibody-mediated pure red cell aplasia: A review of literature and current knowledge. Biotechnol Annu Rev 10:237–250.
11. Birch, J.R. and Racher, A.J. (2006) Antibody production. Adv Drug Deliv Rev 58:671–685.
12. Mirik, G.R., Bradt, B.M., Denardo, S.J. and Denardo, G.L. (2004) A review of human anti-globulin antibody (HAGA, HAMA, HACA, HAHA) responses to monoclonal antibodies. Not four letter words. Q J Nucl Med Mol Imaging 48:251–257.
13. Alete, D.E., Racher, A.J., Birch, J.R., Stansfield, S.H., James, D.C. and Smales, C.M. (2005) Proteomic analysis of enriched microsomal fractions from GS-NS0 murine myeloma cells with varying secreted recombinant monoclonal antibody productivities. Proteomics 18:4689–4704.
14. Woof, J.M. and Burton, D.R. (2004) Human antibody-Fc receptor interactions illuminated by crystal structures. Nat Rev Immunol 4:89–99.
15. Nezlin, R. and Ghetie, V. (2004) Interactions of immunoglobulins outside the antigen-combining site. Adv Immunol 82:155–215.
16. Niwa, R., Natsume, A., Uehara, A., Wakitani, M., Iida, S., Uchidqa, K., Satoh, M. and Shitara, K. (2005) IgG subclass-independent improvement of antibody-dependent cellular cytotoxicity by fucose removal from Asn297-linked oligosaccharides. J Immunol Methods 306:151–160.
17. Deisenhofer, J. (1981) Crystallographic refinement and atomic models of a human Fc fragment and its complex with fragment B of protein A from Staphylococcus aureus at 2.9- and 2.8-Å resolution. Biochemistry 20:2361–2370.
18. van Sorge, N.M., van der Pol, W.L. and van de Winkel, J.G. (2003) FcgammaR polymorphisms: Implications for function, disease susceptibility and immunotherapy. Tissue Antigens 61:189–202.
19. Nimmerjahn, F. and Ravetch, J.V. (2006) Fcgamma receptors: Old friends and new family members. Immunity 24:19–28.
20. Ferrara, C., Brunker, P., Moser, S., Puntener, U. and Umana, P. (2006) Modulation of therapeutic antibody effector functions by glycosylation engineering: Influence of Golgi enzyme localization domain and co-expression of heterologous beta1, 4-N-acetylglucosaminyltransferase III and Golgi alpha-mannosidase II. Biotechnol Bioeng 93:851–861.
21. Mimura, Y., Sondermann, P., Ghirlando, R., Lund, J., Young, S.P., Goodall, M. and Jefferis, R. (2001) The role of oligosaccharide residues of IgG1-Fc in FcγIIb binding. J Biol Chem 276:45539–45547.
22. Krapp, S., Mimura, Y., Jefferis, R., Huber, R. and Sondermann, P. (2003) Structural analysis of human IgG glycoforms reveals a correlation between oligosaccharide content, structural integrity and Fcγ-receptor affinity. J Mol Biol 325:979–989.
23. Jefferis, R., Lund, J., Mizitani, H., Nakagawa, H., Kawazoe, Y., Arata, Y. and Takahashi, N. (1990) A comparative study of the N-linked oligosaccharide structures of human IgG subclass proteins. Biochem J 268:529–537.
24. Walsh, G. and Jefferis, R. (2006) Post-translational modifications in the context of therapeutic proteins. Nat Biotechnol 24:1241–1252.
25. Qian, J., Liu, T., Yang, L., Daus, A., Crowley, R. and Zhou, Q. (2007) Structural characterization of N-linked oligosaccharides on monoclonal antibody cetuximab by the combination of orthogonal matrix-assisted laser desorption/ionization hybrid quadrupole-quadrupole time-of-flight tandem mass spectrometry and sequential enzymatic digestion. Anal Biochem 364:8–18.
26. Huang, L., Biolosi, S., Bales, K.R. and Kuchibhotla, U. (2006) Impact of variable domain glycosylation on antibody clearance: An LC/MS characterization. Anal Biochem 349:197–207.

27. Umana, P., Jean-Mairet, J., Moudry, R., Amstutz, H. and Bailey, J.E. (1999) Engineered glycoforms of an antineuroblastoma IgG1 with optimized antibody-dependent cellular cytotoxic activity. Nat Biotechnol 17:176–180.

28. Davies, J., Jiang, L., LaBarre, M.J., Anderson, D. and Reff, M. (2001) Expression of GTIII in a recombinant anti-CD20 CHO production cell line: Expression of antibodies of altered glycoforms leads to an increase in ADCC thro' higher affinity for FcRIII. Biotechnol Bioeng 74:288–294.

29. Shields, R.L., Lai, J., Keck, R., O'Connell, L.Y., Hong, K., Meng, Y.G., Weikert, S.H. and Presta, L.G. (2002) Lack of fucose on human IgG1 N-linked oligosaccharide improves binding to human Fcgamma RIII and antibody-dependent cellular toxicity. J Biol Chem 277:26733–26740.

30. Ymane-Ohuki, N., Kinoshita, S., Inoue-Urakubo, M., Kusunoki, M., Iida, S., Nakano, P., Wakitani, M., Niwa, R., Sakurada, M., Uchida, K., Shitara, K. and Satoh, M. (2004) Establishment of FUT8 knockout Chinese hamster ovary cells: an ideal host cell line for producing completely defucosylated antibodies with enhanced antibody-dependent cellular cytotoxicity. Biotechnol Bioeng 87:614–622.

31. Sondermann, P., Huber, R., Oosthuizen, V. and Jacob, U. (2000) The 3.2-A crystal structure of the human IgG1 Fc fragment-FcγRIII complex. Nature 406:267–273.

32. Radaev, S., Motyka, S., Fridman, W.H., Sautes-Fridman, C. and Sun, P.D. (2001) The structure of human type III Fcγ receptor in complex with Fc. J Biol Chem 276:16469–16477.

33. Hinton, P.R., Xiong, J.M., Johlfs, M.G., Tang, M.T., Keller, S. and Tsurushita, N. (2006) An engineered human IgG1 antibody with longer serum half-life. J Immunol 176:346–356.

34. Dumont, J.A., Bitonti, A.J., Clark, D., Evans, S., Pickford, M. and Newman, S.P. (2005) Delivery of an erythropoietin-Fc fusion protein by inhalation in humans through an immunoglobulin transport pathway. J Aerosol Med 18:294–303.

35. Lund, J., Takahashi, N., Pound, J., Goodall, M. and Jefferis, R. (1996) Multiple interactions of IgG with its core oligosaccharide can modulate recognition by complement and human FcγRI and influence the synthesis of its oligosaccharide chains. J Immunol 157:4963–4969.

36. Andersen, D.C. and Reilly, D.E. (2000b) Production technologies for monoclonal antibodies and their fragments. Curr Opin Biotechnol 15:456–462.

37. Saint-Jore-Dupas, C., Faye, L. and Gomorod, V. (2007) From planta to pharma with glycosylation in the toolbox. Trends Biotechnol 25:317–323.

38. Cox, K.M., Sterling, J.D., Regan, J.T., Gadaska, J.R., Frantz, K.K., Peele, C.G., Black, A., Passmore, D., Moldovan-Loomis, C., Srinivasan, M., Cuison, S., Cardarelli, P.M. and Dickey, L. (2006) Glycan optimization of a human monoclonal antibody in the aquatic plant Lemna minor. Nat Biotechnol 24:1591–1597.

39. Kaprivova, A., Stemmer, C., Altmann, F., Hoffmann, A., Kopriva, S., Gorr, G., Reski, R. and Decker, F.L. (2004) Targeted knockouts of Physcomitrella lacking plant-specific immunogenic N-glycans. Plant Biotechnol J 2:517–523.

40. Strasser, R., Altmann, F., Mach, L., Glossl, J. and Steinkellner, H. (2004) Generation of Arabidopsis thaliana plants with complex N-glycans lacking beta1,2-linked xylose and core alpha1,3-linked fucose. FEBS Lett 561:132–136.

41. Hill, D.R., Aumiller, J.J., Shi, X. and Jarvis, D.L. (2006) Isolation and analysis of a baculovirus vector that supports recombinant glycoprotein sialylation by SfSWT-1 cells cultured in serum-free medium. Biotechnol Bioeng 95:37–47.

42. Hamilton, S.R., Davidson, R.C., Sethuraman, N., Nett, J.H., Jiang, Y., Rios, S., Bobrowicz, P., Stadheim, T.A., Li, H., Choi, B.K., Hopkins, D., Wischenewski, H., Roser, J., Mitchell, T., Strawbridge, R.R., Hoopes, J., Wildt, S. and Gerngross, T.U. (2006) Humanization of yeast to produce complex terminally sialylated glycoproteins. Science 313:1441–1443.

43. Georgiou, G. and Segatori, L. (2005) Preparative expression of secreted proteins in bacteria: status report and future prospects. Curr Opin Biotechnol 16:538–545.

44. Wacker, M., Feldman, M.F., Callwaert, N., Kowanik, M., Clarke, B.R., Pohl, N.L., Hernandez, M., Vines, E.D., Valvano, M.A., Whitfield, C. and Aebi, M. (2006) Substrate specificity of bacterial oligosaccharyltransferase suggests a common transfer mechanism for the bacterial and eukaryotic systems. Proc Natl Acad Sci USA 103:7088–7093.

45. Combe, C., Tredree, R.L. and Schellekens, H. (2005) Biosimilar epoetins: An analysis based on recently implemented European

medicines evaluation agency guidelines on comparability of biopharmaceutical proteins. Pharmacotherapy 25:954–962.

46. Jefferis, R. (2007) Antibody therapeutics: Isotype and glycoform selection. Expert Opin Biol Ther 7:1401–1413.

47. Mimura, Y., Ashton, P.R., Takahashi, N., Harvey, D.J. and Jefferis, R. (2007) Contrasting glycosylation profiles between Fab and Fc of a human IgG protein studied by electrospray ionization mass spectrometry. J Immunol Methods 326:116–126.

Chapter 14

N-Glycosylation of Plant Recombinant Pharmaceuticals

Muriel Bardor[1], Gleysin Cabrera[1], Johannes Stadlmann, Patrice Lerouge, José A. Cremata, Véronique Gomord and Anne-Catherine Fitchette

Summary

N-glycosylation is a maturation event necessary for the correct function, efficiency, and stability of a high number of biopharmaceuticals. This chapter presented here proposes various methods to determine whether, how, and where a plant pharmaceutical is *N*-glycosylated. These methods rely on blot detection with glycan-specific probes, specific deglycosylation of glycoproteins followed by mass spectrometry, *N*-glycan profile analysis, and glycopeptide identification by LC-MS.

Key words: Plant pharmaceuticals, *N*-glycosylation, Western Blotting, Immunodetection, Affinodetection, Glycosidase, Mass spectrometry, *N*-glycan profiling, HPLC, Glycopeptide, LC-MS, MALDI-TOF.

14.1. Introduction

Glycosylation is the most widespread posttranslational modification encountered on biopharmaceutical proteins. Indeed, more than one third of biopharmaceuticals are glycoproteins *(1)*, for which functions, efficiency, and stability are modulated by the presence and/or composition of glycans (*see* **Chapter 13, Jeffris**). Having the capacity of glycosylation is an advantage for any system used for biopharmaceutical production.

Among the different production systems available today, whole plants and cultured plant cells are becoming more and

[1] The two authors contributed equally to the work.

Loïc Faye and Véronique Gomord (eds.), *Methods in Molecular Biology, Recombinant Proteins From Plants, vol. 483*

DOI: 10.1007/978-1-59745-407-0_14

more attractive because of their low cost and large-scale production capacity, their low risks of contaminating/infecting a patient (*see* **chapter 1**), and their ability to perform most posttranslational modifications including glycosylation *(1)*. However, like the other production systems, such as mammals, cultured mammalian cells, insect cells, or yeast, plants and plant cells are not able to perform a perfect human-type glycosylation on pharmaceutical recombinant proteins.

In eukaryotes, glycosylation can be of two types: *N*- or *O*-glycosylation, depending on the linkage involved between the oligosaccharide and the protein backbone. Although *N*-glycosylation begins to be well documented for pharmaceutical recombinant proteins, data available for their *O*-glycosylation are scarse *(1)*. Therefore, this maturation event will not be detailed here.

N-glycosylation occurs exclusively on proteins that enter the secretory pathway of eukaryotic cells. It starts with the cotranslational transfer in the endoplasmic reticulum of an oligosaccharide precursor, $Glc_3Man_9GlcNAc_2$, onto specific asparagine residues contitutive of *N*-glycosylation sequences present on the nascent protein (Asn-*X*-Ser/Thr, where *X* can be any amino acid except proline and aspartic acid). Then, processing of the $Glc_3Man_9GlcNAc_2$ precursor occurs along the secretory pathway as the glycoprotein moves from the endoplasmic reticulum through the Golgi apparatus to its final destination. Glycosidases and glycosyltransferases convert successively the precursor into high-mannose-type *N*-glycans, ranging from $Man_9GlcNAc_2$ to $Man_5GlcNAc_2$, and to complex-type *N*-glycans (**Fig. 14.1**). This maturation pathway is common to all eukaryotic cells, including plant and human cells, up to the $Man_5GlcNAc_2$ structure (**Fig. 14.1**). Contrastingly, complex-type *N*-glycans differ according to the cell system in which they are matured. In plants, for example, the proximal GlcNAc of the core can be substituted with an α1,3-fucose, instead of an α1,6-fucose for human *N*-glycans. The β-mannose of the core can be substituted with a β1,2-xylose or a β1,4-GlcNAc in plants and humans, respectively. Moreover, glycan antennae are composed of a terminal β1,3-galactose and an α1,4-fucose linked to the terminal GlcNAc, forming the Lewis a structure, in plants (**Fig. 14.1**) *(2)*. In humans, terminal GlcNAc are substituted with β1,4-galactose to which sialic acid can be linked (**Fig. 14.1**).

This discrepancy between human and plant complex *N*-glycans raises problems when a recombinant plant glycoprotein is to be used as a pharmaceutical in human. The plant β1,2-xylose and the α1,3-fucose have been shown to be immunogenic in all laboratory mammals, with the exception of the BALB/c mouse. Their presence on a plant pharmaceutical will render this therapeutical protein insecure for injection to humans. Moreover, the differences existing between plant and human complex *N*-glycan

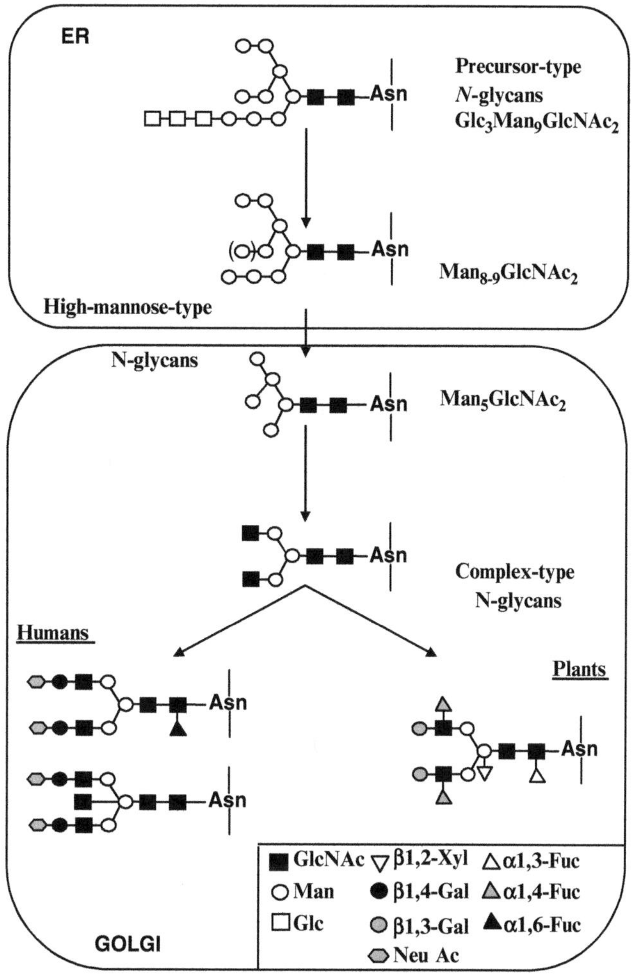

Fig. 14.1. Maturation pathway of *N*-glycans in humans and plants. ***GlcNAc N***-acetylglucosamine, ***Man*** mannose, ***Glc*** glucose, ***Xyl*** xylose, ***Gal*** galactose, ***NeuAc N***-acetylneuraminic acid, ***Fuc*** fucose.

structures can provoke a loss in biological activity and/or half-life of the plant pharmaceutical, once administrated. To circumvent these problems, several *in vitro* or *in vivo* strategies are being developed currently to humanize the *N*-glycans from plant pharmaceuticals (*see* **chapter 13**). In any case, a scrupulous study of the *N*-glycosylation, present on a plant recombinant protein, will be inevitable to consider using this protein as a pharmaceutical for human therapy.

In this chapter, the authors propose various methodologies, already developed in their respective laboratories, to define whether, how, and where a pharmaceutically relevant protein expressed in plant is glycosylated.

14.2. Materials

14.2.1. Is My Protein a Glycoprotein?

1. DIG Glycan Detection Kit (Roche Applied Science, Mannheim, Germany) (store at 2–8°C).

14.2.2. How Is My Protein Glycosylated?

1. TTBS buffer: 20 mM Tris-HCl, pH 7.4, containing 0.5 M NaCl and 0.1% Tween-20.

14.2.2.1. Detection of N-Glycans by Western Blotting

14.2.2.1.1. Affinodetection with Lectins

2. Lectin-biotin: GNA (*Galanthus nivalis* agglutinin)-biotin (Vector Labs/Abcys, France) (store at 2–8°C), and WGA (*Triticum vulgaris* agglutinin)-biotin (Sigma-Aldrich, Lyon, France) (store at 2–8°C).
3. Streptavidin–peroxidase conjugate (Amersham Biosciences/GE Healthcare, Uppsala, Sweden) (store at 2–8°C).
4. TBS buffer: 20 mM Tris-HCl, pH 7.4, containing 0.5 M NaCl.
5. 4-Chloro-1-naphtol (Bio-Rad, Hercules, CA) (store at –20°C).
6. H_2O_2 37%.
7. Ribonuclease B from bovine pancreas (Sigma-Aldrich) (store at 2–8°C).
8. Methyl α-D-mannopyranoside (Sigma-Aldrich) (store at room temperature).
9. Ovalbumin (Sigma-Aldrich) (store at 2–8°C).
10. Tris Buffered Saline Containing Tween 20 (TTBS)* buffer: 20 mM Tris-HCl, pH 7.4, containing 0.5 M NaCl, 1 mM $CaCl_2$, 1 mM $MgCl_2$, and 0.1% Tween-20.
11. Concanavalin A (Sigma-Aldrich) (store at 5–8°C).
12. Horseradish peroxidase (HRP, Sigma-Aldrich) (store at 2–8°C).
13. Soybean agglutinin (Sigma-Aldrich) (store at 2–8°C).

14.2.2.1.2. Immunodetection with Specific Antibodies

1. Tris Buffered Saline (TBS) buffer: 20 mM Tris-HCl, pH 7.4, containing 0.5 M NaCl.
2. Gelatin (Bio-Rad).
3. Anti-HRP rabbit immunserum (Sigma-Aldrich) (store at 2–8°C).
4. Anti-*Apis mellifera* venom (honeybee venom; Sigma-Aldrich) (store at 2–8°C).
5. Mouse antihuman Lewis a monoclonal antibody (Calbiochem, Meudon, France) (store at –20°C).
6. TTBS buffer: 20 mM Tris-HCl, pH 7.4, containing 0.5 M NaCl and 0.1% Tween-20.
7. Horseradish peroxidase (HRP) conjugated goat antibodies directed at mouse polyvalent immunoglobulins (Sigma-Aldrich) (store at –20°C).

8. HRP-conjugated goat antibodies directed at rabbit immunoglobulins (Bio-Rad) (store at 2–8°C).
9. Phospholipase A2 from honeybee venom (Sigma-Aldrich) (store at 2–8°C).
10. Bean phytohemagglutinin L (PHA-L, Sigma-Aldrich) (store at 2–8°C).
11. Recombinant avidin expressed in maize (Sigma-Aldrich) (store at 2–8°C).

14.2.2.2. Glycan Analysis by Its Monosaccharide Composition

1. Teflon-faced screw cap: kit flacons 2 mL (Interchim, Montluçon, France).
2. 2 mM Inositol solution in water (store at 4°C).
3. Methanolic-HCl 3 *N* Kit (Supelco, Bellefonte, PA) (store at 2–8°C). Open one ampule of 3 N methanolic-HCl and dilute it with methanol, to have a 1 N methanol-HCl solution (*see* **Note 1**).
4. Acetic acid, anhydrous 98% A.C.S. reagent (Sigma-Aldrich, Aldrich product) (store at room temperature).
5. Pyridine, anhydrous 99.8% (Sigma-Aldrich) (store at room temperature).
6. Silylation reagent: HMDS + TMCS + pyridine, 3:1:9 (Sylon HTP) kit (Supelco) (store at 2–8°C; *see* **Note 1**).
7. Cyclohexane Rectapur (VWR, Fontenay sous Bois, France) (store at room temperature).
8. Gas chromatography apparatus equipped with a silica column CP-SIL 5 CB (25 m × ×0.25 mm).

14.2.2.3. Analysis of the Protein After Glycan Release by N-glycosidases

1. 1% Sodium dodecyl sulfate (SDS).
2. 150 mM Sodium acetate buffer, pH 5.7.
3. Endo H: endoglycosidase H from *Streptomyces plicatus,* recombinant from *Escherichia coli* (Roche Applied Science) (store at 2–8°C; *see* **Note 2**).
4. 0.1 M Tris-HCl buffer, pH 7.5, containing 0.1% SDS.
5. 0.1 M Tris-HCl buffer, pH 7.5, containing 0.5% Nonidet P40.
6. Nonidet P40 (called now IGEPAL; Sigma-Aldrich).
7. PNGase F: peptide-N-glycosidase F from *Flavobacterium meningosepticum* (Roche Applied Science) (store between −15°C and −25°C).
8. 10 mM HCl solution, pH 2.2.
9. C18 column: Bond Elut C18 column (Varian, Palo Alto, CA) (store at room temperature).
10. Acetonitrile, minimum 99.5% (Sigma-Aldrich).

11. 100 mM Sodium acetate buffer, pH 5.5.
12. Trypsin from bovine pancreas (Sigma-Aldrich) (store at –20°C).
13. 50 mM ammonium bicarbonate buffer, pH 8.0.
14. PNGase A: peptide-*N*-glycosidase A from sweet almond (Roche Applied Science) (store between –15°C and –25°C).
15. α-Cyano-4-hydroxycinnamic acid (Applied Biosystems, Foster City, CA).
16. Water/acetonitrile 70/30 (v/v) containing 0.1% of trifluoroacetic acid (TFA).
17. Sequazyme™ Peptide Mass Standards Kit (Applied Biosystem).
18. Enzymatic deglycosylation kit (Prozyme, San Leandro, CA) (store at 4°C).

14.2.2.4. Identification of the N-Glycans After Their Release by N-Glycosidases

1. AG50-WX2 resin (Dowex, Dow Chemical Company, Michigan, USA).
2. C18 column: Bond Elut C18 column (Varian) (store at room temperature).
3. Acetonitrile, minimum 99.5% (Sigma-Aldrich).
4. GlycoClean H cartridges (Prozyme).
5. 2, 5-dihydroxybenzoic acid (Gentisic acid, Sigma-Aldrich).
6. Water/acetonitrile 70/30 (v/v).
7. Standard peptides: des-Arg^1-Bradykinin (904.4681 Da), Angiotensin I (1296.6853), Glu^1-Fibrinopeptide B (1570.6774 Da), ACTH clip 18–39 (2465.1989), and bovine insulin (5730.6087) (Applied Biosystems).

14.2.2.5. N-Glycan Profiling by HPLC

1. TSK Amide-80 normal-phase HPLC column, 4.6 mm × 250 mm (Tosoh Biosep LLC, Montgomeryville, PA).
2. Acetonitrile 212, of low fluorescence grade for HPLC (Caledon, Georgetown, Canada).
3. 2-Aminobenzamide (2AB) solution: 0.35 M 2AB solution in dimethyl sulfoxide/glacial acetic acid (7/3, v/v) containing 1 M sodium cyanoborohydride solution (*see* **Note 3**).
4. 2AB-labeled dextran ladder (Ludger Ltd., Oxford, UK).
5. 3MM chromatographic paper.
6. GlycoClean H cartridges (Prozyme).
7. Solvent A: 50 mM ammonium formate, pH 4.4 (*see* **Note 4**).
8. Solvent B: Acetonitrile
9. $Man_3GlcNAc_2$ standard isolated from Phospholipase A_2 from bee venom as the most abundant *N*-glycan, and further characterized by ESI-MS.

14.2.3. Where Is My Protein Glycosylated?

14.2.3.1. Identification of N-Glycosylation Sites Using MALDI-TOF MS

1. 50 mM Ammonium bicarbonate, pH 8.0.
2. Trypsin: Sequencing grade trypsin from bovine pancreas (Sigma-Aldrich, Fluka product) (store at –20°C).
3. Chymotrypsin: Sequencing grade chymotrypsin from bovine pancreas (Sigma-Aldrich) (store at –20°C).
4. C18 column: Merck C18 Spherisorb ODS 2 column (5 μm, 4.6 × ×250 mm, VWR).
5. Solvent A: 90:10 water/acetonitrile containing 0.1% Trifluoroacetic acid.
6. Solvent B: 10:90 water/acetonitrile containing 0.1% TFA.

14.2.3.2. Glycopeptide Identifications by LC-MS

1. 50% acetonitrile.
2. 0.1 M NH_4HCO_3.
3. 10 mM Dithiothreitol in 0.1 M NH_4HCO_3.
4. Centrifugal evaporator (e.g., SpeedVac).
5. 55 mM iodoacetamide in 0.1 M NH_4HCO_3 (prepare it freshly and protect from light).
6. trypsin stock: 50 ng/μL in 1 mM HCl of sequencing grade trypsin (Roche Applied Sciences) (Store at 4°C).
7. 25 mM NH_4HCO_3.
8. 5% formic acid.
9. ESI-MS with resolution of at least 5,000 (e.g., Waters Q-TOF Ultima, Waters, St Quentin en Yvelines, France).
10. Capillary LC with gradient capability.
11. Nanoflow solvent switch module (e.g., Rheodyne 10-port valve, Varian).
12. Metal spray needle for capillary flow rates.
13. Solvent A: 0.1% formic acid in HPLC grade water.
14. Solvent B: HPLC grade acetonitrile.
15. Aquastar C18 precolumn (30 × 0.32 mm, Thermo Electron Whaltman, MA, USA).
16. Biobasic C18 column (100 × 0.18 mm, Thermo Electron).

14.3. Methods

14.3.1. Is My Protein a Glycoprotein?

Only one approach is available to answer the question globally: is my protein a glycoprotein? It necessitates the use of a general sugar detection kit that allows the detection and quantification of glycoproteins on blot. All the reagents necessary for such a determination are provided in a detection kit, the DIG Glycan

Detection Kit commercialized by Roche Applied Science, which also includes the manufacturer's instructions. It should be noted that the only information the experimenter will get using this kit is that the proteins do bear at least one oligosaccharide moiety. No information will be provided about the type of glycan involved or its linkage to the protein backbone.

14.3.2. How Is My Protein Glycosylated?

Several strategies can be initiated to determine the nature of the glycan borne by a glycoprotein. The first one involves the use of probes specific for glycoprotein *N*-glycan moieties in Western blotting experiments. This strategy requires neither glycoprotein sample purification nor glycan cleavage or separation (*see* **Subheading 14.3.2.1**). Another strategy is to analyze the N-linked glycan directly by its monosaccharide composition, after the glycoprotein isolation (*see* **Subheading 14.3.2.2**). Alternatively, the *N*-glycan can be selectively cleaved from the purified glycoprotein by enzymatical treatment. The glycoprotein is then analyzed by electrophoresis or mass spectrometry before and after deglycosylation (*see* **Subheading 14.3.2.3**). The protein mass difference observed by mass spectrometry will give information on the structure of the N-linked glycans. The determination of the N-linked glycans obtained from deglycosylation can also be investigated by MALDI-TOF mass spectrometry (*see* **Subheading 14.3.2.4**). Eventually, an *N*-glycosylation profiling can be performed by HPLC after the *N*-glycan release from the glycoprotein (*see* **Subheading 14.3.2.5**). For high sensitivity detection and quantification, oligosaccharides have been coupled to various amino fluorescent tags by reductive amination. HPLC and Mass Spectrometry are widely used for unlabeled and 2AB-labeled *N*-glycans identification and their sequencing. These analytical procedures have been successfully carried out for the profiling of natural and recombinant plant glycoproteins, particularly plantibodies (reviewed in **ref.** *3*).

14.3.2.1. Detection of N-Glycans by Western Blotting

The detection of *N*-glycans borne by glycoproteins can be performed after Western blotting of either 1D or 2D electrophoresis gels. The probes used for this detection are of two types: lectins or antibodies.

14.3.2.1.1. Affinodetection with Lectins

Lectins are plant proteins able to bind oligosaccharide moieties specifically. They have been characterized on the basis of their affinity for mammalian glycoproteins. Only a few of them are available for the detection of plant glycans. They are presented in **Table 14.1** (*see* **Note 5**).

These lectins are commercially available in a biotinylated form. The recognition event between the lectin and the oligosaccharide is visualized on a blot using a streptavidin–peroxidase conjugate. An exception is the lectin concanavalin A, which is

Table 14.1
Lectins used for N-linked glycan detection in plant cell extracts in Western blotting

Lectin	Specificity	Detected oligosaccharides in plants	Suggested positive controls	Sugar inhibitor	Ref.
Concanavalin A from *Canavalia ensiformis*	αD – mannose αD – glucose	*Precursor structures:* $Glc_3Man_9GlcNAc_2$ to $Glc_1Man_9GlcNAc_2$ *High-mannose structures:* $Man_9GlcNAc_2$ to $Man_5GlcNAc_2$	Soybean agglutinin Ovalbumin Ribonuclease B	Methyl α-D-mannopyranoside	(4)
Galanthus nivalis agglutinin (GNA)	Terminal α1, 3-mannose	*High-mannose structures*: $Man_8GlcNAc_2$ to Man_5-$GlcNAc_2$	Ribonuclease B	Methyl α-D-mannopyranoside	(*5, 6*)
Wheat germ agglutinin (WGA) from *Triticum aestivum*	Terminal GlcNAc Internal chitobiose units	All *N*-linked glycans	Ovalbumin	GlcNAc (*see* **Note 22**) Chitotriose Chitobiose	(7)

able to bind directly to horseradish peroxidase. The two protocols are detailed here.

The lectin–biotin/streptavidin–peroxidase procedure

1. Separate the proteins to be studied by 1D or 2D electrophoresis and transfer onto a nitrocellulose membrane.
2. Saturate the blot with TTBS buffer for at least 1 h (*see* **Note 6**).
3. Incubate the blot in TTBS containing the lectin–biotin complex (0.1 mg/20 mL) for 2 h at room temperature.
4. Wash four times in TTBS for 15 min each.
5. Incubate the blot with a streptavidin–peroxidase conjugate diluted 1:1,000 in TTBS for 1 h at room temperature.
6. Wash the blot in TTBS (4 × 15 min) and once in TBS before development.
7. Prepare extemporaneously the peroxidase development mixture by dissolving, in a beaker, 30 mg of 4-chloro-1-naphtol in 10 mL cold methanol (–20°C), and mixing in another beaker 30μL H_2O_2 with 50 mL TBS. Mix the two solutions just before pouring onto the membrane, and shake gently to optimize the development reaction.

8. Stop the reaction by discarding the development mixture and rinsing the blot several times with distilled water. Dry the membrane and store between two sheets of filter paper.
9. Suggested controls for the experiment are presented in **Note 7**.

The concanavalin A (ConA)-peroxidase procedure (modified from **ref.** *4*)

1. Separate the proteins to be studied by 1D or 2D electrophoresis, and transfer onto a nitrocellulose membrane.
2. Saturate the blot for at least 1 h in TTBS at room temperature.
3. Incubate the membrane in TTBS* containing ConA (25 μg/mL) for 2 h at room temperature.
4. Wash the membrane four times for 15 min each in TTBS.
5. Incubate the membrane in TTBS containing horseradish peroxidase (HRP, 50 μg/mL) for 1 h at room temperature.
6. Rinse the blot four times for 15 min each in TTBS and once in TBS for 15 min.
7. Develop the peroxidase reaction as described in the lectin–biotin/streptavidin-peroxidase procedure.
8. Controls have to be performed as presented in **Note 7**.

14.3.2.1.2. Immunodetection with Specific Antibodies

We have shown that the β1,2-xylose and the α1,3-fucose epitopes of plant N-linked complex glycans are highly immunogenic in rabbits *(8)*. As a consequence, sera prepared against glycoproteins containing complex glycans generally contain antibodies directed at the β1,2-xylose and/or the α1,3-fucose. Some commercially available immunsera present the same characteristics as our home-made probes. For instance, immunsera directed against horseradish peroxidase (HRP) can be used as a probe specific for plant complex *N*-glycans with α1,3-fucose and β1,2-xylose residues. More specific is the immunserum raised against honeybee venom proteins, which can be used as a specific probe for α1,3-fucose-containing plant complex *N*-glycans *(8)*. We also raised in our laboratory rabbit antibodies specific for the terminal antennae of complex *N*-glycans, the trisaccharide Lewis a *(2)*. Anti-Lewis a monoclonal antibodies are commercially available, although they are very expensive.

1. Separate the proteins by 1D or 2D electrophoresis and transfer onto a nitrocellulose membrane.
2. Saturate the blot with 3% gelatin prepared in TBS for at least 1 h at room temperature.
3. Incubate the blot in TBS containing 1% gelatin and the immunserum at a convenient dilution for 2 h at room temperature. We used the commercial immunsera described above in **Subheading 14.2.2.1.2** at the following dilutions:

a. Anti-HRP rabbit immunserum, 1:300.

b. Antihoneybee venom rabbit immunserum, 1:200.

c. Anti-Lewis a mouse monoclonal antibody, 1:100.

4. Wash the blot with TTBS buffer four times for 15 min each.
5. Incubate the blot in TBS containing 1% gelatin and the suitable conjugate, that is, a goat antirabbit IgG conjugate coupled to HRP at a dilution of 1:2,000, or HRP-conjugated goat antibodies directed at mouse polyvalent immunoglobulins at a dilution of 1:500 for 1 h at room temperature.
6. Wash the blot four times with TTBS for 15 min each and once in TBS before development.
7. Develop the peroxidase reaction as described for the lectin–biotin/streptavidin–peroxidase procedure (*see* **Subheading 14.3.2.1.1**).
8. For controls, *see* **Notes 8** and **9**.

14.3.2.2. N-Glycan Analysis by Its Monosaccharide Composition

The analysis of the *N*-glycan monosaccharide composition is carried out by hydrolysis with methanol-HCl, followed by derivatization and analysis of the resulting monosaccharide(s) using gas chromatography (*see* **Note 10**).

1. Freeze-dry the sample containing 10 μg of purified protein into a glass reaction tube equipped with a Teflon-faced screw cap.
2. Add 5 to 10 μL of a 2 mM inositol stock solution to the protein under analysis (*see* **Note 11**). Then freeze-dry the sample again before methanolysis.
3. Add 500 μL of 1 N methanol-HCl to the sample, close the tube tightly with the Teflon-faced screw cap, and place the tube overnight at 80°C.
4. Dry the sample under nitrogen atmosphere after methanolysis while heating the sample at 40°C. **Caution:** Work in a hood because of the methanol toxicity.
5. Wash with 250 μL of methanol. Dry under nitrogen atmosphere as in **step 4** and repeat this washing step twice.
6. Resuspend the sample in 250 μL of methanol. Then add 25 μL of acetic acid, 25 μL of pyridine, homogenize, and incubate for 6 h at room temperature. Dry under nitrogen atmosphere as in **step 4**.
7. Add 250 μL of silylation reagent and incubate at 80°C during 20 min (*see* **Note 1**). Air-dry.
8. Wash with 1 mL of cyclohexane. Air-dry and resuspend in 200 μL of cyclohexane, homogenize, and centrifuge before transferring 100 μL of the derivatized sample in a reaction vial with screw cap.

9. Set the gas chromatography apparatus to 1.4 bars with helium. Before injection, heat the injector and FID detector to 250°C and 280°C, respectively. Set the helium pressure to 20 psi and its flow rate to 3 mL/min. Equilibrate the column temperature to 120°C before the injection of 5 μL of the derivatized sample. Eluate the derivatized monosaccharides by a temperature gradient: (1) +10°C/min until 160°C; (2) +1.5°C/min until 235°C; (3) 235°C for 4 min; (4) +20°C/min until 280°C, and, finally, (5) 280°C for 3 min.

14.3.2.3. Analysis of the Protein After Glycan Release by N-glycosidase

Further information can be deduced by treating the purified glycoprotein with glycosidases. Endoglycosidase H (Endo H) is only able to release high-mannose-type *N*-glycans from plant glycoproteins by hydrolyzing the glycosidic bond between the two GlcNAc residues on the core of the *N*-glycan (**Fig. 14.2**). Peptide-N-glycosidases (PNGases) hydrolyze the bond between the Asn of the peptide backbone and the proximal GlcNAc of the oligosaccharide part, for both high-mannose-type *N*-glycans and complex-type *N*-glycans. However, PNGase F, which is widely used in the analysis of mammalian glycoproteins, is active on high-mannose and complex plant *N*-glycans, except those presenting an α1,3-fucose residue linked to the proximal GlcNAc. PNGase A is able to release all types of plant *N*-glycans (**Fig. 14.2**), but it is almost only efficient on glycopeptides and necessitates the proteolytic digestion of the glycoprotein before deglycosylation *(9, 10)*. The deglycosylated protein or peptides can be analyzed for (1) an increased electrophoretic mobility, (2) the loss of glycoprotein reactivity on blots with glycan-specific probes after Endo H or PNGase F treatment, or (3) mass spectrometry analysis after Endo H, PNGase F, or PNGase A treatment.

14.3.2.3.1. Deglycosylation with Endoglycosidase H

1. Before enzyme deglycosylation with Endo H, denature the purified protein by heating it to 100°C for 5 min in the presence of 1% SDS (w/v).
2. Dilute the sample fivefold with 150 mM sodium acetate, pH 5.7, and add then 10 mU Endo H per mg of protein to be analyzed. Incubate the mixture at 37°C for at least 6 h.
3. After the deglycosylation reaction, add an equal volume of twice concentrated electrophoresis sample buffer if the result of the digestion is followed by electrophoresis gel, affino-, or immunodetection (*see* **Subheading 14.3.2.1**). The sample can also be desalted and analyzed by mass spectrometry.

14.3.2.3.2. Deglycosylation with PNGase F

1. Dissolve the purified protein to be digested in 0.1 mM Tris-HCl, pH 7.5, containing 0.1% SDS.
2. Heat the sample for 5 min at 100°C to denature the protein. Let it cool down at room temperature and add an equal

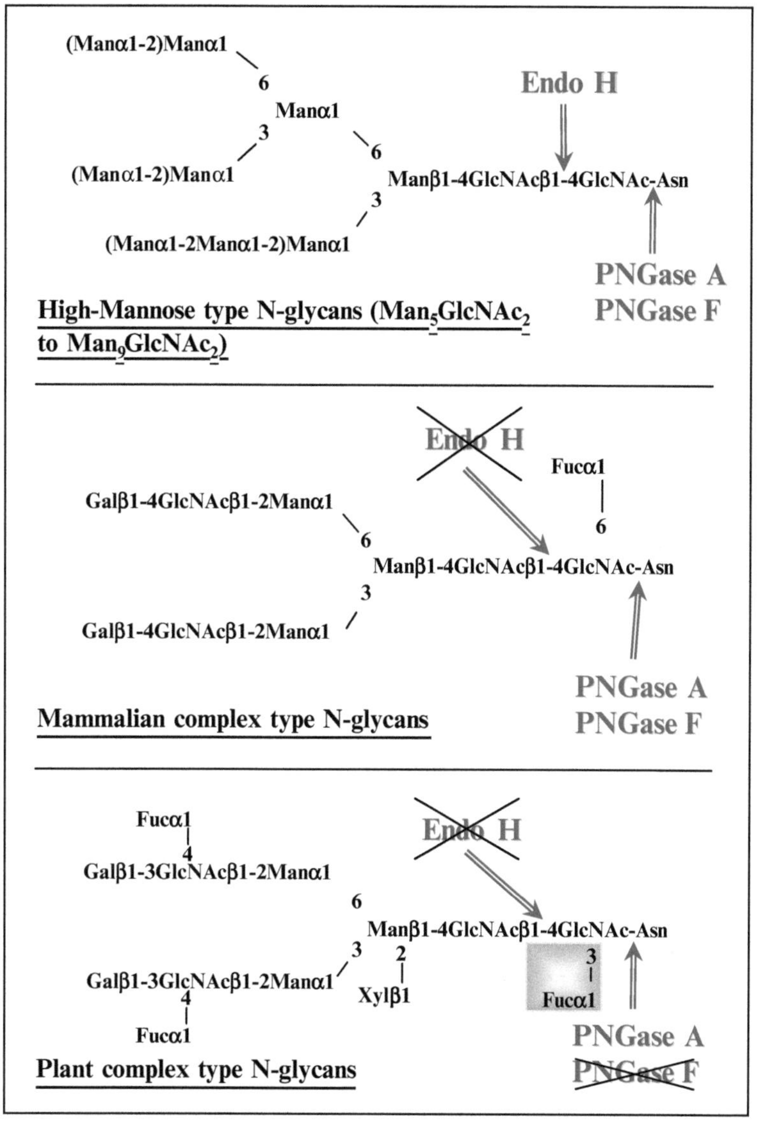

Fig. 14.2. Enzymatic deglycosylation of proteins bearing N-linked glycans. High-mannose-type *N*-glycans can be cleaved by either endoglycosidase H (Endo H), peptide-*N*-glycosidase A (PNGase A), or peptide-N-glycosidase F (PNGase F). Mammalian complex-type *N*-glycans can be removed using PNGase A and PNGase F. Plant complex-type *N*-glycans can only be released by PNGase A, owing to the presence of α1,3-fucose linked to the core. It should be noted that PNGase A mostly acts on glycopeptides and is not efficient on a complete glycoprotein.

volume of 0.1 mM Tris-HCl, pH 7.5, containing 0.5% Nonidet P-40 (*see* **Note 12**).

3. Incubate the sample with PNGase F (1 U enzyme for 100 μg protein) at 37°C during 24 h.
4. After digestion, precipitate, by 4 vols of cold ethanol, the deglycosylated protein overnight at –20°C or twice with shorter incubation time.

5. Wash the pellet two to three times with cold ethanol.
6. Centrifuge the sample. Recover the deglycosylated protein in the pellet and dissolve it in a suitable buffer for gel electrophoresis or mass spectrometry analysis.

14.3.2.3.3. Deglycosylation with PNGase A

1. Dissolve 100 µg of purified protein in 500 µL of 50 mM ammonium bicarbonate, pH 8.0, and heat for 3 min at 100°C. Add, then, two aliquots of 5 µg each of trypsin successively. After addition of each aliquot of enzyme, incubate the sample at 37°C for 2 h and overnight, respectively. Finally, stop the digestion by heating for 5 min at 100°C *(11)* (*see* **Note 13**).
2. Cool the sample down and take 10% of the solution to purify the peptide and glycopeptide mixture on a C18 column according to the following steps.
3. First rinse the C18 column with 5 mL of acetonitrile followed by 5 mL of water. Apply the diluted sample to 1 mL final volume with distilled water. Rinse the column with 5 mL of water to remove salts from the sample, and elute the peptides that bind to the column by using 5 mL of acetonitrile. Concentrate the peptide sample by evaporating the acetonitrile. Analyze the mass of the peptides and glycopeptides by MALDI-TOF mass spectrometry, to get the glycopeptide masses before PNGase A treatment.
4. Freeze-dry the sample left over and dissolve it in 500 µL of 100 mM sodium acetate, pH 5.5.
5. Incubate at 37°C for 18 h with 0.1 mU PNGase A. After digestion, freeze-dry the sample and separate the peptides from the released oligosaccharides using a C18 column, as explained above in **steps 2** and **3**. Analyze the deglycosylated peptides by MALDI-TOF mass spectrometry.
6. Dissolve the deglycosylated peptides in water and add the same volume of a fresh solution of α-cyano-4-hydroxycinnamic acid, used as matrix. The matrix is dissolved at 10 mg/mL in a water/acetonitrile 70/30 (V:V) containing 0.1% TFA. Drop 2 µL of the sample–matrix mixture onto the MALDI plate and let them co-crystallize at room temperature or using a vacuum room.
7. When the sample–matrix mixture is dry, analyze it by MALDI-TOF mass spectrometry using, for example, a Voyager DE-Pro MALDI-TOF instrument (Applied Biosystems) equipped with a 337-nm nitrogen laser. Using this instrument, deglycosylated peptide mass spectra are acquired in the reflector, delayed positive mode with an accelerating voltage of 20,000V, a grid voltage of 80%, a mirror voltage ratio 1.12, a guide wire of 0.002%, an extraction delay time

of 150 ns and laser intensity around 900. Laser shots are accumulated for each spectrum to obtain an acceptable signal to noise ratio. The spectra are externally calibrated using Sequazyme™ Peptide Mass Standards Kit.

The comparison of the two spectra, obtained before and after the deglycosylation treatment, provides a mass difference for some ions owing to the removal of N-linked glycans. This mass difference, in addition to the enzyme used for the deglycosylation procedure and the knowledge of plant glycosylation, indicates which type of glycan is N-linked to the protein and also gives an idea of the structure of this glycan (**Table 14.2**). All the reagents necessary for such a determination are included in a detection kit, the enzymatic deglycosylation Kit commercialized by Prozyme.

14.3.2.4. Analysis of N-Glycans by MALDI-TOF After Their Release by N-Glycosidase

Before *N*-glycan profiling by MALDI-TOF mass spectrometry, the *N*-glycans need to be removed from the glycoprotein of interest (*see* **Subheading 14.3.2.3**, deglycosylation by Endo H, PNGase F or A). The *N*-glycans are then purified using several steps of chromatography.

1. Deglycosylate of the glycoprotein to remove *N*-glycans.
2. Desalt then the sample using a cation exchange column. Therefore, prepare a home-made 2 mL column using AG50-WX2 resin. Wash abundantly the column with water. Dissolve the sample in 0.5 mL of water and run it through the column. Wash the column with water and keep the nonretained fraction which does contain the *N*-glycans. Neutralize this fraction by addition of 150 µL of ammonium hydroxide 4 M. Freeze-dry the sample.
3. Separate the *N*-glycans from peptides using a C18 column as previously described (*see* **Subheading 14.3.2.3**). Keep the nonretained fraction containing the *N*-glycans.
4. Purify the *N*-glycans using a GlycoClean H cartridge. Equilibrate the GlycoClean H cartridge with 3 mL of 50% (v/v) acetonitrile containing 0.1% TFA, and 6 mL of 5% (v/v) acetonitrile containing 0.1% TFA. Apply the sample and wash the column, first with 3 mL of distilled water and, then, with 3 mL of 5% (v/v) acetonitrile containing 0.1% TFA. Elute the *N*-glycans with 4 × 0.5 mL of 50% (v/v) acetonitrile containing 0.1% TFA. Mix the four fractions and lyophilize.
5. Acquire then the *N*-glycan profiles on a Voyager DE-Pro MALDI-TOF instrument (Applied Biosystems) equipped with a 337-nm nitrogen laser. Mix the matrix, 2,5-dihydroxybenzoic acid (also called gentisic acid), freshly dissolved at 5 mg/mL in a 70/30 (v/v) water/acetonitrile, with water containing 0.1% trifluoroacetic acid and the

Table 14.2

Structures and theoretical (M + Na)⁺ values of high-mannose-type *N*-glycans ($Man_5GlcNAc_2$ to $Man_9GlcNAc_2$) and complex-type *N*-glycans isolated from plants

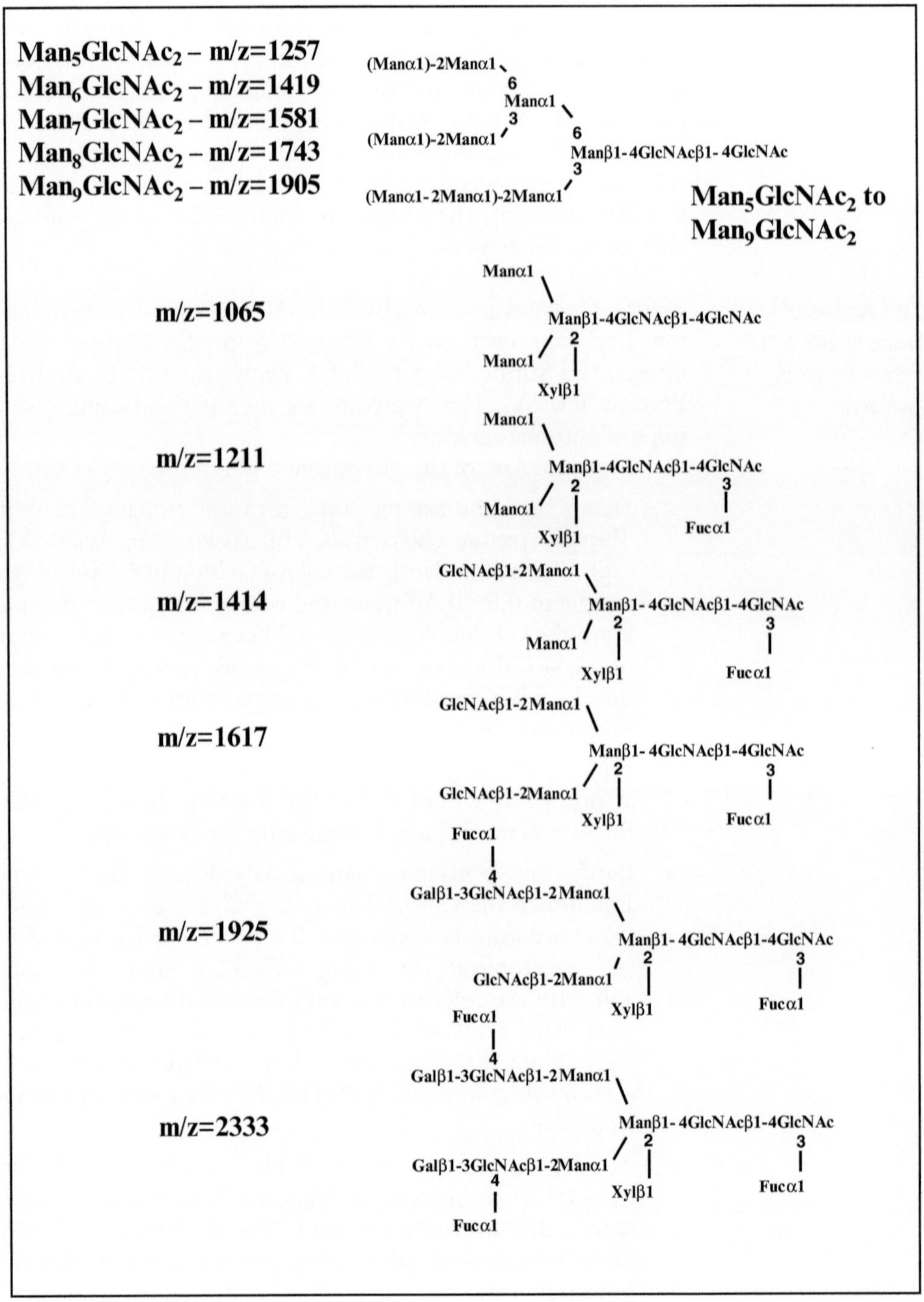

Values indicated on the left correspond to the *m/z* (mass/charge) ratio calculated for the (M + Na)⁺ adducts that can be observed when *N*-glycans are analyzed by MALDI-TOF mass spectrometry.

solubilized oligosaccharides, in a ratio 1:1 (v/v). Drop then 1–2 μL of this mixture onto the MALDI plate before crystallization, which can be done at room temperature or under a vacuum room. Record mass spectra in the reflector delayed positive extraction mode, using an acceleration voltage of 20,000 V with a delay time of 100 ns, a grid voltage of 80%, a mirror voltage ratio 1.12, a guide wire of 0.002%, and laser intensity around 1,300. Accumulate laser shots for each spectrum to obtain an acceptable signal to noise ratio. Smooth spectra once and externally calibrate them using commercially available mixtures of peptides and proteins (Applied Biosystems). The MALDI-TOF mass spectra of *N*-glycans can be calibrated using des-Arg^1-Bradykinin (904.4681 Da), Angiotensin I (1296.6853), Glu^1-Fibrinopeptide B (1570.6774 Da), ACTH clip 18–39 (2465.1989), and bovine insulin (5730.6087), which cover the entire mass range of plant *N*-glycans.

6. Refer **Table 14.2** to associate determined *N*-glycan masses to *N*-glycan-known structures.

Confirmation of *N*-glycan structures can be obtained by coupling MALDI-TOF *N*-glycan profiling and exoglycosidase sequencing *(12)*.

14.3.2.5. N-Glycan Profiling by HPLC

Several aromatic amines have been used as fluorescent tags for labeling *N*-glycans enzymatically released (either by PNGase F or A) from glycoproteins of different origins and applied with success for the characterization of plant glycoprotein *N*-glycans, for example, 2-aminopyridine (2AP) *(13, 14)*, 8-amino-1,3, 6-naphthalene trisulfonic acid (ANTS) *(15)*, and 2AB *(16, 17)*. 2AB-labeled *N*-glycan profiling is carried out by normal-phase HPLC. Retention times are expressed as Glucose Units (GU) referred to a series of isomalto-oligosaccharides derived from Dextran hydrolysis, called Dextran ladder *(18)*. The order of elution of each *N*-glycan is related to the number of monosaccharide residues, thus the larger the oligosaccharide, the higher the retention time and the GU value. GU values are suitable for structure prediction taking into consideration that each monosaccharide and specific linkage contribute with a given amount to the entire *N*-glycan GU. Normal-phase HPLC profile could be sufficient when comparing *N*-glycosylation patterns of IgG produced from different sources (from mammalian cell cultures or even plantibodies) since they have been well characterized in several reports. If further confirmation of any unknown structure is required, exoglycosidase sequencing and mass spectrometry is recommended *(19, 20)*. Moreover, glycoprotein *N*-glycan profiling can be done in SDS-PAGE gel bands using HPLC combined with exoglycosidase array digestions *(21)*.

14.3.2.5.1. Isolation of N-linked Glycans

It is advisable to purify the *N*-glycans after deglycosylation reaction to eliminate salts and other compounds presents. For unlabeled *N*-glycans good results has been achieved by sequential separation in C18 (*see* **Subheading 14.3.2.3**) and GlycoClean H cartridges (*see* **Subheading 14.3.2.4**) (*see* **Note 14**).

14.3.2.5.2. Preparation of fluorescently –labeled oligosaccharides

1. Dissolve *N*-glycans in 5 μL of 2AB solution. Vortex and spin down.
2. Incubate the labeling mixture at 60°C for 2 h.
3. The reaction mixture is spotted onto a strip of 3MM chromatographic paper 10 cm length and 3 cm width (*see* **Note 15**).
4. Place the paper strip in a small chromatographic vessel assuring the solvent does not touch the application point. Perform the ascending paper chromatography using acetonitrile at room temperature for about 4 h.
5. Remove the paper strip from the vessel and let it air dry. Use a UV lamp to check if separation of reagent excess from labeled *N*-glycans was successfully accomplished.
6. Elute the labeled *N*-glycans from the paper by cutting the application point, placing it in a PCR tube, and adding 500 μL distilled water twice and mixing.
7. Spin down, remove the solution and lyophilize or dry the labeled *N*-glycans in speed vac.

14.3.2.5.3. Normal-Phase HPLC of the N-glycan 2AB-derivatives

This step should be carried out in an HPLC system equipped with fluorescence detector. TSK-GEL Amide-80 column (4.6 × 250 mm) is suitable for normal-phase separation of 2AB-labeled *N*-glycans.

1. Prepare the sample for injection by dissolving the lyophilized mixture of 2AB-labeled *N*-glycans or isomalto-oligosaccharides (dextran ladder) in 80% aqueous acetonitrile (*see* **Note 16**). The sample volume depends on the amount of *N*-glycans in it, but 100 pmoles of labeled isomalto-oligosaccharides gives a nice chromatogram.
2. Set the fluorescence detector excitation and emission wavelength to 330 nm and 420 nm, respectively.
3. Set the initial gradient conditions to 20% solvent A in solvent B at a flow rate of 0.4 mL/min. Run a blank (no sample injection) in a linear gradient of 20–53% solvent A in solvent B in 132 min. Pass then from 53% to 100% solvent A in the next 3 min. Immediately after the second gradient program, increase the flow rate to 1.0 mL/min in 2 min. Wash then the column with 100% solvent A for 5 min before returning to the initial conditions within a minute. Solvants are A: 50 mM ammonium formate, pH 4.4, and B: acetonitrile.

4. Run the samples using the same conditions as described in **step 3** (*see* **Notes 17** and **18**).

14.3.2.5.4. Calculating the Retention Times as GU Values

Structure assignment is done by comparison of the experimental GU values with those reported in the GlycoBase from the Dublin-Oxford GlycoBiology Laboratory (http://glycobase.ucd.ie/cgi-bin/public/glycobase.cgi).

1. Calibrate daily your HPLC system by running the Dextran ladder and a standard of known GU value. This avoids problems due to column deterioration, solvent preparation, or any troubles related to pump functioning.
2. The most reliable GU values are from 1 to 15 that corresponds to peak 1 to peak 15 in the chromatogram of the dextran ladder.
3. Verify your system by determining the experimental GU value of the standard, for example, $Man_3GlcNAc_2$-2AB (GU = 4.40). The allowed standard deviation of this *N*-glycan is 0.02 GU.
4. Determine the experimental GU values of your sample by interpolation of its retention time with respect to the Dextran ladder.
5. For structure prediction, compare the experimental GU values of the working sample with the GlycoBase (*see* **Note 19**). An example of the *N*-glycan analysis of a plantibody fused to endoplasmic retention signal tetrapeptide (KDEL) is shown in **Fig. 14.3**.

14.3.3. Where Is My Protein Glycosylated?

This approach will allow the experimenter to get information about both the distribution of the *N*-glycans on the protein backbone and the structures of the glycans themselves. These experiments can be performed on a purified glycoprotein or on a protein isolated from a 2D or 1D electrophoresis gel. In a first approach, the protein is digested by endoproteases, and the resulting peptide and glycopeptide mixture is separated by HPLC. The presence of glycans within the collected fractions is analyzed by sugar composition. The glycopeptide fractions are then analyzed by MALDI-TOF mass spectrometry before and after PNGase A digestion (*see* **Subheading 14.3.3.1**) *(22, 23)*. In a second method, the peptides and glycopeptides are detected by ESI-MS on-line with HPLC. If a capillary or nano-flow LC is used, this approach provides very high sensitivity (*see* **Subheading 14.3.3.2**).

It should be noted that these procedures are only applicable to glycoproteins for which the primary amino acid structure is known, as it is always the case for plant pharmaceuticals. The structure of the *N*-glycans released by PNGase A can also be analyzed by MALDI-TOF using specific procedures described before (*see* **Subheading 14.3.2.4**) *(22, 23)*.

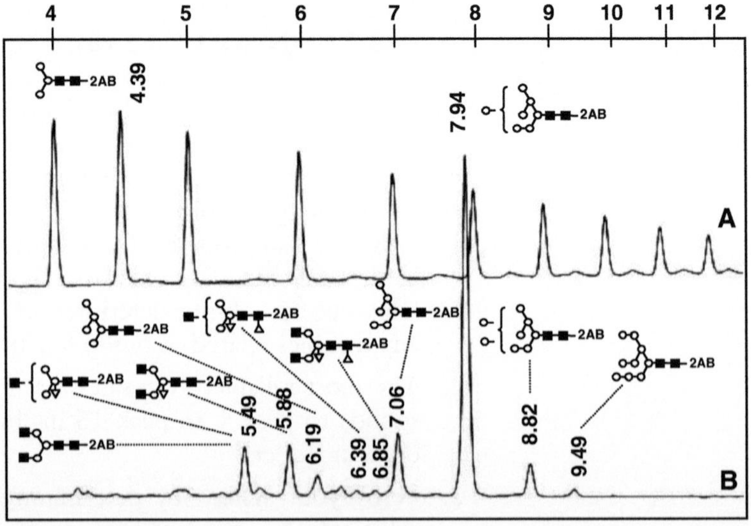

Fig. 14.3. NP-HPLC profile of the N-glycan pool from Plantibody fused to endoplasmic reticulum-retention signal expressed in tobacco plants. A Dextran ladder + $Man_3GlcNAc_2$-2AB standard (internal standard, GU = 4.39). B Plantibody *N*-glycan profile with preliminary structure assignment from the experimental GU values. ***GU*** Glucose Units.

14.3.3.1. Identification of N-Glycosylation Sites by MALDI-TOF Mass Spectrometry

1. Dissolve 100 µg of purified protein in 500 µL of 50 mM ammonium bicarbonate, pH 8.0, and heat for 3 min at 100°C.
2. Add 5 µg of TPCK-treated trypsin from bovine pancreas and carry out the digestion for 2 h, at 37°C.
3. Add a second aliquot of 5 µg of the enzyme and incubate for additional 2 h at 37°C.
4. To perform a double digestion, dissolve 5 µg of TLCK-treated chymotrypsin in 50 mM ammonium bicarbonate, pH 8.0, and add them to the trypsin-digested sample. Incubate the sample at 37°C for 2 h.
5. Incubate for 5 min at 100°C to stop the protease digestion.
6. Separate peptides and glycopeptides generated by the trypsin/chymotrypsin digestion by reverse-phase HPLC (C18 column) using a 60-min linear gradient from 0% to 60% of solvent B and a flow rate of 1 mL/min with eluant peaks monitored by UV at 214 nm. Solvents are A: 90:10 water/acetonitrile containing 0.1% TFA, and B: 10/90 water/acetonitrile containing 0.1% TFA. Collect 2-mL fractions and freeze-dry them.
7. Analyze an aliquot (10%) of each collected fraction for sugar composition (*see* **Subheading 14.3.2.2**) and select the fractions containing oligosaccharides.

8. Analyze the HPLC fractions containing the glycopeptides by MALDI-TOF to obtain the glycopeptide masses.
9. Digest the glycopeptides with PNGase A as previously described (*see* **Subheading 14.3.2.3.2**). Separate the peptides from their oligosaccharide moiety on a C18 column as described above (*see* **Subheading 14.3.2.3.2**).
10. Analyze the resulting de-glycosylated peptides by MALDI-TOF mass spectrometry to determine their masses (*see* **Subheading 14.3.2.3, step 7**, de-glycosylation with PNGase A). These masses will allow the experimenter to identify the occupied glycosylation site(s) of the studied protein. The mass difference observed between glycosylated and de-glycosylated peptides will provide structural information about the glycans borne by the protein (**Table 14.2**).

14.3.3.2. Glycopeptide Identification by LC-MS

1. Subject 5 μg of glycoprotein to standard SDS-PAGE followed by Coomassie blue staining (*see* **Note 20**). Excise the band corresponding to the glycoprotein of interest and transfer it to a 1.2 mL reaction vial.
2. Add 50 μL of 50% acetonitrile, incubate 15 min at room temperature and discard liquid.
3. Repeat **step 2**.
4. Add 50 μL acetonitrile, gently shake the preparation (*see* **Note 21**), and discard liquid.
5. Add 30 μL 0.1 M NH_4HCO_3, and allow the piece of gel to re-swell for 5 min.
6. Add 30 μL acetonitrile (eventually yielding a 50% acetonitrile-solution), incubate for 15 min at room temperature, and discard liquid.
7. Dry the piece of gel in a speed vac evaporator.
8. Add 50 μL 10 mM DTT in 0.1 M NH_4HCO_3. Verify that the piece of gel is covered with buffer and let it re-swell for 5 min. Incubate for 45 min at 56°C and discard liquid.
9. Add 50 μL 55 mM iodoacetamide in 0.1 M NH_4HCO_3, incubate for 30 min at room temperature in the dark, and discard liquid.
10. Repeat the procedure from **step 2** to **step 7**.
11. Add 15 μL of 15 ng trypsin/μL in 20 mM NH_4HCO_3 to the dried gel piece and let it re-swell for 5 min. Add 30 μL of 20 mM NH_4HCO_3 and incubate at 37°C, overnight.
12. After tryptic digestion, add 30 μL of 20 mM NH_4HCO_3 and extensively shake for 15 min at room temperature.

13. Add 30 μL of acetonitrile and continue shaking for 15 min at room temperature.
14. Carefully transfer the resulting supernatant into a fresh vial. Add 30 μL of 5% formic acid to the gel piece and continue shaking for 15 min at room temperature.
15. Carefully combine the resulting supernatant with the previous one. Add again 30 μL of 5% formic acid to the gel piece and continue shaking for 15 min at room temperature.
16. Carefully combine the resulting supernatant with the previous one and dry it in a speed vac evaporator.
17. Dissolve the sample in 8 μL of 0.1% formic acid.
18. Inject 4 μL of the sample onto the pre-column of the ESI-MS apparatus, which has been previously equilibrated with 5% acetonitrile containing 0.1% formic acid, at a flow rate of 2 μL/min.
19. Rinse the pre-column with 0.1% formic acid for 6 min.
20. Switch the flow so that the effluent from the pre-column goes to the analytical column. The analytical column has been already rinsed with 5% acetonitrile containing 0.1% formic acid, at a flow rate of 2 μL/min.
21. Run a gradient from 5% to 50% acetonitrile in 0.1% formic acid at a flow rate of 2 μL/min. The mass spectrometer must be set with an ion source voltage of 3 kV, a desolvatation temperature of 80°C, and a source temperature of 120°C. Choose a mass range between 150 and 1,800 for the data.
22. Calculate theoretical monoisotopic masses of glycopeptides. The monoisotopic masses of theoretical tryptic peptides of a given protein can be calculated using the program PeptideMass (http://www.expasy.org/tools/peptide-mass.html). Glycopeptide masses are then calculated by adding the monoisotopic masses of the respective monosaccharides (e.g., hexoses 162.0528 Da, *N*-acetyl-hexosamines 203.0794 Da, desoxy-hexoses 146.0579 Da, pentoses 132.0423 Da, *N*-acetyl-neuraminic acid 291.0954 Da) to the peptide mass.
23. Search for ions of relevant masses and appropriate charge states in the mass spectrum.
24. Sum the MS spectra in the relevant time frame of the identified peak.
25. Identify further glycoforms of the glycopeptide within the summed spectrum, using masses calculated in **step 22**.

The glycopeptide identification can be facilitated by running the same sample after deglycosylation with PNGase A (*see* **Subheading 14.3.2.3.2**). The nonglycosylated peptide elutes slightly after its glycosylated siblings.

14.4. Notes

1. Use fresh ampules of 3 N methanolic-HCl and silylation reagent for each experiment.
2. Do not freeze Endo H as this inactivates the protein.
3. The solution must be prepared fresh daily.
4. Prepare the ammonium formate buffer by adjusting formic acid solution to the required pH using ammonia solution and diluting to give the appropriated concentration in the final solution.
5. The experimenter should be careful in the blot interpretation, as WGA recognizes GlcNAc in N-linked glycans, as well as *O*-linked GlcNAc.
6. The solution used for blocking binding sites on the nitrocellulose needs to be devoid of glycoproteins. This is why we recommend using Tween-20 to coat the nitrocellulose *(24)* in this procedure.
7. Controls for specificity
   a. It might be advisable to blot the proteins cited in **Table 14.1** on the membrane ("suggested positive controls"), to get positive controls for the affinodetection.
   b. Lectin binding specificity must also be checked by running affinodetection in the presence of 0.3 M inhibitory sugar (**Table 14.1**).
8. Control for *N*-glycan specificity of the immunodetection: it might be wise to check for the specificity of sera toward *N*-glycans attached to the protein studied. This can be done by performing a mild periodate oxidation on the blot before immunodetection. Mild periodate treatment oxidizes glycans and abolishes any recognition of the glycoprotein by the antiglycan antibodies. Any remaining signal will be the consequence of a protein backbone antibody/recognition *(25)*.
   a. After saturation with gelatin, incubate the blot in a 100 mM sodium acetate buffer, pH 4.5, containing 100 mM sodium metaperiodate for 1 h in the dark at room temperature, changing the incubation solution after 30 min.
   b. Incubate the blot in PBS containing 50 mM sodium borohydride for 30 min at room temperature.
   c. Rinse the blot with TBS, saturate for 15 min with TBS containing 1% gelatin, and perform the immunodetection, as described in **Subheading 14.3.2.1.1**.
9. Controls for fucose or xylose specificity: some proteins can be used as positive controls for the *N*-glycan immu-

nodetection. Phospholipase A2 from honeybee venom contains the α1,3-fucose residue, and is devoid of β1,2-xylose *(26)*. PHA-L and recombinant avidin produced in maize are glycoproteins containing both β1,2-xylose and α1,3-fucose *(27, 28)*.

10. Caution: The methanol-HCl hydrolysis cleaves equally *N*- and *O*-linked glycans.
11. Inositol is used as an internal standard.
12. The role of Nonidet P40 is to complex free SDS.
13. Because of possible contamination of the plant *N*-glycans with mammalian-type oligosaccharides coming from the pepsin, which was historically used, pepsin digestion at this step has been substituted by other protease digestion such as trypsin *(11)*.
14. Practical experience has shown that many unexplained peaks in the chromatograms are easily eliminated if a cleaning step is introduced before the labeling reaction is accomplished. Deglycosylation with PNGase A generates a huge amount of peptides that should be removed to get as little interference as possible when running the NP-HPLC chromatogram.
15. The paper strips should be recently prepared by rinsing three times in water and drying in an oven at 60°C. Do not store paper strips. Alternatively some authors recommend using GlycoClean S cartridges instead of 3MM paper chromatography. As low as 25–50 pmoles of *N*-glycans can be seen when viewing with UV lamp.
16. Always use Acetonitrile for HPLC grade of low fluorescence grade. Quantitation of the *N*-glycans is easily carried out by peak integration since only one 2AB molecule (fluorophore) is linked to the *N*-glycan reducing terminal, that is, there is a linear response of the intensity of florescence with the molar concentration of a given *N*-glycan, no matter the size and complexity of the glycan residue.
17. For accurate GU value determination, it is always (daily) advisable to run the dextran ladder, an *N*-glycan standard of known GU ($Man_3GlcNAc_2$-2AB, GU = 4.40) and the samples (**Fig. 14.3**).
18. Taking advantage of PNGase F to cleavage α1,3-fucose-containing *N*-glycans the comparison of the HPLC profiles of the *N*-glycan pools after individual deglycosylations with PNGase F and A will give relevant information about fucosylated and afucosylated *N*-glycans in the glycoprotein under study.
19. GU values of neutral *N*-glycans are very reproducible. However, sialylated oligosaccharides show higher variations; standard deviations of 0.3 are acceptable.

20. It is better to use Colloidal Coomassie blue G-250 for the Coomassie blue staining.
21. The gel pieces shrink in size and become white upon acetonitrile washing.
22. Free GlcNAc is a ligand of WGA and will displace the bound glycoproteins from the immobilized lectin.

Acknowledgements

We thank present and former colleagues from our lab who contributed to the development and optimization of the strategies described in this manuscript. Work on glycobiology at the University of Rouen is supported by the French Ministère de la Recherche and the Centre National de la Recherche Scientifique (CNRS). J.A.C. and G.C. gratefully thank the support and collaboration between Biomedical Research and AgroBiotech Directions from the Center for Genetic Engineering and Biotechnology, Havana, Cuba.

References

1. Saint-Jore-Dupas, C., Faye, L. and Gomord, V. (2007) From planta to pharma with glycosylation in the toolbox. Trends Biotechnol. 25, 317–323.
2. Fitchette-Lainé, A.-C., Gomord, V., Cabanes, M., Michalski, J.-C., Saint Macary, M., Foucher, B., Cavelier, B., Hawes, C., Lerouge, P. and Faye, L. (1997) *N*-glycans harboring the Lewis a epitope are expressed at the surface of plant cells. Plant J. 12, 1411–1417.
3. Bardor, M., Cabrera, G., Rudd, P.M., Dwek, R.A., Cremata, J.A. and Lerouge, P. (2006) Analytical strategies to investigate plant *N*-glycan profiles in the context of plant-made pharmaceuticals. Curr. Opin. Struct. Biol. 16, 576–583.
4. Faye, L. and Chrispeels, M.J. (1985) Characterization of *N*-linked oligosaccharides by affinoblotting with concanavalin A-peroxidase and treatment of the blots with glycosidases. Anal. Biochem. 149, 218–224.
5. Shibuya, N., Goldstein, I.J., van Damme, E.J.M. and Peumans, W.J. (1988) Binding properties of a mannose-specific lectin from the snowdrop (*Galanthus nivealis*) bulb. J. Biol. Chem. 263, 728–734.
6. Animashaun, T., Mahmood, N., Hay, A.J. and Hughes, R.C. (1993) Inhibitory effect of novel mannose-binding lectins on HIV-infectivity and syncitium formation. Antiv. Chem. Chemother. 4, 145–153.
7. Nagata, Y. and Burger, M.M. (1974) Wheat germ agglutinin. Molecular characteristics and specificity for sugar binding. J. Biol. Chem. 249, 3116–3122.
8. Faye, L., Gomord, V., Fitchette-Lainé, A.-C. and Chrispeels, M.J. (1993) Affinity purification of antibodies specific for Asn-linked glycans containing $\alpha 1->3$ fucose or $\beta 1->2$ xylose. Anal. Biochem. 209, 104–108.
9. Tomiya, N., Kurono, M., Ishihara, H., Tejima, S., Endo, S., Arata, Y. and Takahashi, N. (1987) Structural analysis of *N*-linked oligosaccharides by a combination of glycopeptidase, exoglycosidases, and highperformance liquid chromatography. Anal. Biochem. 163, 489–499.
10. Ogawa, H., Hijikata, A., Amano, M., Fukushima, H., Ishizuka, I., Kurihara, Y. and Matsumoto, I. (1996) Structures and contribution to the antigenicity of oligosaccharides of Japanese cedar (*Cryptomeria japonica)* pollen allergen *Cry j* I: relationship between the structures and antigenic epitopes of plant *N*-linked complex-type glycans. Glycoconj. J. 13, 555–566.

11. Séveno, M., Cabrera, G., Trigero, A., Burel, C., Leprince, J., Rihouey, C., Vézina, L.-P., D'aoust, M.-A., Rudd, P.M., Royle, L., Dwek, R.A., Harvey, D.J., Lerouge, P., Cremata, J.A. and Bardor, M. (2008) Plant *N*-glycan profiling of minute amounts of material. Anal. Biochem. 379(1), 66–72.
12. Bakker, H., Bardor, M., Molthoff, J.W., Gomord, V., Elbers, I., Stevens, L.H., Jordi, W., Lommen, A., Faye, L., Lerouge, P. and Bosch, D. (2001) Galactose-extended glycans of antibodies produced by transgenic plants. Proc. Natl. Acad. Sci. USA 98(5), 2899–2904.
13. Rayon, C., Cabanes-Macheteau, M., Loutelier-Bourrhis, C., Saliot-Maire, I., Lemoine, J., Reiter, W.D., Lerouge, P. and Faye, L. (1999) Characterization of *N*-glycans from *Arabidopsis thaliana*. Application to a fucose-deficient mutant. Plant Physiol. 119, 725–733.
14. Cabanes-Macheteau, M., Fitchette-Lainé, A.-C., Loutelier-Bourhis, C., Lange, C., Vine, N.D., Ma, J.K., Lerouge, P. and Faye, L. (1999) *N*-glycosylation of a mouse IgG expressed in transgenic tobacco plants. Glycobiology 9, 365–372.
15. Bardor, M., Cabanes-Macheteau, M., Faye, L. and Lerouge, P. (2000) Monitoring the *N*-glycosylation by fluorophore-assisted carbohydrate electrophoresis. Electrophoresis 21, 2550–2556.
16. Ko, K., Tekoah, Y., Rudd, P.M., Harvey, D.J., Dwek, R.A., Spitsin, S., Hanlon, C.A., Rupprecht, C., Dietzschold, B., Golovkin, M. and Koprowski, H. (2003) Function and glycosylation of plant-derived antiviral monoclonal antibody. Proc. Natl. Acad. Sci. USA 100, 8013–8018.
17. Triguero, A., Cabrera, G., Cremata, J.A., Yuen, C.-T., Wheeler, J. and Ramirez, N.I. (2005) Plant-derived mouse IgG monoclonal antibody fused to KDEL endoplasmic reticulum-retention signal is *N*-glycosylated homogeneously throughout the plant with mostly high-mannose-type *N*-glycans. Plant Biotech. J. 3, 449–457.
18. Guile, G.R., Rudd, P.M., Wing, D.R., Prime, S.B. and Dwek, R.A. (1996) A rapid high-resolution high-performance liquid chromatographic method for separating glycan mixtures and analyzing oligosaccharide profiles. Anal. Biochem. 240, 210–226.
19. Rudd, P.M., Colominas, C., Royle, L., Murphy, N., Hart, E., Merry, A.H., Hebestreit, H.F., Dwek, R.A. (2001) A high-performance liquid chromatography based strategy for rapid, sensitive sequencing of *N*-linked oligosaccharide modifications to proteins in sodium dodecyl sulphate polyacrylamide electrophoresis gel bands. Proteomics 1, 285–294.
20. Harvey, D.J., Dwek, R.A. and Rudd, P.M. (2006) Determining the structure of glycan moieties by mass spectrometry, in "Current Protocols in Protein Science", (Coligan, J.E., Dunn, B.M., Speicher, D.W. and Wingfield, P.T. eds.), John Wiley and Sons, New York, Unit 12, pp. 12.7.1–12.7.18.
21. Royle, L., Radcliffe, C.M., Dwek, R.A. and Rudd, P.M. (2006) Detailed structural analysis of *N*-glycans released from glycoproteins in SDS-PAGE gel bands using HPLC combined with exoglycosidase array digestions, in "Methods in Molecular Biology, Glycobiology Protocols", (Brockhausen-Schutzbach, I. ed.), Humana Press, vol. 347, pp. 125–144.
22. Bardor, M., Loutelier-Bourhis, C., Marvin, L., et-al. (1999) Analysis of plant glycoproteins by matrix-assisted laser desorption ionisation mass spectrometry: application to the *N*-glycosylation of bean phytohemagglutinin. Plant Physiol. Biochem. 37, 319–325.
23. Bardor, M., Faye, L. and Lerouge, P. (1999) Analysis of the *N*-glycosylation of recombinant glycoproteins produced in transgenic plants. Trends Plant Sci. 4, 376–380.
24. Bird, C.R., Gearing, A.J.H. and Thorpe, R. (1988) The use of Tween-20 alone as a blocking agent for the immunoblotting can cause artefactual results. J. Immunol. Methods 106, 175–179.
25. Lainé, A.-C. and Faye, L. (1988) Significant immunological cross-reactivity of plant glycoproteins. Electrophoresis 9, 841–844.
26. Kubelka, V., Altmann, F., Staudacher, E., Trotter, V., März, L., Hard, K., Kamerling, J.P. and Vliegenthert, J.F.G. (1993) Primary structures of the *N*-linked carbohydrate chains from honeybee venom phospholipase A2. Eur. J. Biochem. 213, 1193–1204.
27. Vitale, A., Warner, T.G. and Chrispeels, M.J. (1984) *Phaseolus vulgaris* phytohemagglutinin contains high-mannose and modified oligoasaccharide chains. Planta 160, 256–263.
28. Bardor, M., Loutelier-Bourhis, C., Paccalet, T., Cosette, P., Fitchette, A.-C., Vézina, L.P., Trepanier, S., Dargis, M., Lemieux, R., Lange, C., Faye, L. and Lerouge, P. (2003) Monoclonal C5-1 antibody produced in transgenic alfalfa plants exhibits a *N*-glycosylation that is homogeneous and suitable for glyco-engineering into a human-compatible structure. Plant Biotech. J. 1, 451–462.

Chapter 15

Companion Protease Inhibitors to Protect Recombinant Proteins in Transgenic Plant Extracts

Meriem Benchabane, Daniel Rivard, Cécile Girard, and Dominique Michaud

Summary

We describe a general approach for the use of recombinant protease inhibitors as stabilizing agents for clinically useful proteins extracted from transgenic plant tissues. A procedure is first described to assess the overall (in)stability of heterologous proteins in transgenic plant crude protein extracts. Step-by-step protocols are then presented for the choice and use of companion protease inhibitors inhibiting the host plant proteases during extraction. This strategy, that reproduces the protein-stabilizing effect of low-molecular-weight protease inhibitors often added to protein extraction media, does not require the exogenous addition of such expensive and often toxic compounds. It also presents the advantage of being intrinsically scalable to the amount of biomass processed.

Key words: Clinically useful proteins, Companion protease inhibitors, Protein degradation, Protein stabilization, Recombinant proteins, Transgenic plants.

15.1. Introduction

Higher plants represent versatile expression platforms for the heterologous production of clinically useful proteins *(1, 2)*, but also pose serious challenges to protein developers dealing with naturally unstable proteins *(3–5)*. The well-documented importance of proteolytic enzymes in plant cells *(6)*, in particular, is often a major hurdle to recombinant protein integrity, both in planta at the expression stage and in vitro during protein recovery from plant tissues *(5)*. For any production platform, the postharvest processing of plant biomass is critical not only in determining the economic feasibility of the overall production scheme, but

Loïc Faye and Véronique Gomord (eds.), *Methods in Molecular Biology, Recombinant Proteins From Plants, vol. 483*

DOI: 10.1007/978-1-59745-407-0_15

also in strongly impacting final quality and yield of the recombinant product *(7, 8)*. Downstream procedures for recombinant proteins are in most cases laborious, typically involving multiple extraction, clarification, and purification steps potentially deleterious to the recombinant product *(9)*. Endogenous proteases and other protein-denaturing compounds released in the extraction medium after cell disruption may significantly destabilize the overall structure of protein backbones and lead to irreversible precipitation, partial trimming, or even complete hydrolysis of the extracted proteins *(10)*. As a result, the end-product may show lower solubility, altered integrity and higher heterogeneity, with a negative impact on biological activity and commercial value.

Several general guidelines may be followed to minimize protein loss at the extraction and purification stages, including the processing of plant biomass shortly after harvest, the handling and treatment of plant tissues at low temperature, the inclusion of phenolics-neutralizing compounds in extraction buffers, the use of pH conditions unfavorable to proteolysis, and the use of low-molecular-weight protease inhibitors active against the host plant proteases *(9–12)*. As a complement, we described recently an in-built system for the protection of recombinant proteins in vitro, based on the heterologous coexpression of a companion protease inhibitor acting as a "mouse trap" against host plant proteases during extraction *(13)*. This strategy, that reproduces the protein-stabilizing effect of low-molecular-weight protease inhibitors added to protein extraction media, does not require the addition of such expensive and often toxic compounds, while also presenting the advantage of being intrinsically scalable to the amount of biomass processed. In practice, this strategy first involves the identification of a proteinaceous protease inhibitor effective in protecting the recombinant protein of interest, and then the coexpression of the two proteins in planta. This chapter describes the main steps to perform for adapting this strategy to a particular protein expression platform, from the identification of a proteolytic problem to the choice and use of a companion protease inhibitor.

15.2. Materials

15.2.1. Plant Material

Transgenic plant lines expressing recombinant protease inhibitors and their corresponding [non-transgenic] isoline are used for the tests. Transgenic lines of potato (*Solanum tuberosum* L., cv. Kennebec) expressing the serine protease inhibitor bovine aprotinin or the Kunitz protease inhibitor tomato cathepsin D inhibitor (CDI) are used here for demonstration purposes, along with control transgenic and nontransgenic sister lines (*see* **ref.** *13*).

15.2.2. Special Laboratory Tools and Materials

1. Liquid nitrogen.
2. Mortar and pestle.
3. Miracloth (Calbiochem, La Jolla, CA).
4. Mini-Protean II™ Electrophoresis Unit for protein gel electrophoresis (Bio-Rad, Richmond, CA).
5. Minitransfert Unit™ for protein electrotransfer to nitrocellulose membranes (Bio-Rad).
6. Hybond C nitrocellulose membranes (Amersham Biosciences, Baie d'Urfé QC, Canada).
7. Standard reagents for sodium dodecyl sulphate-polyacrylamide gel electrophoresis (SDS-PAGE) and immunoblotting.
8. Specific polyclonal or monoclonal antibodies for detection of human fibronectin, human serum albumin, human erythropoietin, or human stefin A (Sigma-Aldrich, Oakville ON, Canada).
9. Appropriate secondary antibodies for immunodetection (Sigma-Aldrich).
10. Microtek ScanMaker II scanner (Microteck Laboratory, Torrance, CA).
11. Image analysis software NIH Image 1.6 (National Institutes of Health, Bethesda, MD).

15.2.3. Buffers and Other Solutions

All buffers and solutions are made up as aqueous solutions. Solutions for widely used procedures such as SDS-PAGE and immunodetection are not described here. Detailed, step-by-step protocols for SDS-PAGE and immunoblotting are given in **refs.** *14* and *15*, respectively. For immunodetection, simple and efficient procedures involving alkaline phosphatase- or peroxidase-labeled secondary antibodies and appropriate reagents for color development are usually provided by the suppliers.

1. Leaf extraction buffer: 100 mM Tris-HCl, pH 7.0, containing 0.2 mM polyvinyl-polypyrrolidone (PVPP) to neutralize phenolics. Prepare fresh.
2. Protein preparations: Each protein of interest is dissolved in water or in an appropriate low ionic strength buffer (e.g., 25–50 mM salts) at 1 mg/mL final concentration. Human fibronectin (Sigma-Aldrich), human serum albumin (Sigma-Aldrich), human erythropoietin (Sigma-Aldrich), and a recombinant form of human stefin A expressed in *Escherichia coli (16)* are used here as test proteins for demonstration purposes.
3. Inhibitor preparations: Each potential companion protease inhibitor is dissolved at 1 mg/mL final concentration in leaf extraction buffer. The eight inhibitory unit of tomato multicystatin (TMC-8) *(17)*, tomato CDI *(18)*, and bovine aprotinin (Sigma-Aldrich) are used here as candidate companion inhibitors for demonstration purposes.

4. Trichloroacetic acid (TCA) solution: 10% (w/v) TCA. Keep at 4°C until use. Stable for months.
5. Acetone solution: 80% (v/v) acetone. Keep at –20°C until use. Stable for months.
6. Electrophoresis sample buffer (double strength): 62.5 mM Tris-HCl, pH 6.8, 2% (w/v) SDS, 20% (w/v) glycerol, 10% (v/v) ß-mercaptoethanol, 0.01% (w/v) bromophenol blue. Add ß-mercaptoethanol just before use. Nonreduced buffer stable for weeks at 4°C.

15.3. Methods

15.3.1. *In Vitro* Degradation of the Recombinant Protein

This section describes a simple way to evaluate the overall stability of a given protein in a host plant extract containing proteolytic enzymes (*see* **Note 1**). The procedure consists, in brief, (i) to extract the proteins of a candidate host plant, (ii) to incubate a protein of interest with the plant protein extract; and (iii) to determine the residual amount of protein [and/or the generation of protein fragments] by immunodetection and image analysis after different periods, compared to the amount measured at time 0. Potato leaf extracts and human variants of fibronectin, serum albumin, erythropoietin, and stefin A are used here for demonstration purposes.

15.3.1.1. Extraction of Plant Soluble Proteins

1. Grind the fourth, fifth, and sixth leaves of ~30-cm long potato plants to a fine powder in liquid nitrogen, using a mortar and a pestle (*see* **Note 2**).
2. Transfer 7.5 g of leaf powder to an ice-cold plastic tube, and add 20 mL of cold leaf extraction buffer containing PVPP.
3. After extraction of soluble proteins on ice for 10 min, pass the mixture through one layer of Miracloth tissue to remove leaf debris, and centrifuge the solution at 15,000*g* for 15 min at 4°C. Discard the pellet and use the supernatant (leaf extract) as a source of plant endogenous proteases for further analyses. If not used immediately, the extract can be quick-frozen in liquid nitrogen and stored at –80°C for several weeks without loss of protease activity.

15.3.1.2. Visualization of Protein Degradation *In Vitro*

1. Place 25 µL (~20 µg protein) of leaf extract (*see* **Subheading 15.3.1.1, step 3**) in microcentrifuge tubes, and add 5 µL of protein preparation (5 µg of protein) in each tube.
2. Gently agitate the mixtures and let the tubes standing at 4°C or 20°C for different periods (e.g., 0, 10, or 20 min) to allow proteolytic degradation by leaf proteases.

3. Stop the reaction by adding 36 μL of cold TCA solution [5% (w/v) final concentration]. Keep the tubes standing on ice for 15 min to allow precipitation of the proteins. Centrifuge at 12,000*g* for 10 min at 4°C.
4. Discard the supernatant and wash the protein-containing pellet with 75 μL of acetone solution. Centrifuge at 12,000*g* for 10 min at 4°C, and discard the supernatant.
5. Repeat **step 4** once and dissolve the pellet in 30 μL of SDS-PAGE sample buffer (double strength). This sample contains the host plant proteins and the protein of interest after (possible) proteolytic cleavage.
6. Perform standard SDS-PAGE *(14)* using the Bio-Rad Mini-Protean II Electrophoresis Unit™. Load the samples onto a 0.75-mm thick gel (1–2 μL per well), and perform migration at 200 V until the bromophenol blue tracking dye reaches the bottom of the gel.
7. After electrophoresis, transfer the resolved proteins onto a nitrocellulose (Hybond C) membrane using the Bio-Rad Minitransfer Unit™, and reveal the protein of interest by immunodetection *(15)* using appropriate antibodies and reagents for enzymatic development.
8. Estimate the residual amount of protein in the extract by densitometry, using a ScanMaker II scanner and the image analysis software NIH Image, v. 1.6. Data may be expressed as the relative amount of protein remaining in the extract (%), compared to the amount measured at time 0 (100%) (*see* **Note 3**). **Figure 15.1** illustrates the susceptibility of

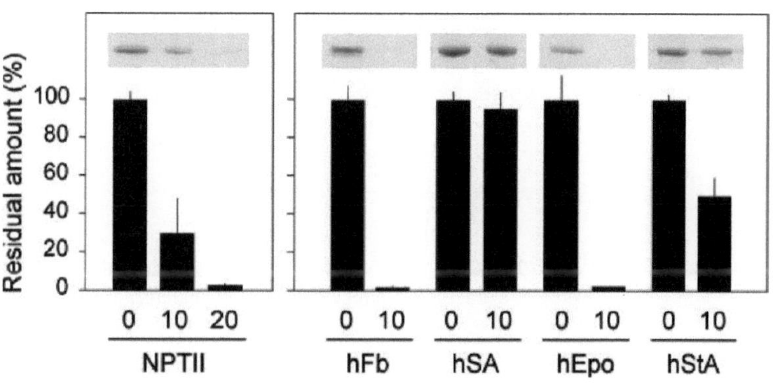

Fig. 15.1. Heterologous protein degradation in potato leaf crude protein extracts. The proteins were immunodetected using appropriate primary and secondary antibodies (inset gels), after incubation at 20°C with a leaf protein extract for 0, 10, or 20 min. The marker protein NPTII (*left panel*) was expressed in planta and detected directly in leaf protein extracts of transgenic potato line SP4 (*see* **ref.** *13*). The test proteins human fibronectin (*hFb*), human serum albumin (*hSA*), human erythropoietin (*hEpo*), and human stefin A (*hStA*) (*right panel*) were added exogenously to a control (nontransgenic) potato leaf extract before incubation and detection. Data are expressed as the residual amount of protein (%) compared to the initial amount at time 0 (100%). Each bar is the mean of three values ± se.

different proteins, including proteins of clinical interest, to in vitro cleavage by potato leaf endogenous proteases.

15.3.2. Companion Protease Inhibitors for Heterologous Protein Stabilization

This section describes a short protocol to identify companion (proteinaceous) protease inhibitors efficient in stabilizing a protein of interest in crude protein extracts of the host plant. The cysteine protease inhibitor TMC-8 and the serine protease inhibitors tomato CDI and bovine aprotinin are used as candidate companion inhibitors to illustrate the procedure.

1. Place 20 μL (~16 μg protein) of leaf extract (*see* **Subheading 15.3.1.1, step 3**) in microcentrifuge tubes. Add 5 μL of protein preparation and 5 μL of inhibitor preparation (5 μg inhibitor) in each tube. Also prepare a positive control by replacing the inhibitor preparation by 5 μL of leaf extraction buffer.
2. Incubate the mixtures for different periods, resolve the proteins by SDS-PAGE, and detect the protein of interest by immunoblotting, as described above (see **Subheading 15.3.1.2, steps 2–7**).
3. Determine the residual amount of protein by densitometry as described above (*see* **Subheading 15.3.1.2, step 8**), and estimate the potential of each inhibitor tested to protect the proteins. **Figure 15.2** illustrates the strong protein-stabilizing effects of TMC-8, tomato CDI, and bovine aprotinin added to potato leaf protein extracts.

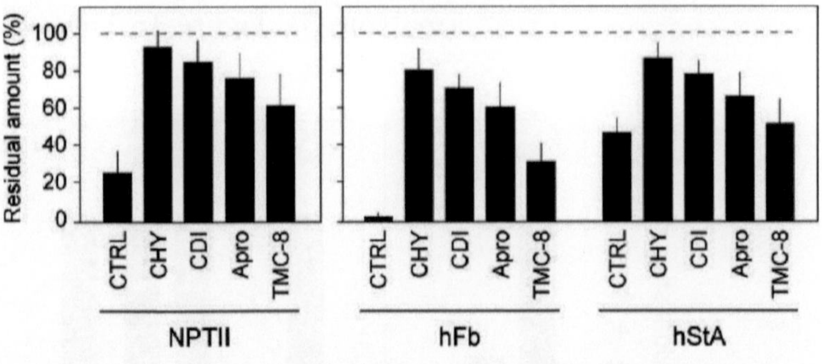

Fig. 15.2. Protease inhibitor-mediated protection of heterologous proteins in potato leaf crude protein extracts. The proteins were immunodetected using appropriate primary and secondary antibodies after incubation for 10 min at 20°C with a leaf protein extract, in the presence or absence of protease inhibitors. The marker protein NPTII (*left panel*) was expressed in planta and detected directly in leaf protein extracts of transgenic potato line SP4 (*see* **ref.** *13*). The test proteins human fibronectin (*hFb*) and human stefin A (*hStA*) (*right panel*) were added exogenously to a control (nontransgenic) potato leaf extract before incubation and detection. The broad-spectrum inhibitor of serine and papain-like proteases, chymostatin (*CHY*), was used as a positive control for plant protease inhibition. Data are expressed as the residual amount of protein (%) compared to the initial amount at time 0 (100%). Each bar is the mean of three values ± se. *CTRL*, control tests with no inhibitor added in the incubation medium.

15.3.3. An In-Built Protease Inhibitor System for Protein Stabilization In Vitro

This section describes a simple procedure to assess the in vitro protein-stabilizing effects of recombinant protease inhibitors expressed in planta. Transgenic potato lines expressing bovine aprotinin or tomato CDI are used as models to illustrate the procedure (*see* **Note 4**).

1. Extract leaf soluble proteins of bovine aprotinin- and tomato CDI-expressing transgenic potato lines *(13)* as described above for nontransgenic (control) plants (*see* **Subheading 15.3.1.1**). Also prepare a control extract with nontransgenic leaf material for comparison purposes.
2. Place 25 μL of transgenic (or control) leaf extract in microcentrifuge tubes, and add 5 μL of protein preparation in each tube.
3. Incubate the mixtures for different periods, resolve the proteins by SDS-PAGE, and detect the protein of interest by immunoblotting, as described above (see **Subheading 15.3.1.2, steps 2–7**).
4. Determine the residual amount of protein by densitometry as described above (*see* **Subheading 15.3.1.2, step 8**), and estimate the potential of each transgenic line to protect the test proteins. **Figure 15.3** illustrates the important protein-stabilizing effects of recombinant bovine aprotinin and tomato CDI in potato leaf crude protein extracts.

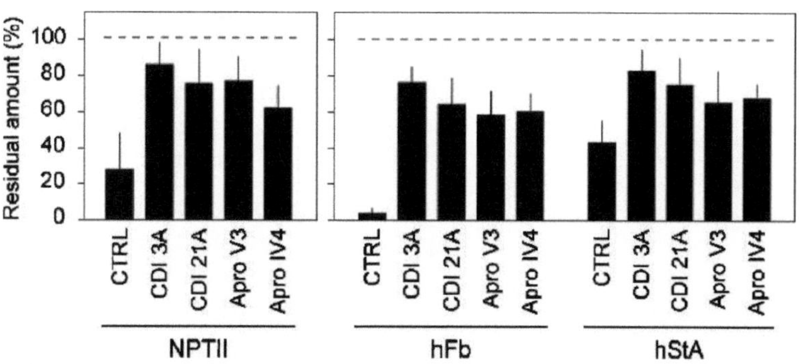

Fig. 15.3. Protection of heterologous proteins in potato leaf crude protein extracts by the broad-spectrum serine protease inhibitors tomato cathepsin D inhibitor (*CDI*) and bovine aprotinin expressed in planta. The proteins were immunodetected using appropriate primary and secondary antibodies after incubation for 10 min at 20°C with a transgenic leaf protein extract containing *CDI* or aprotinin (*see* **ref.** *13*). The marker protein NPTII (*left panel*) was coexpressed in planta and detected directly in leaf extracts of transgenic potato lines expressing or not *CDI* or aprotinin (*see* **ref.** *13* for more details on these lines). The test proteins human fibronectin (*hFb*) and human stefin A (*hStA*) (*right panel*) were added exogenously to the same leaf extracts before incubation and detection. Data are expressed as the residual amount of protein (%) compared to the initial amount at time 0 (100%). Each bar is the mean of three values ± se. *CTRL*, control tests using transgenic potato line SP4 expressing NPTII but no *CDI* or aprotinin.

15.4. Notes

1. The protocols described here [i.e., extraction from green (leaf) tissues, biomass processing and protein extraction at room temperature, no protein-stabilizing agent in the extraction buffer] refer to a "theoretical" protein production scheme. They can be easily adapted to alternative parameters for downstream processing such as tissue grinding at 4°C, protein extraction in acidic conditions, or inclusion of protein-stabilizing agents in the extraction buffer. These protocols also are suitable for the assessment of proteins expressed in nonleaf tissues, provided that efficient protein extraction procedures are available for these tissues.
2. Because the overall cellular complement of proteases in planta depends on various environmental and developmental factors *(6, 19, 20)*, the concurrent analysis of proteases from younger and senescent leaves, or from plants submitted to different cultural conditions, might be useful to obtain a broader picture of the situation, especially when expressing highly labile proteins.
3. The approach described in this chapter is based on the assumption that obtaining the entire protein as the final end-product is essential. Proteolytic events in biological extracts often generate a complex mix of stable, high-molecular-weight intermediates including or not the entire protein product. When maintaining the protein's structural integrity is not mandatory and when product heterogeneity does not alter its commercial value, estimating the total amount of stable intermediates in the extract may also be relevant. This could be the case, for instance, when the goal is to obtain stable antigenic products for vaccine production.
4. Recombinant protease inhibitors may show inhibitory effects against the host plant endogenous proteases and interfere with normal metabolic or developmental processes in planta (e.g., **refs.***21, 22)*. The use of an inhibitor weakly active against the plant's proteases, or its targeting to an alternative cell compartment using appropriate translocation signals may be required in such cases *(18)*.

Acknowledgments

This work was supported by a Discovery grant from the Natural Science and Engineering Research Council (NSERC) of Canada to D. Michaud. M. Benchabane was the recipient of an NSERC doctoral scholarship.

References

1. Twyman, R.M., Stoger, E., Schillberg, S., Christou, P., and Fischer, R. (2003) Molecular farming in plants: host systems and expression technology. Trends Biotechnol. 21, 570–578.
2. Ma, J.K.C., Chikwarnba, R., Sparrow, P., Fischer, R., Mahoney, R., and Twyman, R.M. (2005) Plant-derived pharmaceuticals—the road forward. Trends Plant Sci. 10, 580–585.
3. Doran, P.M. (2006) Foreign protein degradation and instability in plants and plant tissue cultures. Trends Biotechnol. 24, 426–432.
4. Benchabane, M., Goulet, C., Rivard, D., Faye, L., Gomord, V., and Michaud, D. (2008) Preventing unitended proteolysis in plant protein biofactories. Plant Biotechnol. J. 6, 633–648.
5. Faye, L., Boulaflous, A., Benchabane, M., Gomord, V., and Michaud, D. (2005) Protein modifications in the plant secretory pathway: current status and practical implications in molecular pharming. Vaccine 23, 1770–1778.
6. Schaller, A. (2004) A cut above the rest: the regulatory function of plant proteases. Planta 220, 183–197.
7. Daniell, H., Streatfield, S.J., and Wycoff, K. (2001) Medical molecular farming: production of antibodies, biopharmaceuticals and edible vaccines in plants. Trends Plant Sci. 6, 219–226.
8. Nikolov, Z.L., and Woodard, S.L. (2004) Downstream processing of recombinant proteins from transgenic feedstock. Curr. Opin. Biotechnol. 15, 479–486.
9. Menkhaus, T.J., Bai, Y., Zhang, C.M., Nikolov, Z.L., and Glatz, C.E. (2004) Considerations for the recovery of recombinant proteins from plants. Biotechnol. Progr. 20, 1001–1014.
10. Michaud, D., and Asselin, A. (1995) Application to plant proteins of gel electrophoretic methods. J. Chromatogr. A 698, 263–279.
11. Michaud, D. (1998) Gel electrophoresis of proteolytic enzymes. Anal. Biochim. Acta 372, 173–185.
12. Schillberg, S., Twyman, R.M., and Fischer, R. (2005) Opportunities for recombinant antigen and antibody expression in transgenic plants—technology assessment. Vaccine 23, 1764–1769.
13. Rivard, D., Anguenot, R., Brunelle, F., Le, V.Q., Vezina, L.P., Trepanier, S., and Michaud, D. (2006) An in-built proteinase inhibitor system for the protection of recombinant proteins recovered from transgenic plants. Plant Biotechnol. J. 4, 359–368.
14. Smith, B.J. (1984) SDS polyacrylamide gel electrophoresis of proteins, In J.M. Walker (Eds.), Methods in Molecular Biology, vol. 1: Proteins (pp. 165–178). NJ: Clifton,Humana Press
15. Gooderham, K. (1984) Transfer techniques in protein blotting, In J.M. Walker (Eds.), Methods in Molecular Biology, vol. 1: Proteins (pp. 165–178). NJ: Clifton,Humana Press
16 Michaud, D., Nguyen-Quoc, B., Vrain, T.C., Fong, D., and Yelle, S. (1996) Response of digestive cysteine proteinases from the Colorado potato beetle (*Leptinotarsa decemlineata*) and the black vine weevil (*Otiorynchus sulcatus*) to a recombinant form of human stefin A. Arch. Insect Biochem. Physiol. 31, 451–464.
17. Kiggundu, A., Goulet, M.-C., Goulet, C., Dubuc, J.-F., Rivard, D., Benchabane, M., Pépin, G., van der Vyver, C., Kunert, K., and Michaud, D. (2006) Modulating the proteinase inhibitory profile of a plant cystatin by single mutations at positively selected amino acid sites. Plant J. 48, 403–413.
18. Brunelle, F., Girard, C., Cloutier, C., and Michaud, D. (2005) A hybrid, broad-spectrum inhibitor of Colorado potato beetle aspartate and cysteine digestive proteinases. Arch. Insect Biochem. Physiol. 60, 20–31.
19. Stevens, L.H., Stoopen, G.M., Elbers, I.J.W., Molthoff, J.W., Bakker, H.A.C., Lommen, A., Bosch, D., and Jordi, W. (2000) Effect of climate conditions and plant developmental stage on the stability of antibodies expressed in transgenic tobacco. Plant Physiol. 124, 173–182.
20. Martinez, D.E., Bartoli, C.G., Grbic, V., and Guiamet, J.J. (2007) Vacuolar cysteine proteases of wheat (Triticum aestivum L.) are common to leaf senescence induced by different factors. J. Exp. Bot. 58, 1099–1107.
21. Van der Vyver, C., Schneidereit, J., Driscoll, S., Turner, J., Kunert, K., and Foyer, C. (2003) Oryzacystatin I expression in transformed tobacco produces a conditional growth phenotype and enhances chilling tolerance. Plant Biotechnol. J. 1, 101–112.
22. Belenghi, B., Acconcia, F., Trovato, M., Perazzolli, M., Bocedi, A., Polticelli, F., Ascenzi, P., and Delledonne, M. (2003) AtCYS1, a cystatin from *Arabidopsis thaliana*, suppresses hypersensitive cell death. Eur. J. Biochem. 270, 2593–2604.

Chapter 16

Strategies for Improving Vaccine Antigens Expression in Transgenic Plants: Fusion to Carrier Sequences

Jose M. Escribano and Daniel M. Perez-Filgueira

Summary

Transgenic plants are gaining increasing attention from the industry as a natural bioreactor for the production of industrial and chemical products. Optimization of transgene expression in plant cells holds the key to maximizing the potential of plants for producing proteins of commercial interest. This chapter is devoted to the description of the methods utilized for the generation of transgenic plants expressing a canine parvovirus vaccine peptide or virus-like particles from a rabbit calicivirus.

Key words: Recombinant subunit vaccines, Plant-derived vaccines, β-glucuronidase, Ubiquitin; Tetrameric antigens, Protein accumulation, Peptide vaccine, Rabbit haemorrhagic virus, Canine parvovirus.

16.1. Introduction

A very important criterion for an ideal vaccine is low cost. This is especially important in developing countries and even more so for development of vaccines applied to animal species, where vaccines must sell for a few cents a dose. Vaccines produced in transgenic plants constitute a promising alternative to conventional immunogens, presenting the possibility of stimulating secretory and systemic immunity against enteric pathogens when administered orally *(1–4)*. However, expression levels of vaccine antigens in transgenic plants have important consequences in their use as edible vaccines. Previous literature addressing the production of recombinant proteins in transgenic plants and our own experience

Loïc Faye and Véronique Gomord (eds.), *Methods in Molecular Biology, Recombinant Proteins From Plants, vol. 483*

DOI: 10.1007/978-1-59745-407-0_16

has consistently shown that recombinant proteins tend to accumulate when reducing their turnover within cells. This reduction in degradation is frequently found in proteins capable of forming complex structures which, as a result, are more efficiently expressed than monomeric proteins. In previous publications, we have presented different strategies to enhance recombinant protein or peptide accumulation in transgenic plants without affecting their antigenic and immunogenic properties.

Here, we review our experience in enhancing vaccine antigen accumulation in transgenic plants through these different strategies. Methods used for enhancing accumulation of a peptide sequence [2L21 from canine parvovirus (CPV)] are described. Examples include its fusion to (i) other proteins [β-glucuronidase (GUS)] that are stable within the cytoplasmic environment *(5)* or to (ii) multimerization sequences (p53 tetramerization domain) that permit its oligomerization *(6)*. Methods developed to study expression in transgenic plants of other proteins that form complex structures [VP60 from rabbit hemorrhagic disease virus (RHDV)] as single proteins or fused to other sequences are also reviewed *(7)*.

16.2. Materials

1. *Arabidopsis thaliana* plants (Columbia ecotype).
2. *Agrobacterium tumefaciens* (strain C58C0).
3. Oligonucleotide primers.
4. Restriction enzymes, T4 DNA ligase.
5. Agarose gels and DNA sequencing equipments.
6. Protein and DNA electrophoresis equipments.
7. Taq polymerase.
8. Plasmid *pGemT-easy* (Promega, Madison, wisconsin).
9. Binary plasmid pBI121 (Clontech, PaloAlto, California).
10. Plasmid pCMV-p53 (Clontech, Palo Alto, California).
11. Murashige-Skoog (MS) cell culture medium (Sigma, St. Louis, MIssouri).
12. Hybond-N membranes (Amersham Biosciences, Piscataway, New Jersey).
13. Random-Primed DNA Labelling Kit (Roche Diagnostics, Basel) for P32 isotope-labelled probed.
14. Protein extraction buffer: 10 mM 2-(N-morpholino) ethanesulfonic acid (MES), pH 6, 10 mM NaCl, 5 mM EDTA, 0.6% Triton X-100, 0.25 M sucrose, 0.15 mM

spermine, 0.5 mM spermidine, 10 mM DTT and 1 mM phenylmethylsulfonyl fluoride (PMSF).

15. Protein extraction buffer for denaturing SDS-tricine-PAGE: 50 mM Tris-HCl pH 7.4, 10 M EDTA pH 8.0, 0.1% Triton X-100, 1% β-mercaptoethanol and 2 mM PMSF.
16. VLPs extraction buffer: 0.5 M sodium acetate, 0.8 M acetic acid, 0.01 mM magnesium chloride, 0.1 M ascorbic acid, 0.1 mM β-mercaptoethanol and 2 mM PMSF.
17. VLPs suspension buffer: 50 mM sodium acetate, 10 mM acetic acid, 1 mM magnesium chloride and 1 mM EDTA pH8.
18. Protein extraction buffer for native tricine-PAGE: 50 mM Tris-HCl pH 7.4, 0.1% Triton X-100 and 2 mM PMSF.
19. Bradford Protein assay (BioRad, Hercules, California).
20. Monoclonal antibody (MAb) 3C9 (Ingenasa, Madrid) against CPV 2L21 peptide sequence.
21. Anti-mouse IgG conjugated with alkaline phosphatase (Roche Diagnostics, Basel).
22. Nitroblue tetrazolium chloride (Gibco-BRL Lyfe Technology, Rockville, Mary land), used following manufacturer recommendation.
23. Bromochloroindoyl phosphate (Pierce Chemical Company, Rockford, Illinois), used following manufacturer recommendation.

16.3. Methods

The methods described below outline (1) construction of the expression plasmids, (2) plant transformation and (3) analysis of protein expression in the different transgenic plants.

16.3.1. Construction of Expression Plasmids

16.3.1.1. Plasmids for Expression of Peptide 2L21 from Canine Parvovirus Fused to GUS

The DNA sequence corresponding to 2L21 epitope was obtained by Polymerase Chain seaction (PCR) using the following overlapping oligonucleotides: (A) 5′GCGCGGATCCATGTCTGATGGAGCTGTTCAACCAGATGGTGGTCAACCTG 3′ and (B) 5′GCGCGGATCCTCCAGTAGCTCTCTGTTTCTAACAGCAGGTTGACCACCA 3′. Sequence of primers was adapted to the most frequent codon usage in *Arabidopsis (8)*. The annealed oligomers were cloned into the *Bam* HI site of the plasmid pUC19 and then digested with *Xba* I and *Sma* I for subcloning the 2L21 encoding DNA fragment into the binary plasmid pBI121 (Clontech palo Alto, California) following standard procedures *(9)*. The resulting plasmid, named pBI2L21, contained the 2L21 encoding region fused in frame to *GUS* gene under the control of the cauliflower mosaic virus 35S (CaMV 35S) promoter (**Fig. 16.1a**).

16.3.1.2. Plasmids for Expression of Peptide 2L21 from Canine Parvovirus Fused to the Tetramerization Domain of Protein p53

Coding sequences for the 2L21 CPV peptide and p53TD fragment were obtained by PCR amplification using the pBI2L21 *(5)* and pCMV-p53 (Clontech, palo Alto, California) plasmids as templates, respectively. A set of four primers were designed to include appropriate restriction sites at the 5′ and 3′ ends of the 2L21 and p53TD sequences permitting in frame fusion of the two fragments and further cloning into an intermediate plasmid (pGemT-easy, Promega, Madison, wisconsin). In frame start and stop codons were also added upstream and downstream the 2L21 and p53TD sequences, respectively, and the fusion gene was cloned into the pBI-121 p35S/TEV binary vector, under control of the CaMV 35S promoter and tobacco etch virus enhancer (TEVe) leader sequence (pBI2L21-TD; **Fig. 16.1b**).

16.3.1.3. Plasmids for Expression of VP60 Gene from RHDV Fused to Different Sequences

The *VP60* gene was cloned into cassettes either as single open reading frame or as a fusion to other sequences. Expression cassettes were then subcloned into binary plasmids derived from the pBI121 (Clontech palo Alto, California) to produce the above mentioned expression, vectors (**Fig. 16.1c**). Plasmid *pVP60* contains the gene encoding VP60 (Olot/89 strain), obtained by a PCR amplification from plasmid gently provided by Dr. F. Parra (University of Oviedo, Oviedo, Spain). Expression of VP60 in these constructs was driven by the CaMV 35S promoter, followed by a TEV (tobacco etch virus) 5′ leader sequence, inserted downstream of the promoter to increase gene translation (**Fig. 16.1c**). The *prbcSVP60* vector contained the *VP60* gene fused in frame to the *Arabidopsis rbcS* sequence and it was designed to target the synthesized fusion protein to the chloroplasts. *A. thaliana rbcS* cDNA corresponding to small subunit of ribulose-1,5-bis-phosphate carboxylase-oxygenase was obtained by PCR amplification of cDNA from leaves of Columbia ecotype with specific primers 5′-AACTGCAGAGATCTATGGCTTCCT CTATGCTCTC-3′ and 5′-CGTCTAGAACCGGTGAAGCTTGGTGGCTTG-3′. Constructs $pUBQ_{G76}VP60$ and $pUBQ_{A76}VP60$ were designed to express cleavable and non-cleavable UBQ-VP60 fusion proteins, respectively. Original UBQ sequence was obtained from plasmid *pbin19-UBQ-GUS-K48* gently provided by Dr. R. Vierstra (University of Wisconsin, Madison, USA), containing the *ubq11* gene of *A. thaliana (10)*, downstream a 39-bp sequence from the 5′ UTR from alfalfa mosaic virus (AMV) *(11)*.

16.3.2. Plant Transformation

A. thaliana plants (Columbia ecotype) were transformed by infiltration with *A. tumefaciens* as previously described *(12)* with slight modifications *(13)*. Seeds from infiltrated plants (T0) were

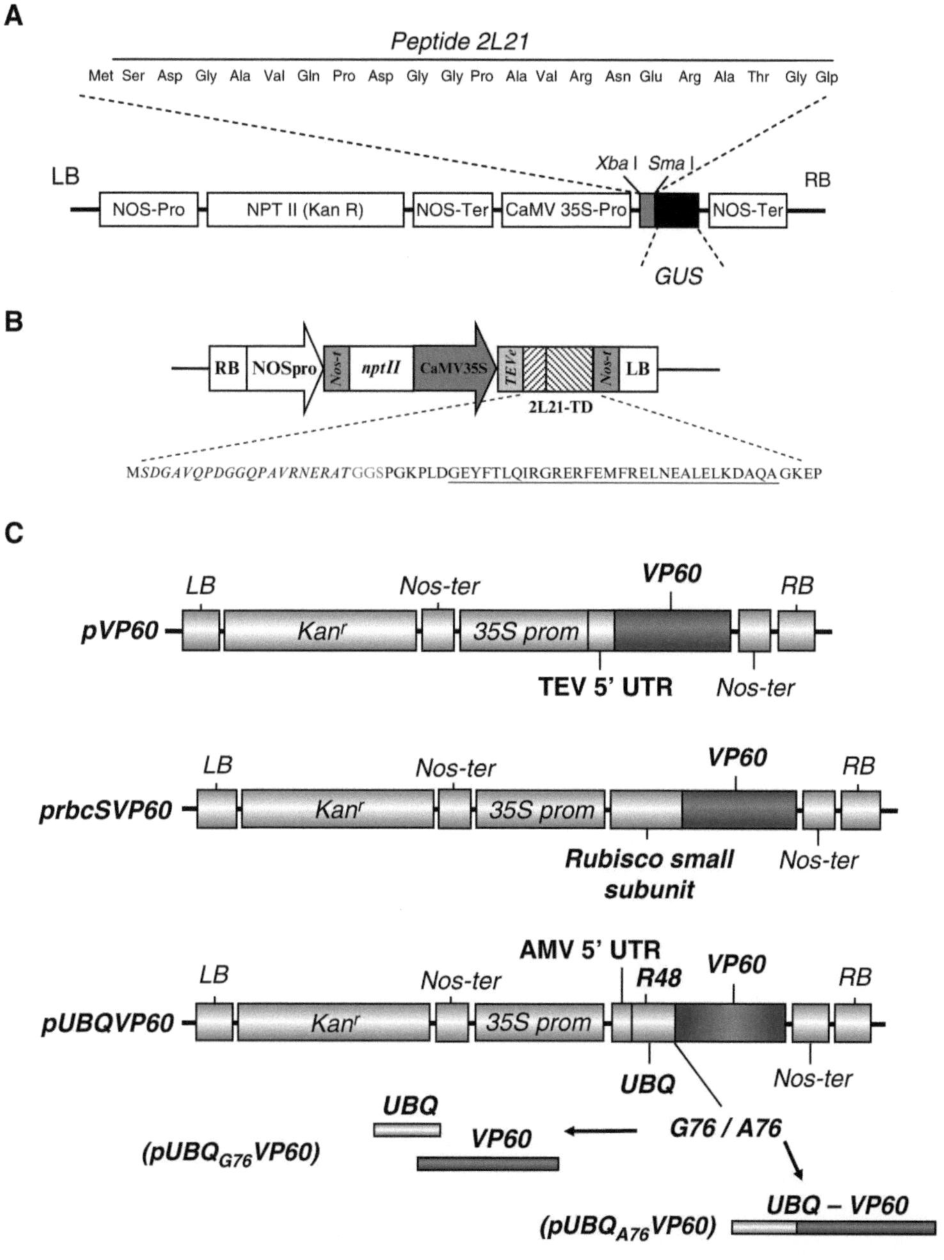

Fig. 16.1. (**A**) Schematic structure of the binary plasmid pBI-2L21 used for Agrobacterium-mediated plant transformation. This panel represents the resulting plasmid after cloning of 2L21 encoding sequence downstream of the CaMV 355 promoter in plasmid pBI121. The amino acid sequence of 2L21 peptide is also shown. (**B**) Schematic representation of the transgene region of the pBI-2L21-TD binary vector. RB and LB, right and left borders of the T-DNA; NOSpro and NOS-t, nopaline synthase gene promoter and polyadenylation signals; nptII, neomycinphosphotransferase gene; CaMV35S, 35S promoter from cauliflower mosaic virus and TEVe from tobacco etch virus enhancer sequence. Amino acid sequence of the 2L21-TD is depicted below, residues for 2L21 peptide are shown in italics, fragment in grey represent a linker sequence and the minimum TD sequence within the p53-derived fragment is underlined. (**C**) Structure of the T-DNA region of different vectors used for expression of VP60 protein of rabbit hemorrhagic disease virus (RHDV) in leaves of transgenic Arabidopsis thaliana plants. The viral gene is inserted into an expression cassette of binary vector pBI121 from which transcription is driven by the cauliflower mosaic virus 35S promoter; Nos-ter element mediates polyadenilation; the tobacco each virus 5′-UTR and AMV (alfalfa mosaic virus) 5′-UTR provide enhancement of translation.

screened onto MS cell culture medium (Sigma, St. Lowis, Missouri) containing 50 μg/ml of kanamycin and antibiotic resistant. T1 plants were analysed for the presence of the foreign DNA sequence by PCR using primers for amplification of a fragment of transgene. Selected T1 plants were self-pollinated to obtain T2 and T3 generations. GUS activities in pBI2L21 transformed plants were determined in plant tissues according to the method of Jefferson et al. *(14)*.

16.3.3. Analysis of Expression of 2L21-GUS Fusion Protein in Transformed Plants

Twenty five independent kanamycin-resistant transgenic lines were obtained appearing similar in morphology to wild-type plants. All lines were positive when screened for the presence of the recombinant genes by PCR analysis. Plant soluble proteins were obtained by homogenization of leaves in a blender with liquid nitrogen, and the resulting powder was re-suspended in protein extraction buffer (0.3 g of fresh wt/ml). The extract was filtered and centrifuged 10 min at 12,000*g*. Plant saps were centrifuged at 12,000*g* at 4?C for 15 min and total soluble protein (TSP) concentrations of the resulting supernatants were measured by Bradford assay *(15)*. Plant extracts (30 μg of TSP) were resolved in 15% SDS-PAGE and transferred to nitrocellulose filters. The presence of 2L21-GUS protein was detected in Western blot using MAb 3C9 at 0.1 μg/ml dilution and an anti-mouse IgG conjugated with alkaline phosphatase (Roche Diagnostics, Basel). Reactions were developed with nitroblue tetrazolium chloride and bromochloroindoyl phosphate.

Fully expanded upper leaves were analysed by Western blot using with MAb 3C9 against CPV-2L21 peptide. A 73 kDa immunoreactive band was detected in most plants, approximately corresponding to the expected electrophoretic mobility of the 2L21-GUS fusion protein (**Fig. 16.2**). No immunoreactive bands could be detected in leaf extracts from plants transformed with pBI121 plasmid or in plants showing low levels of GUS activity. A good correlation between GUS activity and reactivity of fusion protein in Western blot was observed. Homozygous plants from lines showing the highest GUS activity were isolated and used in further experiments. They showed an increase of about 20% of GUS activity with respect to their parental T2 segregating transgenic lines.

Levels of 2L21-GUS fusion protein in leaves from transgenic plants were calculated by densitometry of Western blots using MAb 3C9. Intensity of MAb 3C9-reacting bands in Western blots of known concentrations of baculovirus-expressed CPV VP2 were compared in multiple reactions to that of 2L21-GUS protein contained in 30 μg of leaf extracts. The average integrated densitometry values calculated with every transgenic plant in at least three independent analysis, rendered percentages of 2L21-GUS protein accumulation in the highest expressing plants between

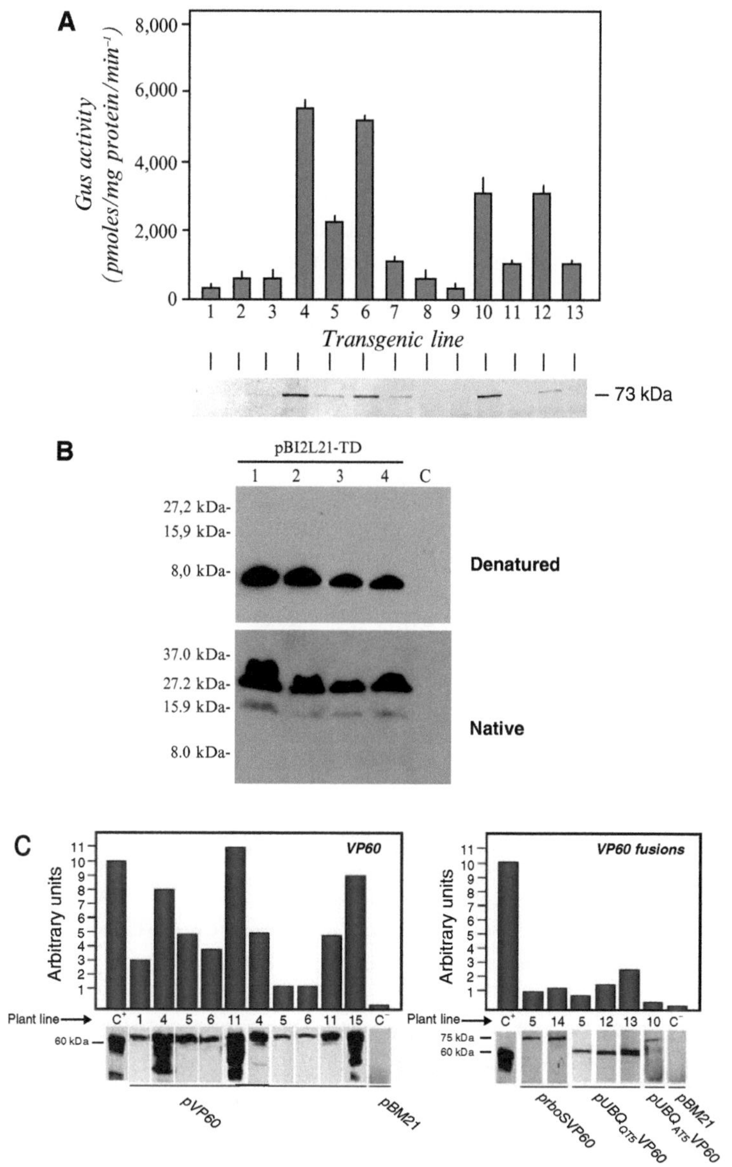

Fig. 16.2. (**A**) Characterization of transgenic plants obtained with plasmid pBI-2L21. This panel represents the levels of GUS activity in transgenic T2 lines transformed with plasmid pBI121 (1) or pBI-2L21 (2-13). GUS activity of leaves in individual transgenic lines was determined and shown as produced amount of 4-methylumbelliferone per min per mg soluble leaf protein at 37?C. Bars represent the mean of GUS activity and standard deviation of different leaves from the same plant. This panel also shows the expression of 2L21-GUS protein in leaves from the same plants (T_1), detected by Western blot using 30 μg of soluble leaf proteins and MAb 3C9. (**B**) Expression of the 2L21-TD polypeptide in transgenic plants. Protein extracts from 4 plant lines were studied by: denaturing PAGE and Western-blot assays using the 3C9 mAb as detector and; non-denaturing PAGE and analysed by Western blot using the same mAb as probe. (**C**) Western-blot analysis for detection of VP60 protein in leaves (30 μg of total leaf soluble proteins) from transgenic Arabidopsis thaliana plant lines (transformed with pBIVP60 or with plasmids expressing VP60 fusions) using specific anti-VP60 monoclonal antibody. VP60 expression levels found in different plants are referred to that obtained with the baculovirus system (15 μg of total soluble proteins from VP60 recombinant baculovirus-infected SF9 insect cells).

0.15% and 3.3% of total soluble leaf protein. It represents yields of 2L21-GUS protein around 75 μg/g of fresh weight of transgenic plant leaves. **Figure 16.2c** shows signals obtained with different concentrations of a recombinant CPV VP2 and compared to those registered with different leaf extracts, as detected by Western blot using the 3C9 MAb as probe. These quantification results were confirmed using the same assay and analysing the extinction signals of serial dilutions of known amounts of CPV VP2, a synthetic 2L21 peptide coupled to KLH protein and leaf extracts. Recombinant protein was histochemically detected by GUS expression in all plant tissues.

16.3.4. Analysis of Expression of 2L21-TD Fusion Protein in Transformed Plants

Twenty five independent kanamycin-resistant transgenic lines were analysed, appearing similar in morphology to wild-type plants and T1 plants were selected based on their 2L21-TD expression levels and sequentially self-pollinated to obtain a T3 generation that was finally used for estimation of yields of the recombinant protein.

Four transgenic lines presenting the highest transcriptional levels for the 2L21-TD gene were analysed for protein expression. Expression of the 2L21-TD fusion in transgenic plants was analysed in both denaturing and native PAGE and Western-blot assays. For denaturing assays, TSPs were obtained by grinding fresh leaves in protein extraction buffer for denaturing SDS-tricine-PAGE. Plant saps were centrifuged at 12,000*g* at 4?C for 15 min and TSP concentrations of the resulting supernatants were measured by Bradford assay *(15)*. Samples containing 40 μg of TSP were resolved in 15% tricine-SDS-PAGE, as previously described *(16)*. Gels were transferred to nitrocellulose membranes and probed using the 3C9 MAb.

A single band with the expected electrophoretic mobility (~7–8 kDa) was readily detected in all tested plants indicating effective expression of the transgene which seemed to roughly correlate with the corresponding transcriptional levels previously registered. To test the ability of the recombinant chimeras to form oligomeric structures, native protein extracts obtained as described above but using an extraction buffer lacking of EDTA and β-mercaptoethanol, were resolved by native 15% tricine-PAGE. Immunoblotting analyses using the 3C9 mAb showed that all plant lines presented a predominant band with a electrophoretic mobility of ~28–30 kDa and a second band with much lower intensity around 16 kDa as expected for 2L21-TD tetramers and dimers, respectively (**Fig. 16.2**). These results indicated that most of the recombinant 2L21-TD expressed in plants was accumulated in stable tetrameric forms, with expression levels that resembled those registered in the corresponding denaturing Western-blot assays. Dimeric species probably represent intermediate forms since it has been described that the native tetramers are actually constituted by dimers of two primary dimers *(17)*. In both assays, no immunoreactive bands could be detected in protein extracts from plants transformed with pBI121 plasmid. Western-blot results presented above showed that tetrameric and monomeric forms of the fusion were immunodetected by the 3C9 mAb with approximately similar efficiency. Based on these observations, the amount of recombinant antigen produced in 2L21-TD transgenic plants was roughly estimated by coating enzyme linked immunosorbent (ELISA) plates with serial dilutions of a 2L21 synthetic peptide and native protein extracts from 2L21-TD transgenic line #6, using the 3C9 mAb to detect the 2L21 peptide. Nanograms of 2L21 synthetic peptide and μg of TSP from 2L21-TD transgenic leaves that produced the same ODs within a linear range (between 0.25 and 0.65 absorbance units) were determined and a correlation between them was established (correlation R2 = 0.9977). Using this approach, we estimated that line #6 of 2L21-TD transgenic plants contained ~12 ng of 2L21 peptide per microgram of TSP (1.2%).

16.3.5. Analysis of Expression of VP60 Fusion Proteins in Transformed Plants

Leaves from 25 independent transgenic lines were analysed by Western blot for VP60 protein expression. Plant soluble proteins were extracted from fresh leaves by homogenization in extraction buffer containing 50 mM TRIS-HCl pH 7.4, 10 mM EDTA pH 8.0, 0.1% TRITON X-100, 1% β-mercaptoethanol, 2 mM PMSF followed by centrifugation at 12,000 *g* for 20 min at 4^{0}C. Thirty microgram of TSP from the resultant supernatants were resolved in 12% SDS-polyacrilamide gels (PAGE) and transferred to nitrocellulose membrane. Western-blot analysis were performed using an specific anti-VP60 MAb (Ingenasa, Modrid) as primary probe followed by and anti-mouse IgG conjugated with horseradish peroxidase (Amersham Biosciences, piscataway, New Jersey). Reactions were developed using ECL-detection system kit (Amersham Biosciences, piscataway, New Jersey) according to manufacturer's instructions.

While control plants transformed with pBI121 plasmid did not show VP60 expression, more than 50% of plants transformed with plasmids *pVP60* showed detectable levels of the VP60 recombinant protein (**Fig. 16.2**). As expected, the amount of VP60 antigen detected in leaves was variable among individual clones. As it was previously reported for the expression of this protein in plants and insect cells, the VP60 protein was detected as a major band of 60 kDa and two additional discrete reactive bands of lower MW, probably arisen by proteolytic degradation of VP60 within the plant cell. Accumulation of the recombinant protein was compared by densitometry in Western blots using baculovirus-expressed VP60 proteins as standard (AST/89; gently provided by Dr. F. Parra, Universidad de Oviedo). Most of the plants expressed between five and ten times lower amounts of recombinant protein per microgram of soluble protein than the baculovirus-infected cell extracts expressing VP60. Nevertheless, some plant lines showed expression levels similar to those of baculovirus-infected cell extracts. On the basis of these results, we estimated that the accumulated recombinant VP60 in these high expressing plants could represent between 0.3% and 0.8% of the total soluble leaf proteins.

Soluble proteins from leaves of independent transgenic lines obtained with plasmids containing the *rbcS*- and *UBQ*-fusion constructs were also analysed by Western blot. Only in two plant lines a protein of predicted molecular weight of 74 kD was detected which corresponded to fusion of *rbcS* and viral VP60 protein (**Fig. 16.2**). In 3 out of 25 transgenic lines obtained after transformation with $pUBQ_{G76}VP60$ construct showed detectable levels of recombinant protein of expected size (60 kD), which confirmed that *UBQ* moiety was rapidly removed from the initial translational product by *UBQ* C-specific hydrolases to release the fused protein in unmodified form as has been shown previously for some proteins *(10)* (**Fig. 16.2**). The expression levels of VP60 in these plants were also compared to those obtained in the baculovirus system, representing expression levels lower than 0.1% of total soluble leaf proteins. We failed to detect the accumulation of VP60 viral protein in any of 25 transgenic lines obtained after infiltration with $pUBQ_{A76}VP60$ which represented the un-cleavable form of the UBQ-fusion. As it was previously shown, non-cleavable variants of UBQ-protein fusion could become substrates for further ubiquitination. The lack of detectable expression of viral antigen from this construct is probably due to rapid polyubiquitination and consequent degradation of fused UBQ-VP60 protein.

For detection of stable oligomeric forms of VP60 in transgenic plant leaves, leaf material from plants transformed with the different constructs were homogenized in VLPs extraction buffer (1 g/ml). After centrifugation for 5 min at 3,500 *g*, the supernatants were precipitated twice with 8% PEG and the resulting pellets were re-suspended in VLPs suspension buffer, resolved in 12% denaturing SDS-PAGE and analysed by Western blot for presence of VP60 related antigens as described above.

The sedimentation pattern of the VP60 produced in *pVP60*-transformed plants was further analysed by subjecting the pellets from the first precipitation to zonal sedimentation through a sucrose gradient (7.5–30%). Sixteen fractions of 0.75 ml, representing different sedimentation densities, were collected and 20 µl of each were analysed in 12% SDS-PAGE electrophoresis and Western blot.

Only leaf extracts from plants transformed with plasmid $pVP60_O$, expressing the non-fused VP60, showed positive reactions in the pellet after sedimentation, indicating the presence of complex structures formed by oligomerization of the recombinant

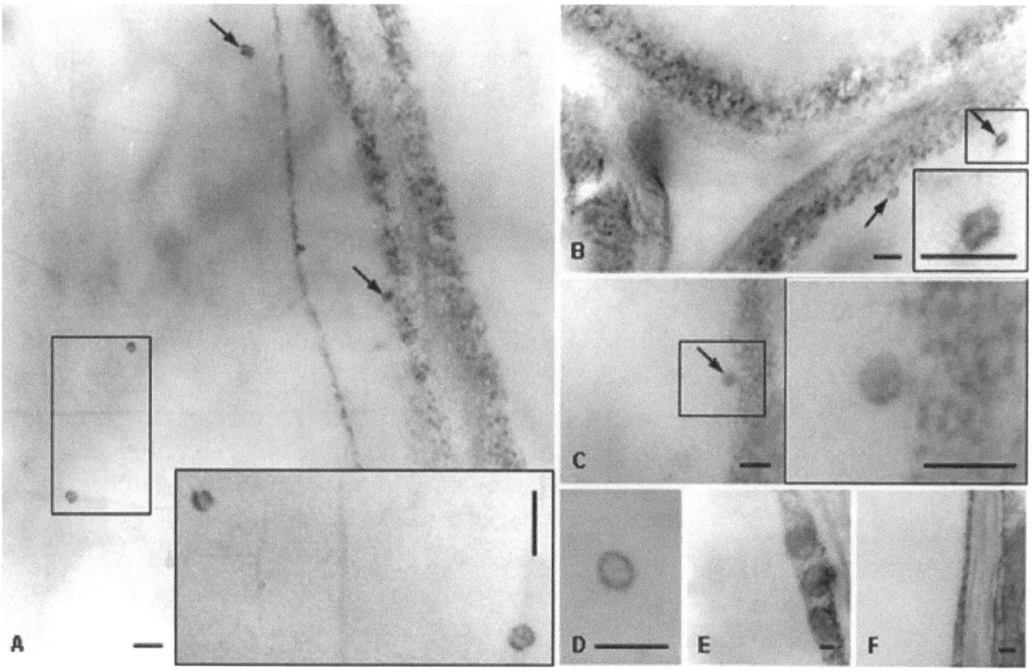

Fig. 16.3. Identification of VLPs in VP60 transgenic plants by electron microscopy. (**A–D**) Tissue sections of pVP60$_0$-transformed plants. See VLPs (arrows). Details from VLPs are enlarged in the insets. (**D, E**) Tissue sections of prbcSVP60 transgenic plants. No particles can be observed. Scale bar = 100 nm.

VP60 antigen produced in this plants (**Fig. 16.3**). The sedimentation pattern of the VP60 produced in *pVP60$_O$*-transformed plants was further analysed by subjecting the pellets from the first precipitation to zonal sedimentation through a sucrose gradient. As it is shown in **Fig. 16.3**, two bands of 60 and 30 kDa were revealed by Western-blot analysis in different fractions, showing different proportions between them along the gradient. These observations may suggest the existence of structures with different sedimentation coefficient produced by recombinant VP60 expressed in these plants.

16.4. Notes

1. The strategies of fusion peptides to GUS or to tetramerization domain of protein p53 facilitate the foreign peptide accumulation in transgenic plants at levels up to 1% of TSP (**Subheadings 16.3.3** and **16.3.4**).

2. Peptide fused to both carriers preserved its antigenic and immunogenic properties, being a good alternative to produce in plants peptide vaccines *(5, 6)*.
3. Other peptide sequences, such as amphipathic proline-rich sequences, have been also reported to enhance accumulation of proteins within plant cells *(18, 19)*.
4. We determined that VP60 protein from RHD virus, expressed in plants as single recombinant protein, resembles a similar conformation of complex structures that found when the protein was expressed in other systems (**Subheading 16.3.5**)
5. Previous papers reported other core proteins, naturally tending to produce oligomeric forms within cells, to accumulate in higher levels than proteins monomers *(20)*. Higher order structures probably makes these protein more resistant to endogenous proteases within plant cells *(21–23)*
6. In contrast to our results obtained with peptides, fusion strategies did not represent an advantage for the VP60 antigen. Our data may indicate that for highly stable self-assembling proteins, such as VP60, translational fusions may disturb protein assembling and did not represent a useful alternative to improve their accumulation (**Subheading 16.3.5**).

References

1. Arakawa, T., Chong, D. K. and Langridge, W. H. (1998). Efficacy of a food plant-based oral cholera toxin B subunit vaccine. Nat Biotechnol 16, 292–297.
2. Modelska, A., Dietzschold, B., Sleysh, N., Fu, Z. F., Steplewski, K., Hooper, D. C., Koprowski, H. and Yusibov, V. (1998). Immunization against rabies with plant-derived antigen. Proc Natl Acad Sci U S A 95, 2481–2485.
3. Tacket, C. O., Mason, H. S., Losonsky, G., Estes, M. K., Levine, M. M. and Arntzen, C. J. (2000). Human immune responses to a novel norwalk virus vaccine delivered in transgenic potatoes. J Infect Dis 182, 302–305.
4. Streatfield, S. J., Lane, J. R., Brooks, C. A., Barker, D. K., Poage, M. L., Mayor, J. M., Lamphear, B. J., Drees, C. F., Jilka, J. M., Hood, E. E. and Howard, J. A. (2003). Corn as a production system for human and animal vaccines. Vaccine 21, 812–815.
5. Gil, F., Brun, A., Wigdorovitz, A., Catala, R., Martinez-Torrecuadrada, J. L., Casal, I., Salinas, J., Borca, M. V. and Escribano, J. M. (2001). High-yield expression of a viral peptide vaccine in transgenic plants. FEBS Lett 488, 13–17.
6. Gil, F., Reytor, E., Perez-Filgueira, D. M. and Escribano, J. M. (2007). Multimerization of peptide antigens for production of stable immunogens in transgenic plants. J Biotechnol 128, 512–518.
7. Gil, F., Titarenko, E., Terrada, E., Arcalis, E. and Escribano, J. M. (2006). Successful oral prime-immunization with VP60 from rabbit haemorrhagic disease virus produced in transgenic plants using different fusion strategies. Plant Biotechnol J 4, 135–143.
8. Chiapello, H., Lisacek, F., Caboche, M. and Henaut, A. (1998). Codon usage and gene function are related in sequences of Arabidopsis thaliana. Gene 209, GC1-GC38.
9. Sambrook, J., Fritsch, E. F. and Maniatis, T. (1989) Molecular Cloning, A Laboratory Manual, 2nd ed. Cold Spring Harbor, New York: Cold Spring Harbor Laboratory Press.

10. Hondred, D., Walker, J. M., Mathews, D. E. and Vierstra, R. D. (1999). Use of ubiquitin fusions to augment protein expression in transgenic plants. Plant Physiol 119, 713–724.
11. Gehrke, L., Auron, P. E., Quigley, G. J., Rich, A. and Sonenberg, N. (1983). 5'-Conformation of capped alfalfa mosaic virus ribonucleic acid 4 may reflect its independence of the cap structure or of cap-binding protein for efficient translation. Biochemistry 22, 5157–5164.
12. Bechtold, N. and Pelletier, G. (1998). In planta Agrobacterium-mediated transformation of adult Arabidopsis thaliana plants by vacuum infiltration. Methods Mol Biol 82, 259–266.
13. Gomez, N., Carrillo, C., Salinas, J., Parra, F., Borca, M. V. and Escribano, J. M. (1998). Expression of immunogenic glycoprotein S polypeptides from transmissible gastroenteritis coronavirus in transgenic plants. Virology 249, 352–358.
14. Jefferson, R. A., Kavanagh, T. A. and Bevan, M. W. (1987). GUS fusions: beta-glucuronidase as a sensitive and versatile gene fusion marker in higher plants. Embo J 6, 3901–3907.
15. Bradford, M. M. (1976). A rapid and sensitive method for the quantitation of microgram quantities of protein utilizing the principle of protein-dye binding. Anal Biochem 72, 248–254.
16. Schagger, H. and von Jagow, G. (1987). Tricine-sodium dodecyl sulfate-polyacrylamide gel electrophoresis for the separation of proteins in the range from 1 to 100 kDa. Anal Biochem 166, 368–379.
17. Sakamoto, H., Lewis, M. S., Kodama, H., Appella, E. and Sakaguchi, K. (1994). Specific sequences from the carboxyl terminus of human p53 gene product form anti-parallel tetramers in solution. Proc Natl Acad Sci U S A 91, 8974–8978.
18. Kogan, M. J., Dalcol, I., Gorostiza, P., Lopez-Iglesias, C., Pons, M., Sanz, F., Ludevid, D. and Giralt, E. (2001). Self-assembly of the amphipathic helix (VHLPPP)8. A mechanism for zein protein body formation. J Mol Biol 312, 907–913.
19. Fernandez-Carneado, J., Kogan, M. J., Castel, S. and Giralt, E. (2004). Potential peptide carriers: amphipathic proline-rich peptides derived from the N-terminal domain of gamma-zein. Angew Chem Int Ed Engl 43, 1811–1814.
20. Perez-Filgueira, D. M., Brayfield, B. P., Phiri, S., Borca, M. V., Wood, C. and Morris, T. J. (2004). Preserved antigenicity of HIV-1 p24 produced and purified in high yields from plants inoculated with a tobacco mosaic virus (TMV)-derived vector. J Virol Methods 121, 201–208.
21. Gross, I., Hohenberg, H. and Krausslich, H. G. (1997). In vitro assembly properties of purified bacterially expressed capsid proteins of human immunodeficiency virus. Eur J Biochem 249, 592–600.
22. Gross, I., Hohenberg, H., Huckhagel, C. and Krausslich, H. G. (1998). N-Terminal extension of human immunodeficiency virus capsid protein converts the in vitro assembly phenotype from tubular to spherical particles. J Virol 72, 4798–4810.
23. Sojikul, P., Buehner, N. and Mason, H. S. (2003). A plant signal peptide-hepatitis B surface antigen fusion protein with enhanced stability and immunogenicity expressed in plant cells. Proc Natl Acad Sci U S A 100, 2209–2214.

Chapter 17

Immunomodulation of Plant Function by In Vitro Selected Single-Chain Fv Intrabodies

Manfred Gahrtz and Udo Conrad

Summary

In this chapter, we discuss and compare the different concepts and examples as well as present the basic protocols for applying intrabody-based approaches in plants for the investigation of cell functions and plant cell–pathogen interactions. The immunomodulation strategy, a molecular technique that allows to interfere with cellular metabolism, signal transduction pathways, or pathogen infectivity, is based on the ectopic expression of genes encoding specific recombinant antibodies. This needs basic prerequisites to be successfully applied as resources and techniques to isolate specific recombinant antibodies with sufficient binding parameters to bind and to block even low-concentrated targets or to compete successfully with substrates and ligands. Also techniques and constructs to efficiently transform plants and to target recombinant antibodies to selected compartments are important requirements. Basic protocols for all these techniques are provided.

Key words: Intrabodies, Immunomodulation, Recombinant antibodies, Phage display, Plant transformation.

17.1. Introduction

Intracellular antibodies or intrabodies can be defined as antibody molecules that are expressed inside of cells. Most commonly, intrabodies were expressed in mammalian cells. They can be specifically directed to different subcellular compartments. After initial reports *(1, 2)* the intrabody concept has been further developed and especially applied for cancer therapy, for the treatment of

Loïc Faye and Véronique Gomord (eds.), *Methods in Molecular Biology, Recombinant Proteins From Plants, vol. 483*

DOI: 10.1007/978-1-59745-407-0_17

HIV-1 infections and for the development of medicals for several other diseases (for review, *see* **refs.** *3, 4)*. A variety of other eukaryotic and prokaryotic cells have also been used for the intracellular expression of recombinant antibodies. In this chapter, we focus to the intracellular expression of recombinant antibodies in plant cells. The specific task is to discuss and to compare the different concepts and examples as well as to present the basic protocols for applying intrabody-based approaches in plants for the investigation of cell functions and plant cell–pathogen interactions. For this general strategy, the specific binding behavior of antibodies is used to affect an intracellular target, for example, a protein, a pathogen-derived structure, or a secondary metabolite as a phytohormone or a toxin. This strategy, called immunomodulation, can be defined as a molecular technique allowing interference with cellular metabolism, signal transduction pathways, or pathogen infectivity caused by the ectopic expression of genes encoding specific recombinant antibodies *(5, 6)*. Basic prerequisites to apply this technology are as follows:

1. Resources and developed techniques to isolate specific recombinant antibodies with sufficient binding parameters to bind and to block even low-concentrated targets or to compete successfully with substrates and ligands.
2. Techniques and constructs to efficiently transform plants and to target recombinant antibodies to selected compartments.

The function of an intrabody in plant cells is to bind to an intracellular target antigen. Therefore, the antigen-binding fragments of antibodies are sufficient for immunomodulation. The most common format used in the studies reviewed here is the single-chain Fv (scFv, *7)*, but also a whole collection of other antibody formats could be used (for review, *see* **ref.** 8*)*. The specific recombinant antibodies could be designed by cloning of the expressed VH and Vκ genes from known hybridomas secreting specific monoclonal antibodies *(9)*. The scFv coding sequences are then designed by PCR assembly *(10)*, by a two-step cloning procedure *(11)*, or by RACE *(12)*. A second and powerful way to select specific recombinant antibodies is the so-called phage display technology. The display of peptides or proteins on phage surfaces has been described for the first time by Smith in 1985 *(13)*. In 1990 McCafferty and coworkers were the first to show the successful display of functional antibody fragments on phage surfaces *(14)*, and in 1991 the antigen-driven selection of specific recombinant antibodies from phage display libraries was described *(15)*. Since this time, several other antibody formats as VH, VHH, and Fab have been displayed and different types of libraries have been developed and used (for review, *see* **ref.** *16)*. Two most critical points in terms of antibody properties have to be taken into account, if the recombinant antibodies will be used for immunomodulation: (a) a sufficient affinity and (b) a structure to be well-folded and stable even at reducing conditions in the cytosol of plant cells. To address the first point, techniques as affinity maturation *(17)* or improved screening techniques as yeast surface display can be applied *(18)*. The second problem was addressed by the generation of scFv scaffolds, which fold into functional molecules in the cytoplasma using genetic engineering, molecular evolution, or molecular chaperones (for review, *see* **ref.** *3)*. Another possibility is the direct in vivo selection of stable fragments by the intracellular antibody capture technology developed in mammalian systems *(19, 20)*. The latter approaches have not been applied for plantibodies, so far.

Since the first report of antibody expression in transgenic plants *(21)*, recombinant antibodies have been expressed in several plant species (for review *see* **refs.** *6, 22)* including dicots and monocots *(23)*. Especially scFvs, for the first time expressed in plants by Owen and coworkers in 1992 *(24)*, have been expressed in several plant cell compartments and plant organs (for review, *see* **ref.** *25)*. Also fusion proteins of scFv have been expressed in plant cells to improve the accumulation in the cytosol *(26)*. To achieve stable and long-time expression of the recombinant antibodies, as necessary for successful immunomodulation, *Agrobacterium tumefaciens*-mediated gene transfer was generally used.

To summarize, methods have been developed to select and to characterize specific and high-affine recombinant antibodies against target antigens in plant cells. Furthermore, the necessary technology to accumulate these specific antibodies in the cell compartment and in the organ of choice in transgenic plants is generally available. In the following, we will shortly discuss, how these technologies are applied to study cellular functions and/or to create pathogen and drug resistance. The first example for immunomodulation of specific functions in plant cells was published already in 1992. Owen and coworkers ubiquitously expressed a scFv against phytochrome in transgenic tobacco and showed aberrant light-controlled germination in homozygous transgenic seeds *(24)*. Jobling and coworkers expressed and targeted enzyme activity inhibiting single-domain antibody fragments (VHH) from camelids against the starch branching enzyme A (SBE A) into potato chloroplasts *(27)*. The authors demonstrated that these antibodies can be correctly targeted to subcellular organelles. The VHHs inhibited enzyme functions in plants more efficiently than antisense approaches. Thus, camelid single-domain antibodies have been demonstrated to be stable even at reducing conditions and could inhibit enzymes quiet efficiently. Other successful experiments describe the modulation of the flavonoid metabolism and the immunomodulation of the polyamine biosynthesis using scFv intrabodies inhibiting enzymes *(28, 29)*. ScFvs against small heat shock proteins (sHSP) have been expressed in the cytosol of transgenic tobacco plants to prevent the assembly of heat stress granula. In this experiments, the formation of heat shock granula by sHSPs and its disintegration was shown to be a prerequisite for survival of plants under continuous stress conditions *(30)*.

The immunomodulation of phytohormone functions was firstly shown for the phytohormone abscisic acid (ABA, *31)*. ABA functions in seed development, during early embryogenesis and in stomata development have been studied using this system *(32–34)*. The creation of an artificial ABA sink in the ER caused by the scFv accumulated in this compartment was found as a potential mechanism *(35, 6)*. A further example of phytohormone immunomodulation is giberellic acid *(36)*. In immunomodulated plants, an increase of the specific phytohormone content was often measured *(6)*. In this connection, Putalun and coworkers found a stimulation of solasodine glycoside synthesis in transgenic hairy roots expressing a specific antibody against this compound, thus representing a new plant breeding strategy to improve secondary metabolite production *(37)*. Plantibody projects were also successful in creating herbicide resistance and/or tolerance *(38–40)*, fungicide tolerance *(41)*, and mycotoxin tolerance *(42)* in transgenic plants.

The creation of plant pathogen resistance via plantibodies was pioneered by the work of Tavladoraki and coworkers *(43)*, who created *Nicotiana benthamiana* plants resistant against the Artichoke Mottled Crinkle Virus (AMCV) by expression of scFv against the AMCV coat protein. In the following, antibody-based resistance to plant pathogens was further developed (for review, *see* **ref.** *44)*. In the last years, several new concepts were developed and applied. Fusion proteins consisting of a Fusarium-specific antibody linked to antifungal peptides were expressed in *Arabidopsis thaliana* and conferred high levels of protection *(45)*. ScFvs against a plant viral RNA-dependent RNA polymerase were expressed in *N. benthamiana* plants and caused multiresistance against four different plant viruses *(46)*. Specific scFv isolated from a stable single-framework phage display library using a scaffold of a cytosolic stable scFv *(43)* caused resistance against the Cucumber Mosaic Virus *(47)*. The fusion of tetrapeptide KDEL even in case of cytosolic accumulation can also cause stable expression and resistance against viruses *(48)*.

Overall, the examples mentioned here demonstrate the high potential of the intrabody concept also in plant science. Methods and strategies developed in the last 15 years could now be applied in cell physiology, developmental biology and in plant pathogen research. We have selected specific protocols describing essential methods for applying the immunomodulation approach.

The protocols presented in this chapter are as follows:

17.3.1. Selection of antigen-specific scFv via phage display

17.3.1.1. Immunoselection

17.3.1.2. Monoclonal phage enzyme-linked immunoabsorbant assay (ELISA)

17.3.1.3. Checking the functionality of the soluble recombinant antibodies by indirect ELISA

17.3.2. Expression of single-chain Fv in tobacco

17.3.2.1. Construction of the expression plasmid

17.3.2.2. Transient expression of single-chain Fv antibodies in *N. tabacum* or *N. benthamiana*

17.3.2.3. Expression of single-chain Fv antibodies in stable transformed *N. tabacum* or *N. benthamiana*

17.3.2.4. Harvesting of leaf tissue and protein extraction

17.3.2.5. Western-blot analysis of leaf extracts

17.2. Materials

17.2.1. Selection of Antigen-Specific scFv via Phage Display

1. Antigen, in phosphate-buffered saline (PBS), usually *Escherichia coli*-expressed recombinant protein, full-length or partial.
2. ScFv-presenting M13 phagemid libraries: Two single framework libraries (A and B) kindly provided by Ian Tomlinsson (MRC, Cambridge, UK) were used *(49)*.
3. *E. coli* strains TG1 (GE Healthcare, Piscataway, NJ), HB2151 (GE Healthcare), XL1-Blue (Stratagene, La Jolla, CA).
4. Helper phage M13KO7 (GE Healthcare).
5. Maxisorb™ microtiter plate (Nunc™ Roskilde, Denmark).
6. BioAssay dish, 243 × 243 mm (Nunc).
7. Media for growth of *E. coli*:

   M9 medium: 0.6% Na_2HPO_4, 0.3% KH_2PO_4, 0.05% NaCl, 0.1% NH_4Cl, adjust pH to 7.4 and autoclave. Then add to 1 l medium 2 ml 1 M $MgSO_4$, 10 ml 20% (w/v) glucose, and 0.1 ml 1 M $CaCl_2$.
   2× TY medium: 1.6% tryptone, 1% yeast extract, 0.5% NaCl.
   TYE medium: 1% tryptone, 0.5% yeast extract, 0.8% NaCl.
   For solid medium add 1.5% agar prior autoclaving.
8. PBS: 8 mM Na_2HPO_4, 2 mM KH_2PO_4, 0.15 M NaCl, pH7; PBST: PBS + 0.1% Tween 20.
9. PEG/NaCl: 20% (w/v) polyethylene glycol 6000, 2.5 M NaCl.

10. Freshly made 100 mM triethylamine.
11. Antibiotics: Ampicillin, kanamycin. Dissolve in water and filter sterilize.
12. Isopropyl β-D-1-thiogalactopyranoside (IPTG, Fluka 59740, Sigma-Aldrich, St. Louis, MO).
13. Bovine serum albumin (BSA).
14. Antibodies for the detection in ELISA: Anti-c-myc monoclonal antibody (9E10), anti-mouse IgG alkaline phosphatase (ALP) coupled (Sigma, A4312), anti-M13 horseradish peroxidase (HRP) monoclonal conjugate (GE Healthcare, 27-9421-01).
15. 3,3',5,5'-Tetramethylbenzidine (TMB, Sigma, T0565), stock solution at 10 mg/ml in DMSO.
16. *p*-nitrophenylphosphate (Sigma, N2640-100Tab).

17.2.2. Expression of Single-Chain Fv in Tobacco

1. Plant material: *N. tabacum* (e.g., cv. Samsun NN), *N. benthamiana*.
2. Plasmids pMG3004, 7P-Nos (DNA Cloning Service, Hamburg, Germany, http://www.dna-cloning-service.com), pCB301-Kan *(50)*.
3. All you need for standard molecular biology techniques, like restriction enzymes, T4 DNA ligase, Shrimp ALP, agarose and agarose gel equipment, DNA Mini prep kit (e.g., QIAprep Spin Miniprep Kit, 27104, Qiagen, Hilden, Germany), oligonucleotide primers, etc.
4. Antibiotics: Ampicillin, spectinomycin, rifampicin, carbenicillin, kanamycin. Dissolve in water and filter sterilize.
5. YEB medium: 0.5% beef extract, 0.1% yeast extract, 0.5% peptone, 0.5% sucrose, pH 7.0, after autoclaving add $MgSO_4$ to a final concentration of 2 mM.
6. SOC medium: 20 g/l tryptone, 5 g/l yeast extract, 0.5 g/l NaCl, and 2.5 ml 1 M KCl/l, pH 7.0, after autoclaving add 1 ml 1 M $MgCl_2$, 1 M $MgSO_4$, and 2 ml 1 M glucose.
7. 1 M acetosyringone in DMSO (3',5'-Dimethoxy-4'-hydroxyacetophenone, Fluka 38766).
8. 1 M MES, pH 5.6.
9. 2× infiltration buffer: Murashige and Skoog (MS) Basal Salt Mixture ready to use powder including vitamins *(51)* (Sigma M5519) 8.8 g/l, glucose 3.6 g/l, and sucrose 100 g/l.
10. Single-use 1-ml syringes, 13-mm cork borer, pestle for 1.5-ml microcentrifuge tube.
11. MS medium *(51)*: Ready to use MS powder including vitamins (Sigma M5519) 4.4 g/l, sucrose 30 g/l (20 g/l for liquid inoculation medium), MES 0.5 g/l, pH 5.6–5.8, agar 9 g/l.

12. MG medium: Ready to use MS powder (Sigma M5519) 4.4 g/l, glucose 16 g/l, MES 0.5 g/l, pH 5.6 to 5.8, agar 9 g/l.
13. Cefotaxim, Benzylaminopurine (BAP), α-Naphthaleneacetic acid (NES).
14. Phosphinothricin (Duchefa, Haarlem, The NetherlandsP0159.0250), BASTA® (Bayer CropScience, Monheim, Germany).
15. Bradford reagent (e.g., Bio-Rad Protein Assay, Bio-Rad, Munich, Germany 500-0006).
16. Sodium dodecyl sulfate-polyacrylamide gel electrophoresis (SDS-PAGE) sample buffer *(52)*: 62.5 mM Tris·Cl, pH 6.8, 25% (v/v) glycerol, 2% (w/v) SDS, 0.01% (w/v) bromophenol blue, 5% (v/v) β-mercaptoethanol.
17. SDS-PAGE equipment (e.g., Bio-Rad Mini PROTEAN® 3) including all necessary reagents to cast and to run the gels according to Laemmli *(51)*.
18. Western-blot equipment (e.g., Bio-Rad Mini Trans-Blot Cell®) including the nitrocellulose membrane (Schleicher & Schuell, Dassel, Germany) and all buffers for blotting and detection as described *(53)*.
19. Antibodies for the detection in Western-blot analysis: Anti-c-myc monoclonal antibody (9E10), anti-mouse Ig HRP coupled (GE Healthcare).
20. Chemiluminescent reagents for Western blotting (e.g., ECL Western Blotting Detection Reagents, GE Healthcare RPN2109).

17.3. Methods

The presented methods require a laboratory equipped for standard molecular biology, microbiology and immunochemistry. It is necessary to have a laminar flow, incubators (28°C, 30°C, 37°C), both for bacterial cultures and for microtiter plates, ELISA reader, ELISA washer, centrifuges also for microtiter plates, autoclaves, and a greenhouse or plant growth chambers. In most countries working with transgenic organisms is regulated. Make sure you fulfill all requirements regarding health and safety regulations.

The focus of the following protocols is on the handling of the phage display technology and the expression of recombinant antibodies in plants and not so much on standard molecular biology techniques (PCR, restriction, ligation, bacterial transformation by heat shock and/or electroporation, miniprep, midiprep). These techniques have not been described in detail here.

17.3.1. Selection of Antigen-Specific scFv via *Phage Display*

Each round of immunoselection spans 4 days; three rounds can be finished within 8 days.

17.3.1.1. Immunoselection

Day 1

1. Coat a microtiter plate (Maxisorb; *see* **Note 1**) with the antigen. The antigen concentration should be 10–100 µg/ml in PBS (*see* **Note 2**). Put 100 µl antigen solution into each of three wells of a 96-well microtiter plate for each library you want to screen and incubate overnight at room temperature.
2. Prepare a BioAssay dish (243 × 243 mm) with TYE (100 µg ampicillin/ml, 1% glucose).

Day 2

1. TG1 (*see* **Note 3**) grown on a M9 minimal medium plate supplemented with 0.5 mg thiamin is used to inoculate 3 ml 2× TY and grown for about 2.5 h at 37°C. Then dilute the culture 1:100 in 20 ml of fresh 2× TY and cultivate to OD_{600} of 0.4.
2. 2. Wash the coated wells three times with PBS.
3. 3. Block the wells with 200 µl 2% BSA in PBS each for 2 h at room temperature.
4. Wash the coated wells three times with PBS.
5. To each three wells give 10^{12} to 10^{13} phages in a volume of 90 µl per well and incubate for 30 min with and 1 h without agitating at room temperature.
6. Waste the free phages in the supernatant. Wash 20 times with PBST (*see* **Note 4**).
7. Elute bound phages from the microtiter plate wells with 100 mM triethylamine per well. Incubate agitating for 10 min at room temperature. During the incubation time prepare a 1.5 ml microcentrifuge tube with 135 µl 1 M Tris, pH 7.4.
8. Transfer the eluted phage from the three wells to the microcentrifuge tube for neutralization.
9. Add all of the neutralized eluted phages to 9 ml of the fresh TG1 culture of OD_{600} of 0.4–0.5. Incubate in a 15 ml Falcon tube for 30 min at 37°C. To harvest the residual phages from the well add 5 µl 1 M Tris pH 7.4 and then 50 µl of the TG1 culture. Incubate the plate at 37°C for 30 min.
10. Combine both cultures. For determining the phage titer dilute the culture 1:10^{-2}, 1:10^{-4} and 1:10^{-6} and plate 50 µl from each dilution on TYE (100 µg ampicillin/ml, 1% glucose). The titer should be 10^4 cfu/ml (*see* **Note 5**).

11. Centrifuge the culture for 10 min at 3,300 *g* and 4°C. Decant all but about 1 ml of the supernatant and resuspend the cells. Plate all on a BioAssay dish (243 × 243 mm) with TYE (100 μg ampicillin/ml, 1% glucose). Grow overnight at 37°C.

Day 3

1. Wash the cells from the BioAssay dish with 7 ml 2× TY, 15% glycerol using a Drigalski spatula.
2. Inoculate 20 ml 2× TY (100 μg ampicillin/ml, 1% glucose) with 50 μl of the cells. Grow the culture to an OD_{600} of about 0.4 (1–2 h). The remaining cells are stored at −80°C.
3. To 10 ml of the culture add 5×10^{10} helper phages (*see* **Note 6**).
4. Incubate for 30 min at 37°C in a water bath without shaking (*see* **Note** 7).
5. Centrifuge for 10 min at 3300 *g* and resuspend the pellet in 50 ml 2× TY (100 μg ampicillin/ml, 50 μg kanamycin/ml).
6. Incubate overnight at 30°C, while shaking.
7. To continue with the next round of selection start with the procedures from day 1 and go on the next days as described.

Day 4

1. Centrifuge the overnight culture for 15 min at 3,300 *g*.
2. Collect the supernatant and add 10 ml PEG/NaCl, mix and incubate for 1 h on ice to precipitate the phages.
3. Centrifuge for 30 min at 3,300 *g*. Decant the supernatant, recentrifuge, and remove the residual supernatant.
4. Resuspend the pellet in 1 ml PBS and centrifuge for 10 min at 11,600 *g* in a microcentrifuge to pellet cellular debris. The supernatant are the phages to be used in the next round of selection.
5. To 400 μl of phages add 200 μl 6% BSA in PBS. Start the next of altogether three rounds of selection.

Days 5–7

The following rounds are performed essentially as described above.

Day 8

1. The microtiter plates of round 3 are used to isolate single phage clones. Pick single colonies from the microtiter plates and put each into 100 μl 2× TY (100 μg ampicillin/ml, 1% glucose) in microtiter plates (Greiner 655182, Greiner Bio-One GmbH Frickenhausen, Germany; *see* **Note 8**). Grow overnight at 37°C and 500 rpm.

Day 9

1. Transfer 2 µl from each well to fresh microtiter plates (well per well using a multichannel pipet) containing 200 µl 2× TY (100 µg ampicillin/ml, 1% glucose) per well and shake 1 h at 37°C at 700 rpm.
2. Add 25 µl 2× TY (100 µg ampicillin/ml, 1% glucose) containing 10^9 helper phages (*see* **Note 6**). Incubate at 37°C (*see* **Note 7**) for 30 min and shake 1 h at 37°C at 300 rpm.
3. Spin down at 1,800 *g* for 10 min and remove the supernatants with a multichannel pipet. Resuspend the pellets in 200 µl 2× TY (100 µg ampicillin/ml, 50 µg kanamycin/ml) and cultivate at 30°C overnight.
4. Grow a culture (10 ml) of HB2151 (*see* **Note 3**) in 2× TY overnight at 37°C.

Day 10

1. Centrifuge the microtiter plate containing the phage solutions at 1,800 *g* for 10 min. Keep the supernatants for the monoclonal phage ELISA (*see* below).
2. Dilute the overnight culture of HB2151 1:100 in 2× TY and cultivate to an OD_{600} of 0.4.
3. Add 200 µl/well of this culture to microtiter plates (*see* **Note 8**) and infect each well with 10 µl monoclonal phage-containing supernatant. Cultivate for 30 min at 37°C without shaking. Remove 2 µl from each well and add glycerol to a concentration of 20% to the rest and freeze at –80°C.
4. Add the removed 2 µl of this cultures to 175 µl 2× TY (100 µg ampicillin/ml, 0.1% glucose) and shake at 500 rpm, 37°C for 3 h.
5. Add 25 µl TY (100 µg ampicillin/ml, 9 mM IPTG) and shake overnight at 30°C (the final concentration of IPTG is 1 mM).
6. Coat a microtiter plate (Maxisorb; *see* **Note 1**) with the antigen (*see* **Note 9**).

Day 11

1. Spin down the cells in the microtiter plates with the cultures for 10 min at 1,800 *g*. The supernatant should contain soluble scFv or VH and can be used for activity tests (ELISA, *see* below).

17.3.1.2. Monoclonal Phage ELISA

This method could be used to identify phages displaying recombinant antibodies specifically binding the respective antigen (*see* **Note 10**). Starting with phage collections after the selection and propagation a polyclonal phage ELISA could be used to monitor the enrichment.

Day 12

1. Take an antigen coated microtiter plate (Maxisorb, *see* day 10), remove the liquid by beating out, add 200 µl with 3% BSA in PBS for blocking, and incubate for 2 h at room temperature.
2. Remove the blocking solution and add 100 µl phages from day 10, diluted 1 + 1 with 2% BSA in PBS and incubate 1 h at room temperature.
3. Remove the phages and wash 5× with PBST.
4. Add 100 µl anti-M13-HRP monoclonal conjugate, 1:5,000 diluted in 1% BSA in PBS, and incubate for 1 h at 37°C.
5. Wash 5× with PBST and add 100 µl substrate solution (100 µg/ml TMB in 100 mM sodium acetate, pH 6.0 with 2 µl 30% hydrogen peroxide added to 10 ml of this solution immediately before use). Leave at room temperature and watch. A blue color should occur after 10 min.
6. Stop the reaction by adding 50 µl 1 M sulphuric acid. The blue color will turn yellow.
7. Read at 650 nm and 450 nm. Subtract OD_{650} from OD_{450} (*see* **Note 11**).

17.3.1.3. Checking the Functionality of the Soluble Recombinant Antibodies by Indirect ELISA

The method described here is based on the c-myc tag and its detection by a specific monoclonal antibody. It could also be used to detect the specific binding of recombinant antibodies expressed in plants to the corresponding antigens.

Day 13

1. Take an antigen coated microtiter plate (Maxisorb, *see* day 10), remove the liquid by beating out, add 200 µl of 3% BSA in PBS for blocking, and incubate for 2 h at room temperature. In parallel, use a microtiter plate coated with BSA as control for every clone.
2. Remove the blocking solution and add 100 µl antibody solution from day 11, diluted 1 + 1 in 2% BSA in PBS and incubate 1 h at room temperature.
3. Remove the antibody solution, wash 5× with PBST and add 100 µl anti-c-myc antibody (9E10) cell culture supernatant, 1:50 diluted in 1% BSA in PBS and incubate 1 h at room temperature.
4. Remove the anti-c-myc antibody, wash 5× with PBST and add 100 µl rabbit anti-mouse IgG-ALP (*see* **Note 12**) diluted 1:2,000 in 1% BSA in PBS for 1 h at room temperature.
5. Remove the conjugate, wash 5× with PBST, 1× with PBS (to avoid bubbles in the last step), and add 100 µl *p*-nitrophenylphosphate solution (1 tablet, 15 mg) diluted in 15 ml

50 mM sodium carbonate buffer, pH 9.6, 0.05 mM magnesium chloride (*see* **Note 13**). Read OD at 405 nm (*see* **Note 14**).

The clones coding for specifically binding recombinant antibodies are selected from the frozen cultures (day 10, **step 3**), plated on solid TYE medium (100 μg ampicillin/ml, 1% glucose), and after growth overnight are regrown in liquid 2× TY (100 μg ampicillin/ml, 1% glucose) again overnight. From these cultures plasmid DNA is prepared by a miniprep method and XL1-Blue (recA⁻) cells are transformed by each of the minipreps and plated. Colonies are selected, cultivated, and high-quality plasmid DNA is prepared and stored as stocks. With these DNAs HB2151 is retransformed. Selected colonies are grown and production of recombinant antibodies is induced by IPTG (*see* above). Supernatants are tested by ELISA again. In the case, the respective activity could be measured, determine the sequence of the corresponding recombinant antibody. In our case, the scFv clones are contained in the phagemid pIT1 (**Fig. 17.1**; *see* **Note 15**). Use the primers LMB3 and PHEN (*see* **Note 16**) for sequencing.

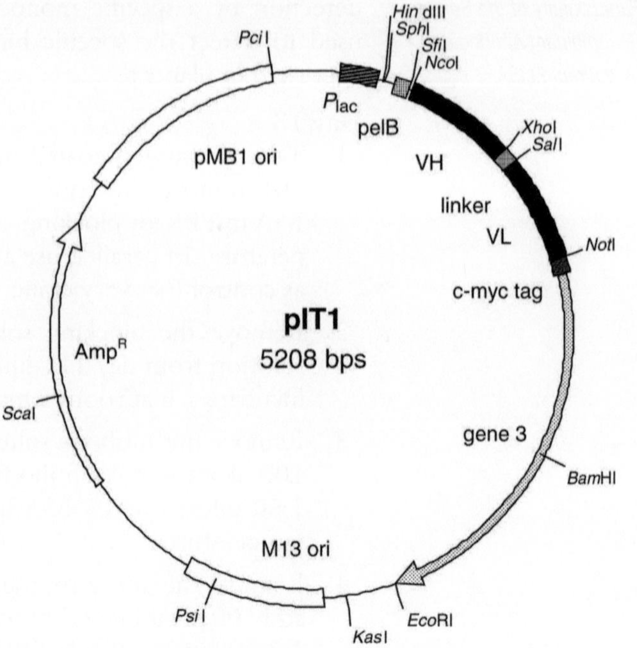

Fig. 17.1. Phagemid pIT1 harboring an scFv encoding insert. P_{lac}: Lac promoter, pelB: pelB signal sequence, VH, VL: Variable domain of the heavy chain and light chain, respectively, linker: $(G_4S)_3$-linker, gene 3: Phage M13 p3 minor coat protein gene, Amp^R: β-lactamase gene for ampicillin resistance.

17.3.2. Expression of Single-Chain Fv in Tobacco

17.3.2.1. Construction of the Expression Plasmid

For the transient and stable transformation of *Nicotiana* species with *A. tumefaciens* a binary expression plasmid for the respective scFv has to be constructed. Thereby the cloning techniques follow the standard procedures described by Sambrook and Russell *(54)*. The scFv encoding sequence can be isolated from the phagemid pIT1 (**Fig. 17.1**) by cutting with the enzymes *Nco*I and *Not*I. Subsequently, it can be ligated into the plant expression cassette vector pMG3004 (**Fig. 17.2**) that is cut with the same enzymes.

Cloning of the scFv sequence this way will result in cytoplasmic expression, since the LeB4 signal sequence will be removed from the plasmid. For targeting the scFv to the ER introduction of *Bam*HI sites at the start and stop codon by PCR using the oligonucleotides sc-Bam-5' and sc-Bam-3' (*see* **Note 17**) and subsequent cloning with *Bam*HI into pMG3004 will be necessary. In this case, the expression has to be targeted to a certain organ or tissue different promoters can be introduced into the construct using the *Eco*RI and *Kpn*I restriction sites. The expression cassette can be cloned with *Hin*dIII into a binary vector like 7P-Nos (**Fig. 17.3a**) or pCB301-Kan (**Fig. 17.3b**; *50*).

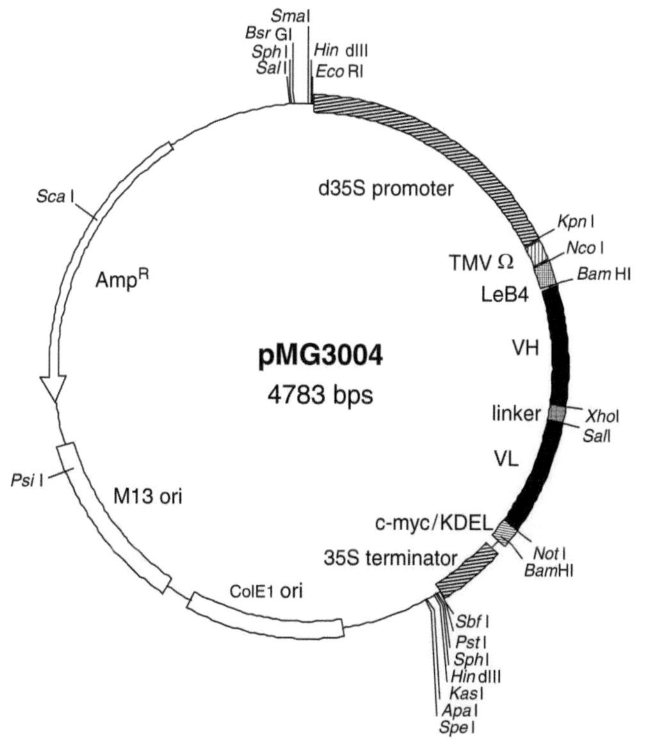

Fig. 17.2. Plasmid pMG3004. d35S promoter: Doubled enhanced CaMV 35S promoter, TMV Ω: Translational enhancer Ω from Tobacco Mosaic Virus (TMV), LeB4: Signal sequence from the *Vicia faba* legumin B4 gene, VH, VL: Variable domain of the heavy chain and light chain, respectively, linker: $(G_4S)_3$-linker, c-myc/KDEL: C-myc tag and KDEL ER retention signal, Amp^R: β-lactamase gene for ampicillin resistance.

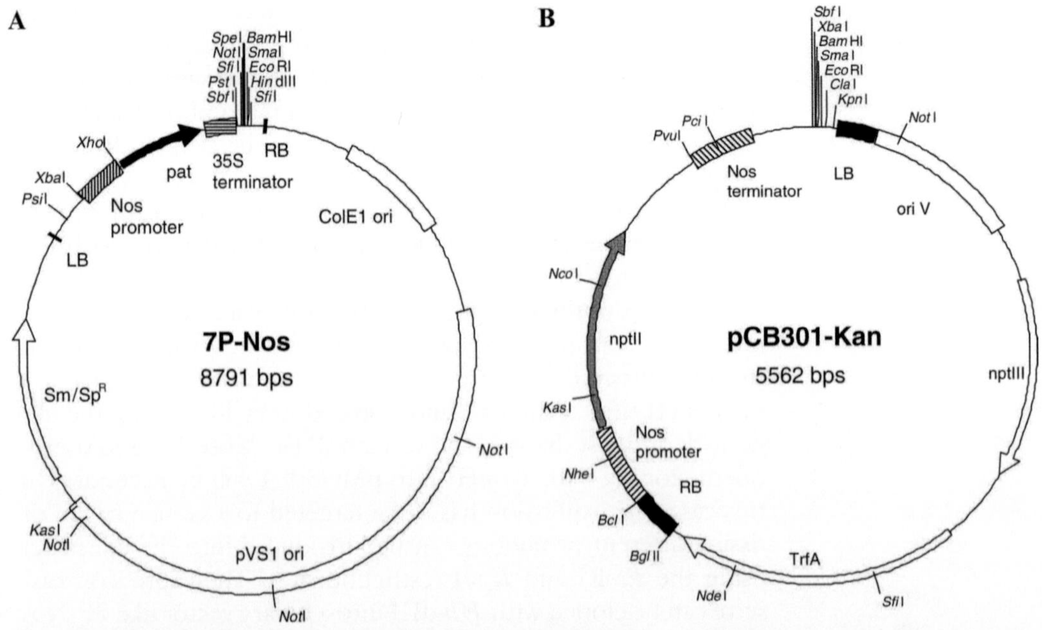

Fig. 17.3. Binary plasmids. A: 7P-Nos, B: pCB301-Kan; LB, RB: T-DNA left and right border, respectively. Nos promoter/terminator: Nopalin synthase promoter/terminator from *Agrobacterium tumefaciens*, Pat: Phosphinothricin acetyltransferase gene from *Streptomyces viridochromogenes* for resistance to the herbicide glufosinate-ammonium, 35S terminator: CaMV 35S terminator, Sm/Sp^R: Streptomycin/spectinomycin adenyltransferase gene *aadA*, nptIII: Aminoglycoside-3'-phosphotransferase and nptII: Neomycin phosphotransferase, both for kanamycin resistance. Genes of interests can be introduced between the right (RB) and left border (LB) using the single cutting restriction enzymes present in the multi-cloning-sites.

The binary plasmid harboring the scFv expression cassette is then transferred to a strain of *A. tumefaciens* like GV2260 *(55)* or LBA4404 *(56)* by electroporation *(57)*. We recommend checking for the integrity of the binary plasmid and the expression cassette in the transformant by a mini DNA preparation and subsequent restriction digest of the plasmid DNA to rule out any recombinations that might have occurred in the Agrobacteria clones.

17.3.2.2. Transient Expression of Single-Chain Fv Antibodies in N. Tabacum or N. Benthamiana

To test the stability and the activity of the scFv, transient expression gives a fast result circumventing the need of stable transformation. This will help to choose between different scFv for stable transformation.

Growth of Agrobacteria and infiltration of *Nicotiana* leaves.

1. Inoculate 10 ml YEB medium containing the respective antibiotics (*see* **Note 18**) with the *Agrobacterium* clones, including an untransformed control, and grow them overnight at 28°C shaking at 170 rpm.
2. Add to the culture acetosyringone up to a final concentration of 20 μM, glucose to a final concentration of 10 mM

and MES from a 1 M stock, pH 5.6 to a final concentration of 10 mM. The cultures are grown overnight as before.

3. The next day determine the OD_{600} and dilute the cultures to an OD_{600} of 1 in 1× infiltration buffer using the double concentrated buffer and water. Add to the diluted cultures acetosyringone to a final concentration of 200 µM.
4. Inject the *Agrobacterium* suspension into tobacco leaves using a single-use 1-ml syringe, by attaching the syringe to the lower leaf surface (**Fig. 17.4**). This is done until about half, or if necessary the whole leaf is infiltrated with the bacterial suspension. For each clone infiltrate at least leaves at two different plants to minimize the risk of low expression due to a nonfit plant. In the case of *N. tabacum*, use a 200-µl pipet tip and make tiny lesions into the lower epidermis, where the syringe is set on to facilitate infiltration.
5. The plants are left in the greenhouse for 4 days, before tissue samples are harvested (*see* **Note 19**).

17.3.2.3. Expression of Single-Chain Fv Antibodies in Stable Transformed N. Tabacum or N. Benthamiana (see Note 20)

1. Ten milliliter of overnight cultures of Agrobacteria clones grown in SOC medium (*see* **Note 18**) are diluted with 10 ml MS medium (2% sucrose).
2. Tobacco leaf squares (about 1-cm edge length), cut from 4- to 5-week-old tissue culture grown *N. tabacum* or *N. benthamiana* (*see* **Note 21**) are incubated with the bacterial suspension for 30 min. Here, it is important that the cut surfaces are in contact with the bacterial suspension.
3. The leaf squares are then shortly put on sterile filter paper to remove excess bacteria.
4. For cocultivation the squares are put upside-down on solid MS medium (3% sucrose) without any growth factors or

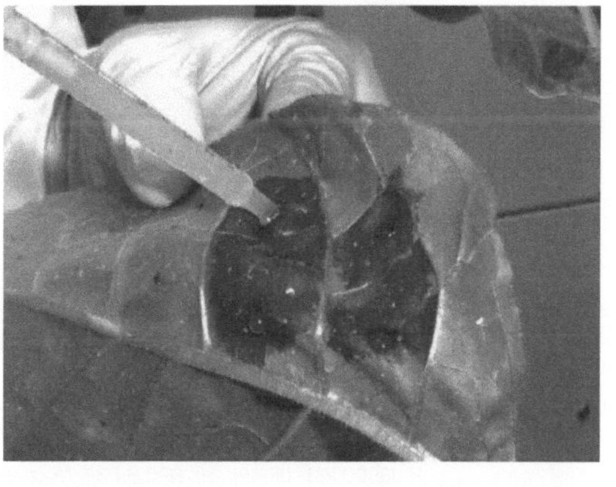

Fig.17.4. Injection of *Agrobacterium tumefaciens* suspension into tobacco leaves.

antibiotics in standard petri dishes and are incubated for 2 days in the dark at 24°C.

5. Then the leaf squares are transferred to solid MG medium (BAP 1 mg/l, NES 0.2 mg/l) with 500-mg cefotaxime/l to suppress the Agrobacteria and either kanamycin (50 mg/l) or phosphinothricin (10 mg/l), depending on the plant selection marker, to start the selection pressure. The plates are incubated under a 16 h light/8 h dark scheme at 24°C.
6. Every 7–10 days the leaf squares are transferred to fresh plates until shoots are emerging from the callus tissue (**Fig. 17.5**). At least four rounds of selection should be carried out.
7. Shoots, which are about 1 cm in size, are then transferred to solid MG medium without growth factors, but with 500 mg cefotaxime per liter and the suitable selecting agent as above in glass or plastic containers that allow further growth of the shoots. This will lead to rooting of the developing plantlets.
8. Rooted plantlets are cut above the root and transferred to solid MS medium (3% sucrose) containing 500 mg cefotaxime/l.
9. Strong plantlets that have rooted also here are then potted on soil and transferred into the greenhouse (*see* **Notes 22** and **23**).

17.3.2.4. Harvesting of Leaf Tissue and Protein Extraction

To test transiently transformed leaves or tissues from stable transformed plants, the tissue samples will be analyzed for the expression of the respective scFv, thereby taking advantage of the c-myc tag at the C-terminus of the protein.

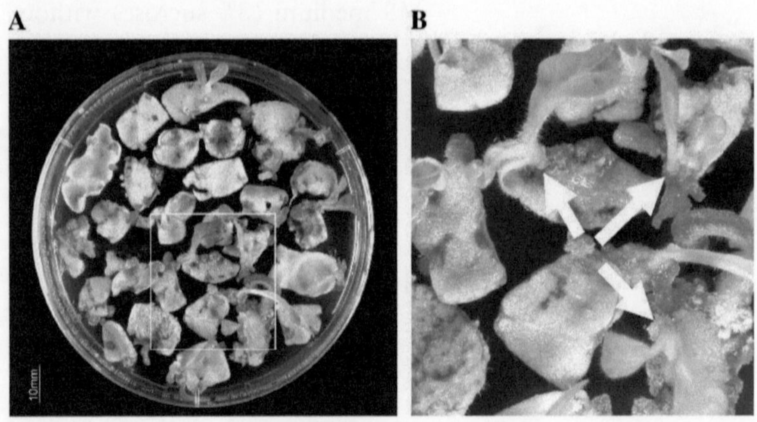

Fig. 17.5. Regeneration of transgenic tobacco shoots on MG medium (1 mg BAP/l, 0.2 mg NES/l, 500 mg cefotaxime/l, 50 mg kanamycin/l). (**A**) Petri dish with regenerating tobacco leaf tissue. Six and a half week after cocultivation still several shoots are developing. (**B**) Inset from A at higher magnification. Arrows indicate emerging shoots.

1. Leaf tissue is sampled with a 13-mm cork borer. Two leaf disks from each infiltrated leaf are transferred to a 1.5-ml microcentrifuge tube and are either frozen at –80°C until further use or directly used for protein extraction.
2. The tissue is homogenized in 100 µl SDS-PAGE sample buffer with a fitting pestle, incubated for 10 min at 95°C and centrifuged at full speed for 15 min (4°C).
3. The supernatant is transferred to a new microcentrifuge tube and the total soluble protein content is determined by the method of Bradford *(58)*.

17.3.2.5. Western-Blot Analysis of Leaf Extracts

A standard SDS-PAGE *(51)* is carried out to separate proteins of leaf extracts on 10–12% polyacrylamide gels. A detailed protocol for the SDS-PAGE and the subsequent Western-blot analysis is described in Conrad et al. *(52)*. About 10–30 µg total soluble protein are loaded per lane. Always include a wild-type negative control and a positive control, which is a defined amount of a c-myc-tagged protein. For the positive control, we use 5 or 10 ng of a purified scFv. After the transfer to nitrocellulose membrane, the c-myc-tagged scFv is detected using an anti-c-myc monoclonal antibody (9E10) from hybridoma cell culture supernatant (1:50) as the primary antibody. Anti-mouse Ig coupled with HRP is used as secondary antibody (1:2,000). Peroxidase activity is detected using the ECL Western blotting detection reagents and subsequent exposure of an X-ray film for 1 min. An example for the detection of a c-myc-tagged scFv is shown in **Fig. 17.6.** An anti-ABA scFv *(31)*, targeted to the endoplasmic reticulum of *A. thaliana* is detected in plant extracts (*see* **Note 24**).

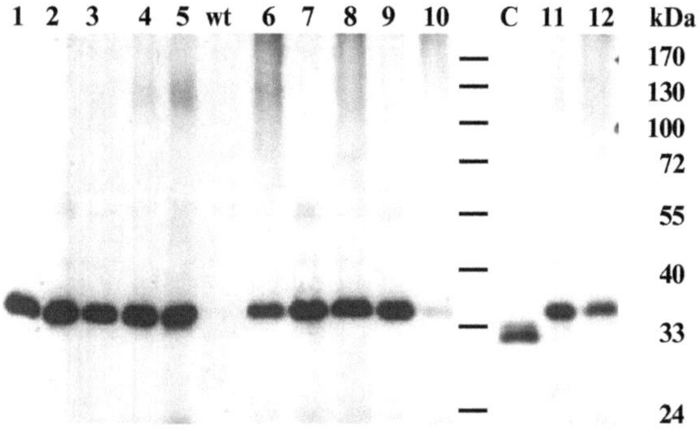

Fig. 17.6. Western-blot analysis of T_3 *Arabidopsis thaliana* plants expressing a c-myc tagged anti-ABA scFv *(31)* in the ER. 1-12: Progeny of one line, tested for the expression of the scFv, wt: wild-type control, C: Five nanogram of a bacterial expressed anti-Fus3 scFv as positive control.

17.4. Notes

1. Other microtiter plates with high binding capacity could also be used. High binding capacity seems to be critical, because this allows minimization of the antigen concentration.
2. The selected concentration depends on the nature of the antigen. Hapten-conjugates with several hapten molecules per BSA or keyhole limpet hemocyanin molecule could be more diluted than large single protein molecules with many different possible epitopes.
3. The libraries used here *(49)* code for scFv-gene 3 fusion proteins separated by an amber stop codon (UAG). To allow the fusion protein to be expressed it is necessary to use a supressor strain like TG1, which translates the amber stop codon as glutamine. Whereas the expression in a nonsupressor strain like HB2151 upon IPTG-induction will give rise to soluble scFv antibodies.
4. The phage containing supernatant and the washes have to be autoclaved.
5. In this first round of selection, at least 10,000 phages should provide a suitable base for the further selection steps. In round 2 and 3, the titers increase often tenfold per round. But in our experience, if the titer increases only slowly or even decreases in round 3, specific scFv with a sufficient affinity can still be isolated.
6. Helper phages can be obtained from GE Healthcare-LifeSciences (M13KO7 27-1524-01). Helper phages (KO7) could be produced by a phage amplification protocol as described. Start with a single plaque from a phage titer plate. It is recommended to select a small or medium-size clearly turbid plaque. During amplification only kanamycin has to be used.
7. It is important not to shake, because otherwise sex pili of the F^+ cells could be sheared and the infection rates can go down dramatically.
8. The properties of the plates are not important here. Any cheap version with flat bottoms can be used.
9. Routinely, the antigen concentrations should be as for the selection. In terms of costs, the concentration could be minimized, of course. ScFv not binding to low-concentrated antigens either have a low affinity or are expressed at a low level. Only in the latter case clones of value are lost.
10. In fact, only a small part of clones resulting in specific binding phages could be expressed as soluble antibodies. In general, around 80% or more of the clones should show sufficient

binding to the antigen. If much lower percentages of the tested phage clones bind, a further round of selection is recommended.

11. A difference to the background of about 0.1 could already represent a suitable binder. Because there are always non-binders, additional controls are not necessary.
12. Any suitable anti-mouse ALP conjugate can be used, as long as it does not cross-react with the antigen (or with plant crude extract). In case of such a cross-reaction, scFv could also be detected directly *via* protein L conjugate. Protein L binding sites occur in many Vκ sequences.
13. Substrate solution could be stored at –20°C in the dark.
14. A difference in optical density to the BSA or unrelated antigen control of about 1.0 represents in our experience a well-binding recombinant antibody.
15. The phagemid pIT1 is similar to pIT2 *(59)*, but at the C-terminal end of the scFv sequence, where pIT2 has a combined His6/c-myc tag pIT1 encodes just the c-myc tag.
16. To obtain the sequence of the encoded scFv within pIT1 or pIT2 the following primers are used:

    PHEN: ctatgcggccccattca, LMB3: gaccatgattacgccaag.
17. Oligonucleotides used for the amplification of *Bam*HI sites-flanked scFv from the phagemid pIT1 or pIT2 (the *Bam*HI sites are underlined):
    sc-Bam-5': acacgtggatccgaggtgcagctgttggagtctggggg,
    sc-Bam-3': gtcacgggatccattcagatcctcttctgagatgagtt.
18. The antibiotics used in the selection of binary plasmids in *E. coli* and Agrobacteria are dependent on the respective marker gene. Here, these are the *aadA* gene conferring resistance to streptomycin/spectinomycin in the binary plasmid 7P-Nos (use 100 μg spectinomycin/ml) and the *nptIII* gene that confers resistance to kanamycin in the binary plasmid pCB301-Kan (use 50 μg kanamycin/ml). Depending on the *Agrobacterium* strain additional antibiotics might be necessary to select for the Ti-plasmid or the strain. So both strains GV2260 and LBA4404 can be selected on 50 μg rifampicin/ml. GV2260 also is resistant to 50 μg carbenicillin/ml due to a resistance gene located on the Ti-plasmid pGV2260.
19. To have a good result, that is, a high expression, it is important that the plants are stressed neither by biotic pests nor by abiotic conditions like high temperatures. In the summer, it can happen that greenhouses get quite hot. In this case, it is recommended to grow the plants in growth chambers that allow controlling the ambient conditions.

20. This protocol is in principle based on the method of Horsch et al. *(60)*. For the manipulation of the tobacco tissue under sterile conditions, the use of a laminar flow is necessary.
21. The whole transformation process for *N. benthamiana* will take some weeks longer than for *N. tabacum*, since regeneration takes much more time.
22. After the transfer to the soil, the plantlets are very sensitive to normal humidity and therefore must be kept at first under high humidity, which is best achieved by using a plastic mini-greenhouse. After 4–5 days, the lid can be partially opened before after some more days it can be removed.
23. Transgenic plants analyzed for phenotypes and expression levels of the recombinant scFvs should display a uniform behavior. Therefore, it is desired to screen for homozygous plants from lines that segregate the transgene 1:3. The segregation pattern is simplest be determined through screening for the resistance by sowing at least 50 T_1-seeds and expose the seedlings to the respective antibiotic/herbicide. In the case of a kanamycin resistance, the seeds should be surface sterilized and then be put on kanamycin-containing MS medium. For the phosphinothricin resistance, it is possible to sow the seeds on soil and spray the seedlings two to three times with the herbicide Basta, which contains the ammonium salt of a racemic DL-phosphinothricin mixture (glufosinate-ammonium). Dilute the glufosinate-ammonium to a concentration of 1 g/l.

Fig. 17.7. ScFv-mediated resistance in *Nicotiana benthamiana*. Contrary to a control plant (right) a transgenic T_1 plant (left), expressing the scFvP55H9, which binds to the RdRp of the Tomato Bushy Stunt Virus (TBSV), shows no symptoms upon TBSV infection 14 d.p.i. (taken from *46)*.

24. When the scFv has been successfully expressed in the plant, the transgenic lines have to be characterized for phenotypical changes. As outlined in **Note 23**, it is best to achieve a more clear and uniform effect, by analyzing homozygous lines. If the phenotypical changes are not obvious under normal growth conditions, the analysis should target features, which are related to the target molecule of the recombinant antibody. So if an enzyme is targeted, the respective enzyme activity or the level of the substrate or product of that enzyme should be determined. In case the target is a RNA-dependent RNA polymerase (RdRp) of a plant virus, plants will be tested for virus resistance, as it is exemplified in **Fig. 17.7**. In parallel, the recombinant antibodies should be tested for their antigen binding. Provided that the expression level is sufficient, the scFv can be purified, by taking advantage of the c-myc tag *(58)*. The purified scFv can be characterized for the specific binding characteristics and these should than be compared to the antigen-binding behavior of the *E. coli* produced variants.

Acknowledgments

The authors thank Doreen Floss for critically reading the manuscript, Silke Krause for helpful comments on phage display technology, and Birgit Schäfer for recording the photographs in **Fig. 17.5**.

References

1. Biocca, S., Neuberger, M. S. and Cattaneo, A. (1990) Expression and targeting of intracellular antibodies in mammalian cells. EMBO J. 9, 101–108.
2. Marasco, W. A., Haseltine, W. A. and Chen, S. Y. (1993) Design, intracellular expression, and activity of a human anti-human immunodeficiency virus type 1 gp120 single-chain antibody. Proc Natl. Acad. Sci. USA 90, 7889–7893.
3. Kontermann, R. E. (2004) Intrabodies as therpeutic agents. Methods 34, 163–170.
4. Stocks, M. (2006) Intracellular antibodies: A revolution waiting to happen? Curr. Opin. Mol. Ther. 8, 17–23.
5. De Jaeger, G., De Wilde, C., Eeckhout, D., Fiers, E. and Depicker, A. (2000) The plantibody approach: expression of antibody genes in plants to modulate plant metabolism or to obtain pathogen resistance. Plant Mol. Biol. 43, 419–428.
6. Conrad, U. and Manteuffel, R. (2001) Immunomodulation of phytohormones and functional proteins in plant cells. Trends Plant Sci. 6, 399–402.
7. Bird, R. E., Hardman, K. D., Jacobson, J. W., Johnson, S., Kaufman, B. M., Lee, S. M., Lee, T., Pope, S. H., Riordan, G. S. and Whitlow, M. (1988) Single-chain antigen-binding proteins. Science 242, 423–426.
8. Dübel, S. and Kontermann, R. E. (2001) Recombinant antibodies, in: *Antibody Engineering* (Kontermann, R. and Dübel, S.,

eds.), Springer-Verlag, Berlin Heidelberg, pp. 3–18.

9. Kohler, G. and Milstein, C. (1975) Continuous cultures of fused cells secreting antibody of predefined specificity. Nature 256, 495–497.
10. Burmester, J. and Plückthun, A. (2001) Construction of scFv fragments from hybridoma or spleen cells by PCR assembly, in *Antibody Engineering* (Kontermann, R. and Dübel, S., eds.), Springer-Verlag, Berlin Heidelberg, pp. 19–40.
11. Breitling, F., Moosmayer, D., Brocks, B. and Dübel, S. (2001) Construction of scFv from hybridoma by two-step cloning of the VH and VL domains, in *Antibody Engineering* (Kontermann, R. and Dübel, S., eds.), Springer-Verlag, Berlin Heidelberg, pp. 41–55.
12. Bradbury, A. (2001) Cloning hybridoma cDNA by RACE, in *Antibody Engineering* (Kontermann, R. and Dübel, S., eds.), Springer-Verlag, Berlin Heidelberg, pp. 56–64.
13. Smith, G. P. (1985) Filamentous fusion phage: Novel expression vectors that display cloned antigens on the virion surface. Science 228, 1315–1317.
14. McCafferty, J., Griffiths, A. D., Winter, G. and Chiswell, D. J. (1990) Phage antibodies: Filamentous phage displaying antibody variable domains. Nature 348, 552–554.
15. Clackson, T., Hoogenboom, H. R., Griffiths, A. D. and Winter, G. (1991) Making antibody fragments using phage display libraries. Nature 352, 624–628.
16. Conrad, U. and Scheller, J. (2005) Considerations on antibody-phage display methodology. Comb. Chem. High Throughput Screen. 8, 117–126.
17. Low, N. M., Holliger, P. H. and Winter, G. (1996) Mimicking somatic hypermutation: Affinity maturation of antibodies displayed on bacteriophage using a bacterial mutator strain. J. Mol. Biol. 260, 359–368.
18. Swers, J. S., Kellogg, B. A. and Wittrup, K. D. (2004) Shuffled antibody libraries created by *in vivo* homologous recombination and yeast surface display. Nucleic Acids Res. 32, e36.
19. Visintin, M., Tse, E., Axelson, H., Rabbitts, T. H. and Cattaneo, A. (1999) Selection of antibodies for intracellular function using a two-hybrid *in vivo* system. Proc. Natl. Acad. Sci. USA 96, 11723–11728.
20. der Maur, A. A., Zahnd, C., Fisher, F., Spinelli, S., Honegger, A., Cambillau, C., Escher, D., Pluckthun, A. and Barberis, A. (2002) Direct in vivo screening of intrabody libraries constructed on a highly stable single-chain framework. J. Biol. Chem. 277, 45075–45085.
21. Hiatt, A., Cafferkey, R. and Bowdish, K. (1989) Production of antibodies in transgenic plants. Nature 342, 76–78.
22. Fischer, R., Stöger, E., Schillberg, S., Christou, P. and Twyman, R. (2004) Plant-based production of biopharmaceuticals. Curr. Opin. Plant Biol. 7, 152–158.
23. Stöger, E., Vaquero, C., Torres, E., Sack, M., Nicholson, L., Drossard, J., Williams, S., Keen, D., Perrin, Y., Christou, P. and Fischer, R. (2000) Cereal crops as viable production and storage systems for pharmaceutical scFv antibodies. Plant Mol. Biol. 42, 583–590.
24. Owen, M., Gandecha, A., Cockburn, B. and Whitelam, G. (1992) Synthesis of a functional anti-phytochrome single-chain Fv protein in transgenic tobacco. Biotechnology (N.Y.) 10, 790–794.
25. Conrad, U. and Fiedler, U. (1998) Compartment-specific accumulation of recombinant immunoglobulins in plant cells: An essential tool for antibody production and immunomodulation of physiological functions and pathogen activity. Plant Mol. Biol. 38, 101–109.
26. Spiegel, H., Schillberg, S., Sack, M., Holzem, A., Nähring, J., Monecke, M., Liao, Y.-C. and Fischer, R. (1999) Accumulation of Antibody fusion proteins in the cytoplasm and ER of Plants. Plant Sci. 149, 63–71.
27. Jobling, S. A., Jarman, C., Teh, M.-M., Holmberg, N., Blake, C. and Verhoeyen, M. E. (2003) Immunomodulation of enzyme function in plants by single-domain antibody fragments. Nat. Biotechnol. 21, 77–80.
28. Santos, M. O., Crosby, W. L. and Winkel, B. S. J. (2004) Modulation of flavonoid metabolism in Arabidopsis using a phage-derived antibody. Mol. Breed. 13, 333–343.
29. Nölke, G., Schneider, B., Fischer, R. and Schillberg, S. (2005) Immunomodulation of polyamine biosynthesis in tobacco plants has a significant impact on polyamine levels and generates a dwarf phenotype. Plant Biotechnol. J. 3, 237–247.
30. Miroshnichenko, S., Tripp, J., zur Nieden, U., Neumann, D., Conrad, U. and Manteuffel, R. (2005) Immunomodulation of function of small heat shock proteins prevents their assembly into heat stress granules

and results in cell death at sublethal temeratures. Plant J. 41, 269–281.

31. Artsaenko, O., Peisker, M., zur Nieden, U., Fiedler, U., Weiler, E. W., Müntz, K. and Conrad, U. (1995) Expression of a single-chain Fv antibody against abscisic acid creates a wilty phenotype in transgenic tobacco. Plant J. 8, 745–750.
32. Phillips, J., Artsaenko, O., Fiedler, U., Horstmann, C., Mock, H.-P., Müntz, K. and Conrad, U. (1997) Seed-specific immunomodulation of abscisic acid activity induces a developmental switch. EMBO J. 16, 4489–4496.
33. Senger, S., Mock, H.-P., Conrad, U. and Manteuffel, R. (2001) Immunomodulation of ABA function affects early events in somatic embryo development. Plant Cell Rep. 20, 112–120.
34. Wigger, J., Phillips, J., Peisker, M., Hartung, W., zur Nieden, U., Artsaenko, O., Fiedler, U. and Conrad, U. (2002) Prevention of stomatal closure by immunomodulation of endogenous abscisic acid and its reversion by abscisic acid treatment: Physiological behaviour and morphological fratures of tobacco stomata. Planta 215, 413–423.
35. Strauss, M., Kauder, F., Peisker, M., Sonnewald, U., Conrad, U. and Heineke, D. (2001) Expression of an abscisic acid-binding single-chain antibody influences the subcellular distribution of abscisic acid and leads to developmental changes in transgenic potato plants. Planta 213, 361–369.
36. Shimada, N., Suzuki, Y., Nakajima, M., Conrad, U., Murofushi, N. and Yamagushi, I. (1999) Expression of a functional single-chain antibody against GA24/19 in transgenic tobacco. Biosci. Biotechnol. Biochem. 63, 779–783.
37. Putalun, W., Taura, F., Qing, W., Matsushita, H., Tanaka, H. and Shoyama, Y. (2003) Anti-solasodine glycoside single-chain Fv antibody stimulates biosynthesis of solasodine glycoside in plants. Plant Cell Rep. 22, 344–349.
38. Eto, J., Suzuki, Y., Ohkawa, H. and Yamaguchi, I. (2003) Anti-herbicide single-chain antibody expression confers herbicide tolerance in transgenic plants. FEBS Lett. 550, 179–184.
39. Almquist, K. C., Niu, Y., Mclean, M. D., Mena, F. L., Yau, K. Y. F., Brown, K., Brandle, J. E. and Hall, J. C. (2004) Immunomodulation confers herbicide resistance in plants. Plant Biotechnol. J. 2, 189–197.
40. Leps, B. (2003) Anti-Quinmerac Einkettenantikörperexpression in transgenen Tabakpflanzen. *Thesis*, University of Halle-Wittenberg, Germany.
41. Leps, M. (2002) Expression von Einkettenantikörpern gegen das Fungizid Kresoxim-methyl in transgenen Tabakpflanzen. *Thesis*, University of Halle-Wittenberg, Germany.
42. Yuan, Q., Hu, W., Pestka, J. J., He, S. Y. and Hart, L. P. (2000) Expression of a functional antizearalenone singel-chain Fv antibody in transgenic *Arabidopsis* plants. Appl. Environ. Microbiol. 66, 3499–3505.
43. Tavladoraki, P., Benvenuto, E., Trinca, S., De Martinis, D., Cattaneo, A. and Galeffi, P. (1993) Transgenic plants expressing a functional single-chain Fv antibody are specifically protected from virus attack. Nature 366, 469–472.
44. Schillberg, S., Zimmermann, S., Zhang, M.-Y. and Fischer, R. (2001) Antibody-based resistance to plant pathogens. Transgenic Res. 10, 1–12.
45. Peschen, D., Li, H.-P., Fischer, R., Kreuzaler, F. and Liao, Y.-C. (2004) Fusion proteins comprising a Fusarium-specific antibody linked to antifungal peptides protect plants against a fungal pathogen. Nat. Biotechnol. 22, 732–738.
46. Boonrod, K., Galetzka, D., Nagy, P. D., Conrad, U. and Krczal, G. (2004) Single-chain antibodies against a plant viral RNA-dependent RNA plymerase confer virus resistance. Nat. Biotechnol. 22, 856–862.
47. Villani, M. E., Roggero, P., Bitti, O., Benvenuto, E. and Franconi, R. (2005) Immunomodulation of cucumber mosaic virus infection by intrabodies selected *in vitro* from a stable single-framework phage display library. Plant Mol. Biol. 58, 305–316.
48. Prins, M., Lohuis, D., Schots, A. and Goldbach, R. (2005) Phage display-selected single-chain antibodies confer hifg levels of resistance against Tomato spotted wilt virus. J. Gen. Virol. 86, 2107–2113.
49. de Wildt, R. M., Mundy, C. R., Gorick, B. D. and Tomlinson, I. M. (2000) Antibody arrays for high-throughput screening of antibody-antigen interactions. Nat. Biotechnol. 18, 989–994.
50. Scheller, J., Henggeler, D., Viviani, A. and Conrad, U. (2004) Purification of spider silk-elastin from transgenic plants and application for human chondrocyte proliferation. Transgenic Res. 13, 51–57.
51. Murashige, T. and Skoog, F. (1962) A revised medium for rapid growth and bioassays with tobacco tissue cultures. Physiol. Plant. 15, 473–497.

52. Laemmli, U. K. (1970) Cleavage of structural proteins during the assembly of the head of bacteriophage T4. Nature 227, 680–685.
53. Conrad, U., Fiedler, U., Artsaenko, O. and Phillips, J. (1998) Single-chain Fv antibodies expressed in plants, in *Methods in Biotechnology*, Vol. 3: Recombinant proteins from plants (Cunningham, C. and Porter, A. J. R., eds.), Humana, Totowa, NJ, pp. 103–127.
54. Sambrook, J. and Russell, D. W. (2001) *Molecular Cloning: A Laboratory Manual.* Cold Spring Harbor Laboratory Press, New York.
55. Deblaere, R., Bytebier, B., De Greve, H., Deboeck, F., Schell, J., Van Montagu, M. and Leemans, J. (1985) Efficient octopine Ti plasmid-derived vectors for *Agrobacterium-mediated* gene transfer to plants. Nucleic Acids Res. 13, 4777–4788.
56. Hoekema, A., Hirsch, P. R., Hooykaas, P. J. J. and Schilperoort, R. A. (1983) A binary plant vector strategy based on separation of *vir-* and T-region of the *Agrobacterium tumefaciens* Ti-plasmid. Nature 303, 179–180.
57. Mattanovich, D., Rüker, F., da Câmara Machado, A., Laimer, M., Regner, F., Steinkellner, H., Himmler, G. and Katinger, H. (1989) Efficient transformation of *Agrobacterium* spp. by electroporation. Nucleic Acids Res. 16, 6747.
58. Bradford, M. M. (1976) A rapid and sensitive method for the quantitation of microgram quantities of protein utilizing the principle of protein-dye binding. Anal. Biochem. 72, 248–254.
59. Holt, L. J., Büssow, K., Walter, G. and Tomlinson, I. M. (2000) By-passing selection: Direct screening for antibody-antigen interactions using protein arrays. Nucleic Acids Res. 28, e72.
60. Horsch, R. B., Fry, J. E., Hoffmann, N. L., Eichholtz, D., Rogers, S. G. and Fraley, R. T. (1985) A simple and general method for transferring genes into plants. Science 227, 1229–1231.

Chapter 18

On-Chip Detection of Low-Molecular-Weight Recombinant Proteins in Plant Crude Extracts by SELDI-TOF MS

Amine M. Badri, Karine Coenen, Louis-Philippe Vaillancourt, Charles Goulet, and Dominique Michaud

Summary

This chapter presents a general procedure for the on-chip detection and quantitation of low-molecular-weight recombinant proteins in transgenic plant crude extracts by surface-enhanced laser desorption ionization time-of-flight mass spectrometry (SELDI-TOF MS). A protocol is first described to detect the protein of interest in crude protein extracts of transgenic plant lines, by differential protein mapping against similar extracts from a control, nontransgenic line. A complementary protocol is then presented to generate a standard curve with the SELDI system, allowing the protein to be quantified in different transgenic lines. Overall, this procedure may be carried out within a few hours, without the need for prior purification or enrichment of the recombinant protein.

Key words: Plant crude protein extracts, Protein detection, Protein quantitation, Recombinant proteins, SELDI-TOF MS, Surface-enhanced laser desorption ionization time-of-flight mass spectrometry.

18.1. Introduction

A key challenge for recombinant protein developers is the cost-effective production of high-quality products. Unlike several nonprotein pharmaceuticals, recombinant proteins present a natural tendency to structure heterogeneity, often giving a complex mixture of protein forms differing in their primary or tertiary structure *(1)*. This natural bias of proteins, which stresses the important issue of product authenticity, also points to the need for efficient analytical tools adapted to protein characterization, from the early steps of expression platform design to the final

Loïc Faye and Véronique Gomord (eds.), *Methods in Molecular Biology, Recombinant Proteins From Plants, vol. 483*

DOI: 10.1007/978-1-59745-407-0_18

steps of product development and approval. Several approaches are currently available to characterize the physicochemical and immunological properties of recombinant proteins *(2)*, including basic tools such as sodium dodecyl sulfate-polyacrylamide gel electrophoresis (SDS-PAGE), isoelectric focusing and immunoblotting, and more sophisticated tools such as circular dichroism, nuclear magnetic resonance, capillary electrophoresis, electrospray ionization mass spectrometry (ESI MS), and matrix-assisted laser desorption ionization time-of-flight MS (MALDI-TOF MS). As a complement, we recently discussed the potential of surface-enhanced laser desorption ionization time-of-flight MS (SELDI-TOF MS) for the rapid detection and quantitation of heterologous proteins in biological (plant) crude extracts *(3)*.

The SELDI-TOF MS technology *(4, 5)*, first described by Hutchens and Yip 15 years ago *(6)*, involves the use of "ProteinChip" arrays for the capture of polypeptides and proteins on solid surfaces, and their subsequent MS analysis using an adapted ProteinChip Reader. Several ProteinChip arrays are available for protein capture, all bearing a specific chromatographic interface for reverse-phase chromatography, cation- or anion-exchange chromatography, or protein-ligand affinity chromatography *(4)*. After derivatization, the arrays are incubated with the proteins or biological extracts of interest, and washed with appropriate buffers to eliminate nonspecific interactions at the membrane-solution interface. An energy-absorbing matrix is then applied to the surface to embed the bound proteins, that are finally laser-desorbed and visualized on MS spectra. In recent years, this analytical platform has been used for applications as diverse as the identification of proteins following proteomic analysis, the discovery of disease-related protein biomarkers, the characterization of protein–protein and protein–DNA interactions, the monitoring of protein post-translational modifications, and the annotation of bacterial proteomes *(5, 7, 8)*. A step-by-step procedure is described here for the rapid detection and quantitation of low-molecular-weight recombinant proteins in plant samples, using as an example leaf crude protein extracts of transgenic potato lines expressing recombinant protease inhibitors of clinical or industrial interest. The whole procedure is carried out in a few hours, and does not require prior purification or enrichment of the recombinant protein.

18.2. Materials

18.2.1. Plant Material

Transgenic plant lines expressing a recombinant protein of interest and their corresponding (non-transgenic) isoline are used for the experiments. Transgenic lines of potato (*Solanum tuberosum* L.)

expressing the clinically useful serine protease inhibitor, bovine aprotinin *(9)* or the papain inhibitor, corn cystatin II *(10)* are used here for demonstration purposes (*see* **Note 1**). Transgenic lines expressing aprotinin secreted in the apoplastic milieu were produced as described recently *(11)*. Lines expressing corn cystatin II in the cytosolic compartment were developed using a procedure described earlier *(12)*.

18.2.2. Special Laboratory Tools and Materials

1. Liquid nitrogen.
2. Mortar and pestle.
3. Branson 450 Digital sonicator (Branson Ultrasonics Corporation, Danbury CT).
4. Multi-Tube Vortexer VX 2500 (VWR, Mississauga ON, Canada).
5. Rocking platform Model 100 (VWR).
6. Humidity chamber (plastic box) containing deionized water (1-mm thick) in the bottom.
7. ProteinChip Reader, model PBSIIC (Bio-Rad, Richmond CA).
8. Protein Chip arrays [CM10 arrays] for weak cationic exchange (Bio-Rad), with active spots coated with carboxylate groups acting as weak cationic exchangers.
9. Normal phase ProteinChip arrays [NP20 arrays] (Bio-Rad) for general protein binding.
10. PBSIIC ProteinChip Reader software, v. 3. 2. 0 or later (Bio-Rad).

18.2.3. Buffers and Other Solutions

1. Leaf extraction buffer: 50 mM Tris-HCl, pH 7.0, 0.1% (v/v) Triton X-100, 1 mM phenylmethylsulfonyl fluoride [to inhibit serine proteases], 0.2 mM polyvinyl-polypyrrolidone [to neutralize phenolics].
2. Protein binding buffer: 50 mM phosphate/citrate, pH 7.0, 0.1% (v/v) Triton X-100. Prepare 50 mM dibasic sodium phosphate and adjust to pH 7.0 with 0.1 M citric acid. Add 0.1% (v/v) Triton X-100.
3. Saturated energy absorbing molecule (EAM) solution: Sinapinic acid matrix (SPA) (Bio-Rad) suspended in 50% (v/v) acetonitrile and 0.5% (v/v) trifluoroacetic acid (TFA). Prepare fresh.
4. External calibration solution: All-in-one protein standard II (Bio-Rad), 20 mM ammonium acetate. Add 10 µL of 20 mM ammonium acetate in one standard protein tube and "flick mix" for 30 s at room temperature. Quick-spin and recover the solution (10 µL) in a 250-µL microtube. Add

another 15 μL of 20 mM ammonium acetate in the standard protein tube and let it rest for 2 min at room temperature. Flick-mix and recover the solution, and then add it to the first 10 μL to obtain a final volume of 25 μL. Aliquots may be prepared and stored at –20°C until use.

5. TFA solution (2% (v/v) TFA): Place 48 mL of Milli-Q water in a 100-mL bottle, and add 1 mL of 98% (v/v) TFA.
6. Insulin and bovine ubiquitin protein standards prepared according to the supplier's instructions (Bio-Rad).
7. Bovine aprotinin dilutions: Dilutions of commercial bovine aprotinin (Sigma-Aldrich, Oakville ON, Canada) suspended in protein binding buffer/leaf soluble protein extract from a nontransgenic [control] line (ratio 4:1). Aprotinin final concentrations are adjusted to 0.0, 0.1, 0.2, 0.3, and 0.5 ng/μL, respectively.

18.3. Methods

18.3.1. Detection of the Recombinant Protein

This section describes a SELDI-TOF MS procedure for the rapid detection of a recombinant protein in a transgenic plant leaf crude extract (*see* **Note 2**). The procedure consists, in brief, (i) to extract leaf proteins of transgenic and control lines in mild conditions, (ii) to capture the extracted proteins on a ProteinChip array, and (iii) to look for the presence of an extra peak of the expected size on MS spectra generated for the transgenic lines, compared to the spectra obtained for the control, nontransgenic line.

18.3.1.1. Extraction of Leaf Soluble Proteins

1. Grind the fourth, fifth, and sixth leaves of 30-cm potato plants to a fine powder in liquid nitrogen, using a mortar and a pestle.
2. Transfer 250 mg of leaf powder to an ice-cold microtube and add 1 mL of leaf extraction buffer. Keep the remaining leaf powder at –80°C for further use.
3. Vortex the mixture briefly and sonicate on ice for 1 min at 0.5-s intervals with 0.5 s-long pulsations, using a Bronsan 450 Digital sonicator at 10% amplitude.
4. Centrifuge the resulting solution at 13,000 × *g* for 15 min at 4°C.
5. Collect the supernatant containing total soluble proteins and keep on ice until performing MS analysis.
6. Protein concentration in the different samples may be uniformized by the addition of leaf extraction buffer, after

assaying the proteins using a standard procedure [e.g., using the Bradford method *(13)*].

18.3.1.2. Calibration of the ProteinChip Reader

An external calibration of the ProteinChip Reader must be performed before each experiment, using appropriate protein standards and NP20 arrays prepared in advance and preserved from humidity.

First External Calibration

1. Gently mix 1 μL of the external calibration solution with 4 μL of fresh saturated EAM solution.
2. Place 3 μL of Milli-Q water on each NP20 spot, and then remove extra water with a clean, folded laboratory wipe gently applied on the side of the spot.
3. Add 1 μL of external calibration/EAM solution mixture on each wet spot, taking care not to touch the spot surface with the pipette tip. Let dry for 15 min at room temperature.
4. Insert the NP20 ProteinChip array in the PBSIIC mass spectrometer, and set laser configuration for low-mass proteins (<30 kDa) [Digitizer rate: 1000 MHz; Detector: 2700 V].
5. Read the NP20 array using the following spot protocol:
   a. Set high mass to 50,000 Da, optimized from 6,000 to 30,000 Da.
   b. Set starting laser intensity to 192.
   c. Set starting detector sensitivity to 10.
   d. Focus by optimization centre.
   e. Set Mass Deflector to 1,000 Da.
   f. Set data acquisition method to SELDI quantitation.
   g. Set SELDI acquisition parameters at positions 20, 25, 30,... up to 75, performing 15 laser shots at each position [SELDI acquisition parameters 20. Delta to 5. Transients per to 15 ending position to 75].
   h. Set warming position with 2 shots at intensity 197. Do not include warming shots.
   i. Process sample.
6. Remove the Baseline.
7. Use peaks for hirudin (6,964 Da), bovine cytochrome C (12,230 Da), equine myoglobin (16,951 Da), and bovine carbonic anhydrase (29,023 Da) to calibrate the mass spectrometer.

Second External Calibration

For more accurate measurements, a second external calibration may be performed by fixing onto NP20 arrays protein standards with a molecular mass similar to the mass of the protein to be detected. Insulin (5,734 Da) and ubiquitin (8,565 Da) can be used, for instance, to look for the presence of bovine aprotinin (6,517 Da) in leaf extracts.

1. Dilute 1 μL of insulin (5,733.6 Da) and 1 μL of ubiquitin (8,564.8 Da) in 20 μL of Milli-Q water.
2. Rinse an NP20 ProteinChip spot with 5 μL of water and allow to air-dry.
3. Apply 1 μL of the protein mix (**step 1**) on the air-dried spot.
4. Add 0.5 μL of saturated EAM solution and allow to air-dry.
5. Read the spot with the Spot protocol described above (*see* **Subheading** First External Calibration, **step 5**), with the following modification for **step 5a**: set high mass to 15,000 Da, optimized from 4,000 to 10,000 Da.

18.3.1.3. Recombinant Protein Detection

1. Pretreat CM10 array spots for 5 min with 5 μL of 10 mM HCl (freshly prepared) (*see* **Note 3**).
2. Remove HCl and rinse three times with 5 μL of Milli-Q water.
3. To equilibrate the spots, apply 5 μL of protein binding buffer and incubate for 5 min in a humidity chamber with gentle agitation, on a VWR rocking platform. Remove the buffer and repeat once.
4. Dilute 2 μL of each leaf protein sample with 8 μL of binding buffer.
5. Apply the mix (10 μL) onto a ProteinChip array spot and incubate for 30 min in a humidity chamber, with gentle agitation on a VWR rocking platform.
6. Remove the sample and apply 5 μL of protein binding buffer on the spot. Incubate for 5 min in a humidity chamber with gentle agitation. Remove the buffer and repeat twice.
7. Apply 5 μL of Milli-Q water on the spot and incubate for 1 min in a humidity chamber. Remove water.
8. Air-dry the array and apply 0.7 μL of saturated EAM solution on each spot. Air-dry and repeat twice with 0.5 μL of saturated EAM solution.
9. Using the appropriate settings (high mass to acquire: 15,000 Da; optimization range: 4,000–10,000 Da; focus mass: 6,500 Da; deflector: 1,000 Da), find the best laser intensity [ex: 192] and sensitivity [ex: 9] to obtain a clear signal with the ProteinChip Reader.
10. Create a spot protocol (*see* **Note 4**):
    a. Set high mass to 15,000 Da, optimized from 4,000 to 10,000 Da.

b. Set starting laser intensity to 192.
c. Set starting detector sensitivity to 9.
d. Focus mass at 6,500 Da.
e. Set mass deflector to 1,000 Da.
f. Set data acquisition method to SELDI quantitation.
g. Set SELDI quantitation parameters at positions 20, 25, 30,... up to 80, performing 10 laser shots at each position [SELDI quantitation parameters 20. Delta to 5. Transients per to 10 ending position to 80].
h. Set warming position with 2 shots at intensity 197. Do not include warming shots.
i. Process sample.

11. Create a Chip protocol using this spot protocol (**step 10**), and analyze all samples in a single batch.
12. After reading, make sure to remove the baseline and normalize the spectra for each spot. Using the ProteinChip Reader software, compare the spectra generated for the test and control lines, looking in particular for the presence of an extra peak of the expected molecular mass (*m/z* ratio) in transgenic lines. **Figure 18.1** illustrates the detection of recombinant bovine aprotinin and corn cystatin II in leaves of transgenic potato lines genetically transformed with the corresponding transgene sequences. The recombinant protein peaks, not detected in leaf extracts of control line K, show molecular masses (*m/z* ratios) closely matching (±2 Da) the theoretical molecular masses of these two proteins.

18.3.2. Quantitation of the Recombinant Protein

This section describes a simple procedure to quantify recombinant proteins detected in leaf crude extracts by SELDI-TOF MS (*see* **Subheading** 18.3.1). The procedure involves (i) the addition [at different concentrations] of the tested protein to a leaf protein extract prepared from a nontransgenic (control) plant; (ii) the generation and analysis of SELDI-TOF MS spectra; and (iii) the calculation of a [linear] standard curve serving as a reference for recombinant protein quantitation. Bovine aprotinin and potato lines expressing this inhibitor *(11)* are used here to illustrate the procedure.

18.3.2.1. Extraction of Leaf Soluble Proteins

Extract leaf soluble proteins from a nontransgenic (control) plant, as described above (*see* **Subheading 18.3.1.1**).

18.3.2.2. Calibration of the Mass Spectrometer

Calibrate the mass spectrometer using the All-in-one protein standard II kit and appropriate protein standards (e.g., insulin plus ubiquitin for aprotinin detection), as described above (*see* **Subheading 18.3.1.2**).

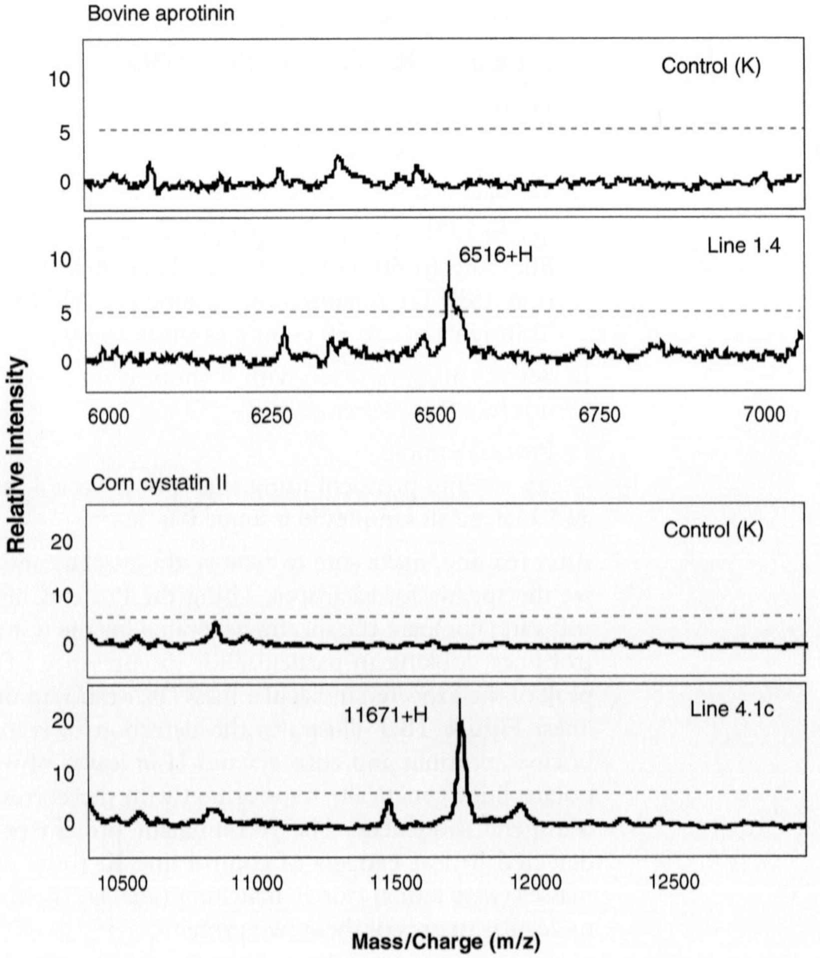

Fig. 18.1. Surface-enhanced laser desorption ionization time-of-flight mass spectrometry (*SELDI-TOF MS*) comparative profiling for the detection of recombinant bovine aprotinin and corn cystatin II in leaf extracts of transgenic potato lines. Spectra are presented for transgenic lines 1.4 (expressing aprotinin) and 4.1c (expressing corn cystatin II) and their corresponding nontransgenic line (line K). Data are presented as relative peak intensities for each protein species. The mean molecular weight of bovine aprotinin is 6,517 Da *(9)*. The mean molecular weight of recombinant corn cystatin II *(14)* bearing here an extra methionine residue at the N terminus is 11,670 Da.

18.3.2.3. Quantitation of the Recombinant Protein

1. Pretreat CM10 array spots for 5 min with 5 μL of 10 mM HCl (prepare fresh).
2. Rinse each spot three times with 5 μL of Milli-Q water.
3. To equilibrate the spots, apply 5 μL of protein binding buffer and incubate for 5 min in a humidity chamber with gentle agitation, using a VWR rocking platform. Remove the buffer and repeat once.
4. Remove the buffer and add immediately 10 μL of each commercial aprotinin dilution. Each dilution is repeated three

times and distributed randomly onto the arrays to avoid position effects in the spectrometer.

5. Generate protein mass spectra using the ProteinChip Reader, as described above for the control and transgenic line samples (*see* **Subheading** 18.3.1.3, **steps 6–12**). **Figure 18.2** illustrates mass spectra obtained for different aprotinin dilutions added to a control potato leaf extract.
6. Infer a standard curve for bovine aprotinin by the least-square method, using the aprotinin peaks generated at **step 5**. The height of each aprotinin peak, as measured from the spectrum baseline, is proportional to the amount of commercial aprotinin applied onto the chip (*see* **Note 5**). **Figure 18.3** illustrates a typical standard curve for bovine aprotinin added in a potato leaf crude protein extract. The concentration of assayed protein may then be expressed on a relative basis, typically as a percentage of total soluble proteins (%TSP), after assaying total soluble proteins in source extracts by a routine procedure (e.g., the Bradford method *13*).

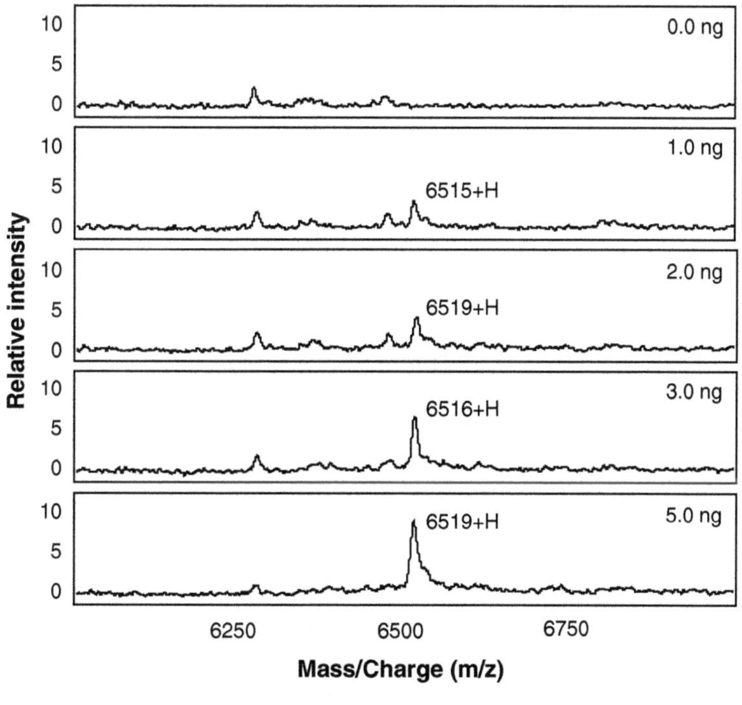

Fig. 18.2. On-chip detection of commercial aprotinin added to a nontransgenic (control) potato leaf soluble protein extract. Different amounts of aprotinin were added to the plant extract before protein capture onto the chip. The height of each peak at m/z = ~6,517 is proportional to the amount of aprotinin applied.

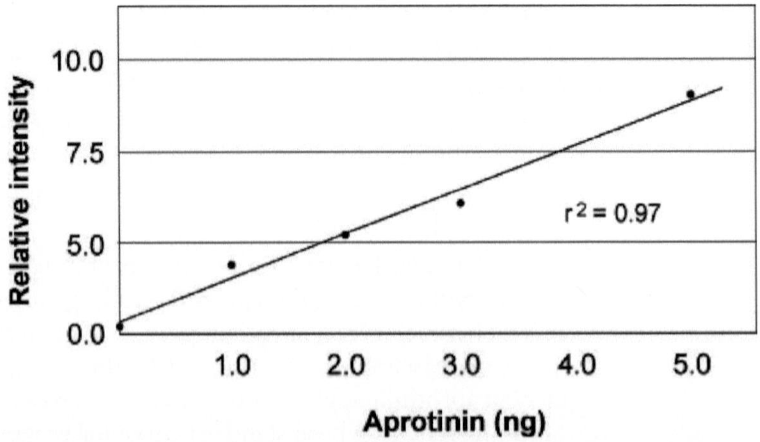

Fig. 18.3. A standard curve for the surface-enhanced laser desorption ionization time-of-flight mass spectrometry (*SELDI-TOF MS*) quantitation of bovine aprotinin in transgenic potato lines. The standard curve was generated by the least-square method based on peak intensities measured for different amounts of aprotinin added to leaf protein extracts (*see* **Fig.18.2**). Commercial bovine aprotinin was used for the tests. Each point is the mean of three independent measurements.

18.4. Notes

1. The approach presented is well suited for the detection of small proteins showing a molecular mass below 15,000 Da. Most proteins with a molecular mass higher than 15–20,000 Da can be fixed onto the chips and easily detected with the ProteinChip Reader under a purified form, but their effective detection in complex protein mixtures depends on their structural and physicochemical characteristics, which in turn determine their ability to be fixed on the chip and laser desorbed in the presence of multiple peptide and polypeptide species.
2. Protocols are described here for leaf extracts, but the protein source has little impact then on SELDI-TOF MS detection. The protocols can be easily adapted to protein extracts from any cell type or tissue, provided that procedures are available for protein extraction in mild (aqueous) conditions.
3. CM10 arrays for weak cationic exchange were chosen based on the basidic nature of the proteins to be detected (e.g., bovine aprotinin). Alternative arrays (e.g., for anionic exchange or mimicking affinity chromatography) may be used pending minor adjustments of the procedure, taking care in particular to the composition of the binding buffer during protein capture.

4. Laser settings can differ depending on the system used and laser age. The settings provided here are for the PBSIIC ProteinChip Reader. Basic starting parameters are suggested by the supplier for the different ProteinChip models.
5. Measuring peak intensities is the most common way of assaying protein levels with the SELDI system. The calculation of peak areas (e.g. ref *3*) represents an interesting and sometimes preferable alternative, especially for non-singleton, overlapping peaks (*see* ref. *15*).

Acknowledgments

This work was supported by a Discovery grant from the Natural Science and Engineering Research Council (NSERC) of Canada to D. Michaud. A.M. Badri was the recipient of a doctoral fellowship from the Government of Tunisia. C. Goulet was the recipient of an NSERC doctoral scholarship.

References

1. Faye, L., Boulaflous, A., Benchabane, M., Gomord, V., and Michaud, D. (2005) Protein modifications in the plant secretory pathway: current status and practical implications in molecular pharming. Vaccine 23, 1770–1778.
2. Chirino, A.J., and Mire-Sluis, A. (2004) Characterizing biological products and assessing comparability following manufacturing changes. Nat. Biotechnol. 22, 1383–1391.
3. Badri, A.M., Rivard, D., Goulet, C., Vaillancourt, L-P., Coenen, K., and Milchaud, D. (2008) A SELDI TOF MS procedure for the detection, quantitation and preliminary characterization of low-molecular-weight recombinant proteins expressed in transgenic plants. Proteomics (in press).
4. Merchant, M., and Weinberger, S.R. (2000) Recent advancements in surface-enhanced laser desorption/ionization-time of flight-mass spectrometry. Electrophoresis 21, 1164–1167.
5. Tang, N., Tornatore, P., and Weinberger, S.R. (2004) Current developments in SELDI affinity technology. Mass Spectrom. Rev. 23, 34–44.
6. Hutchens, T.W., and Yip, T.T. (1993) New desorption strategies for the mass spectrometric analysis of macromolecules. Rapid Commun. Mass Spectrom. 7, 576–580.
7. Barzaghi, D., Isbister, J.D., Lauer, K.P., and Born, T.L. (2004) Use of surface-enhanced laser desorption/ionization time of flight to explore bacterial proteomes. Proteomics 4, 2624–2628.
8. Haleem, J.I., Veenstra, T.D., Conrads, T.P., and Felschow, D. (2002) The SELDI-TOF MS approach to proteomics: protein profiling and biomarker identification. Biochem. Biophys. Res. Commun. 292, 587–592.
9. Ascenzi, P., Bocedi, A., Bolognesi, M., Spallarossa, A., Coletta, M., De Cristofaro, R., and Menegatti, E. (2003) The bovine basic pancreatic trypsin inhibitor (Kunitz inhibitor): a milestone protein. Curr. Protein Pept. Sci. 4, 231–251.
10. Abe, M., Abe, K., Domoto, C., and Arai, S. (1995) Two distinct species of corn cystatins in corn kernels. Biosci. Biotechnol. Biochem. 59, 756–758.
11. Rivard, D., Anguenot, R., Brunelle, F., Le, V.Q., Vézina, L.-P., Trépanier, S., and Michaud, D. (2006) An in-built proteinase inhibitor system for the protection

of recombinant proteins recovered from transgenic plants. Plant Biotechnol. J. 4, 359–368.

12. Michaud, D., and Vrain, T.C. (1998). Expression of recombinant proteinase inhibitors in plants. Meth. Biotechnol. 3, 49–64.
13. Bradford, M.M. (1976) A rapid and sensitive method for the quantitation of microgram quantitites of protein utilizing the principle of protein-dye binding. Anal. Biochem. 72, 248–254.
14. Brunelle, F., Girard, C., Cloutier, C., and Michaud, D. (2005) A hybrid, broad-spectrum inhibitor of Colorado potato beetle aspartate and cysteine digestive proteinases. Arch. Insect Biochem. Physiol. 60, 20–31.
15. Dijkstra, M., Roelofsen, H., Vonk, R.J., and Jansen, R.C. (2006) Peak quantification in surface-enhanced laser desroption/ionization by using mixture models. Proteomics 6, 5106–5116.

Chapter 19

Assessing the Risk of Undesirable Immunogenicity/ Allergenicity of Plant-Derived Therapeutic Proteins

Paul D. Chamberlain

Summary

Therapeutic proteins have an intrinsic potential to induce undesirable immune and allergic responses. The nature of the expression system and the control of the manufacturing process represent extrinsic factors that could modify this potential. Accordingly, regulatory agencies require sponsors to assess the risk of clinical immune and allergic responses that could be associated with the production of therapeutic proteins in transgenic plants. Since factors related to the clinical use of the product – including the immune status and genetic background of subjects – are also relevant, the risk assessment needs to balance the probability of a response induced by a plant-specific factor relative to the likely consequences for the specific product and therapeutic indication.

Key words: Immunogenicity, Allergenicity, Risk assessment, Plant-derived biopharmaceuticals, Plant-derived therapeutic proteins, Host anti-product antibodies.

19.1. Introduction

The subject of this chapter is the production of recombinant protein-based medicinal products that are associated with a potential risk of inducing an *undesirable* host anti-product immune response. This excludes vaccines from the scope, although plant-specific glycosylation might well modify the immunogenic capacity of a vaccine – with implications for the benefit to risk profile.

The assessment of the risk of such undesirable immunogenicity is a major aspect of the regulatory assessment of therapeutic proteins, commencing in the pre-clinical phase, continuing

Loïc Faye and Véronique Gomord (eds.), *Methods in Molecular Biology, Recombinant Proteins From Plants, vol. 483*

DOI: 10.1007/978-1-59745-407-0_19

throughout clinical development, and then further qualified in the post-marketing setting. The most obvious clinical risk is the induction of allergic reactions mediated by IgE that is specific for plant glycans or plant-derived product impurities. But other risks might be associated with indirect effects on molecular structure and antigen presentation caused by differential (non-mammalian) processing.

It is recognised that modifications to the manufacturing process, including a change in the cell line used to express the product, may impact on the potential immunogenicity of a therapeutic protein *(1)*. This reflects the potential of these complex molecules to undergo variable degrees of post-translational structural modification that may alter their capacity to interact with elements of the immune system.

Given the qualitative and quantitative differences that exist between plant and mammalian protein expression systems, it is considered that plant-derived therapeutic proteins could present altered glycoprotein structural motifs to induce an undesirable immune response *(2)*. In addition, since there is less experience in the control of the manufacturing process for plant-derived proteins, particular attention will have to be given to the characterisation of product- and process-related contaminants, and their potential impact on the immunogenicity of the product.

Accordingly, the evaluation of the probability of inducing a humoural *and/or* cell-mediated immune response will represent an essential aspect of the overall risk assessment of a novel plant-derived therapeutic protein. The probability must then be balanced relative to the potential clinical consequences of an anti-product immune or allergic response, as well as the intended conditions of use (patient population, duration of exposure, route of administration, etc.) of the product and its advantages over alternative therapies.

As a cautionary note, it is important to be aware that most biopharmaceutical products have the potential to be immunogenic in at least a small population of human subjects *(3, 4)*.

19.2. Legal Basis for Medicinal Product Registration

The legal bases in the European Union for the registration of medicinal products derived from transgenic higher plants are Regulation 726/2004 and Directive 2001/83/EC as amended. This implies a pan-European review of a marketing authorisation application via the Centralised procedure, although clinical trials remain within the jurisdiction of individual Member States within the scope of Directive 2001/20/EC. The containment provisions of Directive 2001/18/EC also apply to commercial plant-derived medicinal products.

In the USA, medicinal products derived from bioengineered plants are regulated by Food and Drug Administration (FDA), according to the legal basis of *either* the Public Health Service Act *or* the Federal Food, Drug, and Cosmetic Act.

It is anticipated that many of the plant-derived candidate products will be new versions of already marketed biopharmaceutical products that are manufactured using mammalian cell substrates. In such cases, the principles of establishing comparability with the originator product, within the legal provisions for "biosimilar" or "follow-on protein products", could apply *(5)*.

These legal instruments are supported by quality, safety and efficacy guidelines that define the general scope of the data to be submitted in support of an application to market a product. Because product- *and/or* process-specific considerations often apply, actual data requirements should be discussed – typically before commencing clinical trials and then again before submission of an application for marketing authorisation – with concerned regulatory agencies.

19.3. Regulatory Guidance on Immunogenicity/ Allergenicity Testing

19.3.1. Plant-Derived Products

Draft FDA guidance pertaining to plant-derived therapeutic proteins indicate that the potential risk of immunogenic and allergic responses *must* form part of the development programme for these products.

FDA Draft Guidance for Industry: *Drugs, biologics, and medical devices derived from bioengineered plants for use in humans and animals*; September 2002

This guidance explicitly directs sponsors to assess potential immunogenicity and allergenicity of bioengineered plants, including evaluation of the final product for allergenic determinants such as N-glycans.

19.3.2. Biopharmaceutical Products

The ICH S6 guideline on "*Preclinical testing of biotechnology-derived pharmaceuticals*" *(6)* contains helpful information on the scope of testing required. It is acknowledged that such testing is unlikely to be predictive of a clinical response, due to the inter-species differences in immune recognition. Nevertheless, it is necessary to measure the host anti-product immune response to enable the impact on product exposure to be estimated. This should be performed as an integral part of the repeat-dose toxicology studies, using appropriately qualified analytical methods. Unfortunately, routine animal toxicology studies are not very sensitive for predicting allergic responses that occur in humans.

Guidance was recently finalished by the EU regulatory agencies to provide sponsors with a description of the risk factors to be considered as well as the methods to be applied to measure humoural and cell-mediated immune responses *(7)*. This guidance is consistent with the multi-factorial risk approach developed by FDA *(8, 9, 10)*. Product-specific guidelines, for example, those for monoclonal antibodies *(11)*, should also be consulted.

An important aspect of the regulatory guidance in relation to a product comparability exercise – as would be required in the case of "biosimilar" or "follow-on protein products" – is that a *direct* comparison of the immunogenicity of the test *versus* reference product in an animal model might be required; according to the EU guideline *(7)*, "*a clear difference in immunogenicity in the animal model would indicate non-comparability*".

Industry "best practice" for bioanalytical methodology to measure host anti-product antibody responses is also available *(12, 13)*. Since regulatory agencies place particular emphasis on the suitability of bioanalytical methods used to measure immunogenicity, sponsors should ensure that assay qualification and validation standards are strictly adhered to.

19.4. Strategy

An iterative immunogenicity risk assessment is an integral part of the development programme of most, if not all, biopharmaceutical products. This assessment begins in the pre-clinical phase and then continues into the post-marketing phase. The experience accumulated with the particular product, and with other products in the same class, is then used to refine the ongoing risk management plan.

During the development of a plant-derived, recombinant protein-based medicinal product, the sponsor would need to implement a non-clinical and clinical testing strategy that could effectively address a series of critical questions:

- Does the use of a plant expression system alter the intrinsic immunogenicity of the therapeutic protein relative to that expressed in a mammalian cell substrate?

- Are the manufacturing conditions adequately controlled to maintain product- and process-specific impurities within acceptable limits?
- Are the bioanalytical methods suitable to detect immune responses should these occur?
- Is the incidence and magnitude of host anti-product immune response in non-clinical models *comparable* for the plant-derived *versus* mammalian cell substrate-derived product?
- If differences in non-clinical immunogenicity are detected, could such differences impact on pharmacokinetic *and/or* pharmacodynamic parameters, and how might this affect the interpretation of the toxicology studies?
- What precautions should be incorporated into the design of clinical studies to minimise the risk of immunogenic *and/or* allergic reactions?
- Can the pivotal (Phase III) clinical studies be adequately powered to detect clinically significant immune or allergic responses, or will the risk qualification have to be deferred into the post-marketing phase?

These points are considered in the sections that follow.

19.5. Intrinsic Immunogenicity

The intrinsic immunogenicity of a biopharmaceutical product will depend on the presence of "self" and "foreign" structural motifs that are recognised by the host immune system *(1, 3, 14, 15, 16)*.

In the case of plant-derived therapeutic proteins, non-mammalian glycosylation and the presence of contaminants that contain xenodeterminants introduces additional *theoretical* risk factors: (1) provocation of immune/allergic responses in individuals sensitised to the foreign motifs; (2) an adjuvant effect that could increase the probability that immune tolerance to a co-presenting self-antigen might be broken, resulting in an autoimmune response. It is important to emphasise that these potential risks remain unproven with respect to the parenteral administration of plant-made biopharmaceutical products. Experience with marketed biopharmaceutical products, manufactured using mammalian cell substrates that contain non-human glycosylation profiles, has revealed immune responses directed against carbohydrate motifs that have been associated with clinically-significant adverse reactions including anaphylaxis *(17)*. In addition, there is clear evidence that contaminating proteins can increase the immunogenicity of products such as human growth hormone *(18)*.

The impact of plant-specific glycosylation needs to be understood in terms of the interdependence of the glycan and the peptide components of a glycoprotein; intra-molecular interactions between the respective components can affect the conformation of both glycan and peptide *(19)*. Thus, the nature of the glycan may modify the conformation of a peptide that is presented by MHC Class II molecules to T-helper cells; possibly; carbohydrate residues themselves may be directly recognised by T-cell receptors.

In the context of potential modulation of T-cell recognition of glycopeptidic epitopes according to the nature of the glycan moiety, it could be instructive to compare the potential of a glycoprotein, produced in a plant expression system and in a mammalian system, to stimulate antigen-specific human T-cells in vitro *(20)*; such assays can be performed in the presence of antigen presenting cells to mimic the effect of antigen processing. A comparable stimulation index for the plant- *versus* mammalian-derived glycoprotein might then provide reassurance that intrinsic immunogenicity had not been significantly altered, although this would not preclude the need for in vivo testing to qualify the relative risk. Even in the case that the plant-derived glycoprotein generated a higher stimulation index compared with the mammalian-derived protein, central and peripheral tolerance mechanisms may render the clinical risk as negligible.

Should an increased risk of immunogenicity be confirmed by in vivo studies, and the nature of the glycosylation suspected as being the causal factor, consideration could be given to the use of an alternate plant expression system. For example, it was found that alfalfa plants can generate a monoclonal antibody bearing N-glycans that are more "human compatible" than those produced in tobacco *(21)*; although it is recognised that the actual glycosylation pattern may be product-specific.

19.6. Control of Manufacturing Process

Even if intrinsic molecular features of a particular plant-derived therapeutic protein were not to appear to present a higher risk of immunogenicity relative to the same protein expressed in a mammalian cell substrate, a number of extrinsic factors associated with the conditions of the manufacturing process could have an important influence. These include the presence of:

- Aggregates that could cross-link/stimulate B-cell receptors via presentation of repeating motifs
- Degradation products, for example, truncated variants, or other structural modifications that present neo-epitopes

- Incorrectly folded protein that contains modified conformational epitopes
- Process-related contaminants that might act as adjuvants
- Changes in the extent of post-translational modification that modify the tendency for protein aggregation.

Accordingly, regulatory agencies will expect thorough characterisation of the product, supported by validation of the manufacturing process, to demonstrate adequate control of these extrinsic factors.

Particular emphasis will be placed on the identification and minimisation of plant-derived protein contaminants, since these could be associated with a higher risk of inducing immune or allergic responses than the therapeutic protein itself *(22)*. There might be an expectation for sponsors to develop bioanalytical methods to detect antibody responses to potential contaminating plant-derived proteins, as well as to the therapeutic protein, for application during non-clinical and clinical testing.

19.7. Bioanalytical Methods

During the pre-clinical development phase, the sponsor will need to develop bioanalytical methods of adequate sensitivity to:

- Measure host anti-product antibody responses in the relevant non-clinical and clinical sample matrices; *and* to
- Estimate the capacity of any detected antibodies to alter the pharmacokinetic behaviour *and/or* functional activity of the biopharmaceutical product.

Anti-product antibody responses should be described in terms of:

- Incidence
- Magnitude of binding response (corresponding to the titre at the assay cut point)
- Neutralisation capacity for product and endogenous counterparts
- Timing of onset and duration
- Identity of immunoglobulin class/sub-class.

A combination of methods (primary screen, confirmatory assay, cell-based assay for neutralisation, isotyping assay, etc.) will be required to generate the requisite data *(7)*. Alternate ligand-binding formats, for example, enzyme-linked immunosorbent assay (ELISA), multiplex assays, radioimmunoprecipitation *and/or* BIAcore, may be used.

The suitability of these bioanalytical methods must be justified by presentation of data that demonstrate:

- Specificity for the authentic drug product, related proteins and endogenous counterparts
- Benchmarking of limit of quantitation relative to an appropriate positive control antibody
- Optimisation of sample dilution factor relative to sensitivity, matrix effects and interference by residual drug product
- Validity of assay cut point to exclude false negatives whilst minimising false positives.

The most important aspect is the need to demonstrate that the primary screening assay has adequate sensitivity to enable earliest possible detection of binding antibodies that might represent a risk for longer-term administration of the drug product.

Competitive assay formats should be used to confirm that binding responses are effectively inhibited by a modest excess of the solution-phase antigen. Such assay formats can provide a rank-order estimation of binding affinity to related antigens.

Care needs to be exercised to distinguish *antigenicity* from immunogenicity: the observation that antibodies can bind to N-glycans expressed on one protein does not imply that these N-glycans will be immunogenic when present on a therapeutic protein that is administered parenterally at modest dose levels.

A major challenge for understanding the antigenicity of plant-derived glycans is the practical challenge of isolating highly purified core glycans that can be used in such competitive assays; this can seriously limit the interpretation of the specificity of anti-glycoprotein antibodies, that is, whether the antibodies recognise a carbohydrate, peptide or glycopeptide epitope.

The potential cross-reactivity of IgE antibodies demands particular attention to avoid misinterpretation of potential for induction of clinically significant allergic responses *(23)*, as does the analysis of pre-existing antibodies arising from environmental exposure. Since product-specific IgE is likely to be present at much lower levels – possibly, by a factor of several orders of magnitude – than IgG, depletion of the IgG fraction might be necessary to eliminate competition for binding to the antigen *(24)*.

19.8. In Vivo Immunogenicity Evaluation in Non-Clinical Models

Although the evaluation of immunogenicity in non-clinical species is unlikely to be predictive of the response of the response in humans, there could well be merit in comparing the incidence and magnitude of the host anti-product antibody response to a given therapeutic protein that has been manufacturing using alternate manufacturing methods/expression systems. Essentially, the sensitivity of the immune system to detect structural modifications in the biopharmaceutical product enables the host anti-product antibody response to serve as a surrogate “product quality” parameter.

Another approach to understand the contribution of plant N-glycans to the immunogenicity of a glycoprotein is to test the immunogenicity of homologous proteins manufactured using different expression systems. In this way, it was demonstrated that a murine monoclonal antibody produced in transgenic plants was not immunogenic in mice following parenteral administration in the presence of an adjuvant *(22)*. Of course, this does not imply a lack of immunogenicity in other species, but such does help to scale the probability of an undesirable immune response associated with particular plant N-glycans.

Transgenic mouse models *(25)* might enable comparison of the probability of different product forms, for example, a cytokine manufactured using alternate expression systems, inducing a breaking of tolerance to a self-protein, although species differences in the antigen processing/presentation machinery limit absolute predictive power for the clinical scenario.

Mice bearing functional elements of the human immune repertoire, created by engraftment of human CD34+ stem cells into the tissue of new-born immunodeficient mice, offer the potential of an immunogenicity screening model that reflects human-like immune stimulation and tolerance mechanisms *(26)*.

Allergenic potential of therapeutic proteins is not straightforward to assess using animal models, but comparison of the IgE and IgG responses to a protein in wild-type mice might offer the potential to signal an increased allergenic risk for some products *(27)*. The limitation of this approach is that the humoural response is studied in isolation of symptomology.

19.9. Impact on Pharmacokinetics *and/or* Pharmacodynamics

Understanding the potential of a host anti-product antibody response to alter the pharmacokinetics of a biopharmaceutical product is an essential aspect of the interpretation of non-clinical toxicology studies *(6)*, since such a response could substantially limit the systemic exposure to the product during chronic administration. If neutralising anti-product antibodies are generated, these could diminish the primary pharmacodynamic response to the product, thereby resulting in a potential underestimation of the dose-limiting toxicity.

In the case of relatively high sequence homology between a human therapeutic protein and its counterpart in the animal species, host anti-product antibodies could neutralise the activity of the endogenous protein *(28)*. Since this would represent a high-risk factor, suggestive of a potential for inducing a clinical autoimmune response, measurement of the onset and reversibility of a reduction in relevant biomarkers of the physiologic response could be important to reduce concerns about long-lasting neutralisation of an endogenous counterpart of the drug product *(7)*.

Moving into the clinical development stage, definition of the relationship of the timing of an antibody response to an impact on biomarkers represents an important safety consideration. The early detection of an antibody response could allow withdrawal of treatment to preclude adverse reactions; although the relationship between levels of antigen-specific IgE and clinical manifestations of Type I hypersensitivity is uncertain *(29)*. For plant-derived therapeutic proteins, particular attention will need to be given to the specificity of an immune response, that is, to determine whether this is directed to human or plant-related immunogens. Even if there are detectable responses directed toward peptidoglycan or glycan responses, these might be of much lower affinity – and therefore, of negligible clinical relevance – compared with antibodies that recognise B- or T-cell epitopes related to the human protein sequence *(30)*. In the case of therapeutic antibodies, the differential glycosylation could have much greater significance for the Fc functionality relative to potential for immunogenicity *(31)*.

Ultimately, the immunogenicity risk will need to be addressed by clinical evaluation.

19.10. Minimising Risk for Clinical Study Subjects

A number of extrinsic factors have been identified that may influence the probability of induction of a host anti-product antibody response *(7, 16)*. These include:

- Route of administration
- Concomitant medication/immune competence
- Dose level and frequency
- Patient factors, for example,
  - Human leukocyte antigen (HLA) type
  - Nature of underlying disease
  - $Fc_{\gamma}R$ polymorphism.

It is important to remember that these risk factors are not independent variables.

Because biopharmaceutical products may induce immune responses that have life-threatening consequences *(32)*, albeit in rare circumstances, the sponsor will need to design clinical studies with reference to the following considerations:

- Appropriate timing and duration of monitoring of immune response, that is, before administration, and at suitable intervals post-doing to (1) minimise interference by residual circulating drug; *and* (2) detect antibodies formed after cessation of treatment; *and* (3) monitor decline of detected antibodies
- Incorporation of validated bioanalytical methods to enable correlation between pharmacokinetic parameters, pharmacodynamic markers and antibody levels
- Treatment strategy for potential immune-mediated adverse events (infusion reactions and hypersensitivity response)
- Precautions against potential allergenic responses, for example, skin-prick testing using relevant antigens before administration of plant-derived therapeutic proteins.

It may be worthwhile comparing the potential of the product to stimulate histamine release from passively sensitised basophils collected from patients with IgE to relevant plant proteins or carbohydrates as a test of allergenic potential *(33)*.

A number of strategies can be effective in managing adverse events arising from host antibody responses to chronic administration of biopharmaceutical products *(34)* in non-life-threatening indications. These include:

- Exclusion of subjects due to HLA type
- Reduced rate of infusion (or stop completely)
- Shortened time interval between infusions
- Use of pre-medication (e.g. steroid or non-sedating antihistamine)
- Concurrent immunomodulation (e.g. azathioprine or methotrexate)
- Infusion of a combination of IV fluids, antihistamines and steroids in response to adverse reactions
- Switching to another agent [e.g. an alternate anti-TNF (tumour necrosis factor) monoclonal antibody]
- Monitoring of vital signs during administration to provide early detection of infusion reactions
- Use of a dose regimen with maintenance infusions scheduled to induce tolerance.

Certainly, atopic individuals should be excluded from clinical investigational studies of plant-derived biopharmaceutical products, and the decision to administer immunogenic therapeutic proteins to healthy volunteers made after very careful reflection.

19.11. Pre-Authorisation Studies *Versus* Post-Marketing Risk Management

Given the multi-factorial nature of the immunogenic response to biopharmaceutical products, and the relatively low incidence of immune-function-related adverse events for currently marketed products, the power of pre-authorisation clinical studies might not be adequate to identify or enumerate the risks for the broader target population. Even the incidence of immunogenicity may not be estimable on the basis of the population size included in the pre-authorisation programme; for example, anti-rhEPO (recombinant human erythropoietin) antibodies were detected in only 8 out of 5718 (0.14%) subjects that were asymptomatic for Pure Red Cell Aplasia *(35)*.

Consequently, regulatory agencies will require, as part of a marketing authorisation application, the presentation of an effective risk management plan that provides appropriate surveillance of the probability and consequences of the immunogenicity of biopharmaceutical products. Most commonly, this would involve the establishment of a patient registry to identify "at risk" patients/populations. In a number of cases, regulatory agencies have imposed a post-authorisation commitment to undertake an interventional study to generate additional data on the incidence of antibody responses and adverse events in a suitably sized cohort (which may depend on the incidence detected in pre-authorisation studies) for an adequate duration (e.g. 12–18 months). As associations with patient genotype become more evident, it may be logical to consider HLA typing as part of such post-authorisation studies.

For plant-derived therapeutic proteins, the major regulatory concern is likely to be the incidence of allergic responses relative to the same protein manufactured using a mammalian expression system. In the case of a "biosimilar" or "follow-on protein product", directly comparative clinical studies may be required to demonstrate therapeutic equivalence; an increased incidence of immunogenicity or allergic reactions could represent a "major" concern that might prejudice approvability. Nevertheless, the risk that such observations represent would always need to be assessed relative to the therapeutic benefits of the product; an increased incidence of anti-product immune responses would not necessarily affect the licensing of a plant-derived therapeutic protein if the overall benefit to risk profile was acceptable.

19.12. Conclusion: Balancing Probability Versus Consequences

At the present time, the probability that plant-derived therapeutic proteins for parenteral administration, prepared using manufacturing process that are subject to the rigorous controls applied to marketed biopharmaceutical products, are associated with a real increased risk of provoking undesirable immune or allergic reactions is unknown; there simply is no data from controlled clinical studies available in the public domain that would enable an estimation of the risk. Until a number of different (non-vaccine) products have completed their clinical development and registration processes, the risk remains purely theoretical.

Nevertheless, because consequences of immune-mediated responses have been documented for various biopharmaceutical products manufactured using mammalian or bacterial expression systems *(3, 17, 36, 37)*, it is necessary to apply the cautionary principle for plant-derived products.

The prevalence of food allergies attests to the capacity of plant-derived allergens to provoke serious hypersensitivity reactions, albeit in a restricted number of individuals. But, it is questionable whether highly purified, well-characterised, plant-derived protein therapeutic agents could be associated with a similar level of risk.

The methods developed for managing the risks associated with the expression of recombinant proteins in mammalian, fungal or bacterial expression systems will be equally suitable for plant-derived therapeutic proteins.

The most important consideration is that, having accumulated an immunogenicity/allergenicity database of adequate size and relevance to the proposed use of a product to enable an estimation of the "probability" vector, the risk for public health is then moderated by a stratification of the potential and demonstrated consequences *(32)*. If there is no significant change in the overall benefit to risk profile, economic factors might then have the overriding influence on choice of expression system.

References

1. Schellekens H. (2002) Bioequivalence and the immunogenicity of biopharmaceuticals. Nat. Rev. Drug Discov. 1, 457–462.
2. Bardor M., Laye L. and Lerouge P. (1999) Analysis of N-glycosylation of recombinant glycoproteins produced in transgenic plants. Trends Plant Sci. 4, 376–380.
3. Koren E., Zuckerman L.A. and Mire-Sluis A.R. (2002) Immune responses to therapeutic proteins in humans—Clinical significance, assessment and prediction. *Curr.* Pharm. Biotechnol. 3(4), 349–360.
4. Patten P.A. and Schellekens H. (2003) The immunogenicity of biopharmaceuticals. In '*Immunogenicity of Therapeutic Biological Products*', eds. Brown F. and Mire-Sluis A.R. *Dev. Biol.* Basel, Karger, 112, 81–97.
5. The European Agency for the Evaluation of Medicinal Products (EMEA). (2006) Guideline on Similar Biological Medicinal

Products containing biotechnology-derived proteins as active substance: Non-clinical and clinical issues, EMEA/CHMP/BMWP/42832/2005. (http://www.emea.europa.eu/pdfs/human/biosimilar/4283205en.pdf)

6. International conference on harmonisation (ICH) of technical requirements for registration of pharmaceuticals for human use. (1995) ICH harmonised tripartite guideline S6 on preclinical safety evaluation of biotechnology-derived pharmaceuticals. CPMP/ICH/302/95 (http://www.emea.europa.eu/pdfs/human/ich/030295en.pdf)
7. The European Agency for the Evaluation of Medicinal Products (EMEA). (2007) Guideline on immunogenicity assessment of biotechnology-derived therapeutic proteins, EMEA/CHMP/BMWP/14327/2006 (http://www.emea.europa.eu/pdfs/human/biosimilar/1432706enfin.pdf)
8. Rosenberg A.S. and Worobec A. (2004). A risk-based approach to immunogenicity of therapeutic protein products, Part 1: Considering consequences of the immune response to a protein. Biopharm. Int. November 2004, 22–26 (www.biophaminternational.com/biopharm/article/articleDetail.jsp?id=134110)
9. Rosenberg A.S. and Worobec A. (2004). A risk-based approach to immunogenicity of therapeutic protein products, Part 2: Considering host-specific and product-specific factors impacting immunogenicity. Biopharm. Int. December 2004, 34–42. (www.biopharminternational.com/biopharm/article/articleDetail.jsp?id=140695)
10. Rosenberg A.S. and Worobec A. (2005). A risk-based approach to immunogenicity of therapeutic protein products, Part 3: Effects of manufacturing changes in immunogenicity and the utility of animal immunogenicity studies. Biopharm. Int. January 2005, 32–36. (www.biopharminternational.com/biopharm/article/articleDetail.jsp?id=146216)
11. US Department of Health and Human Services Food and Drug administration (FDA). (1997) Points to consider in the manufacture and testing of monoclonal antibody products for human use (http://www.fda.gov/Cber/gdlns/ptc_mab.pdf).
12. Mire-Sluis A.R., Barrett Y.C., Devanarayan V., Koren E., Liu H., Maia M., Parish T., Scott G., Shankar G., Shores E., Swanson S.J., Taniguchi G., Wierda D. and Zuckerman L.A. (2004) Recommendations for the design and optimization of immunoassays used in the detection of host antibodies against biotechnology products. J. Immunol. Methods 289 (1–2), 1–16.
13. Gupta S., Indelicato S.R., Jethwa V., Kawabata T., Kelley M., Mire-Sluis A.R., Richards S.M., Rup B., Shores E., Swanson S.J. and Wakshull E. (2007) Recommendations for the design, optimization, and qualification of cell-based assays used for detection of neutralizing antibody responses elicited to biological therapeutics. J. Immunol. Methods 321 (1–2), 1–18.
14. Chirino A.J., Ary M.L. and Marshall S.A. (2004) Minimizing the immunogenicity of protein therapeutics.. Drug Discov. Today 9(2), 82–90.
15. Rosenberg A.S. (2006) Effects of protein aggregates: An immunologic perspective. AAPS J. 8 (3), Article 59, E501-E507 (http://www.aapsj.org/articles/aapsj080359/aapsj080359.pdf)
16. Strand V., Kimberly R. and Isaacs J.D. (2007) Biologic therapies in rheumatology: Lessons learned, future directions. Nat. Rev. Drug Discov. 6 (1), 75–92.
17. Chung C.H., Mirakhur B., Chan E., Le Q.-T., Berlin J., Morse M., Murphy B.A., Satinover S.M., Hosen J., Mauro D., Slebos R.J., Zhou Q., Gold D., Hatley T., Hicklin D.J. and Platts-Mills T.A.E. (2008) Cetuximab-induced anaphylaxis and IgE specific for galactose-alfa-1,3-galactose. N. Eng. J. Med. 358, 1109–1117.
18. Jones A.J.S. (2002) The use of an animal immunogenicity model in the development of Protropin ® Somatrem (methionyl human growth hormone). In '*Biologics 2000 – Comparability of Biotechnology Products*', eds. Brown F., Lubiniecki A. and Murano G. Dev. Biol. Basel, Karger, 109, 107–118.
19. Lisowska E. (2002) The role of glycosylation in protein antigenic properties. *Cell.* Mol. Life Sci. 59, 445–455.
20. Baker M.P. and Jones T.D. (2007) Identification and removal of immunogenicity in therapeutic proteins. Curr. Opin. Drug Discov. Devel. 10 (2), 219–227.
21. Bardor M., Loutelier-Bourhis C., Paccalet T., Cosette P., Fitchette A.-C., Vézina L.-P., Trépanier S., Dargis M., Lemieux R., Lange C., Faye L. and Lerouge P. (2003) Monoclonal C5–1 antibody produced in transgenic alfalfa plants exhibits a N-glycosylation that is homogeneous and suitable for glyco-engineering into human-compatible structures. Plant Biotech. J. 1, 451–462.
22. Chargelegue D., Vine N.D., van Dolleweerd C.J., Drake P.M.W. and Ma J.K.-C. (2000)

A murine monoclonal antibody produced in transgenic plants with plant-specific glycans is not immunogenic in mice. Transgenic Res. 9, 187–194.

23. Mari A., Iacovacci P., Afferni C., Barletta B., Tinghino R., Di Felice G., Pini C. (1999) Specific IgE to cross-reactive carbohydrate determinants strongly affect the *in vitro* diagnosis of allergic diseases. *J.* Allergy Clin. Immunol. 103 (6) 1005–1011.
24. Wachholz P.A., Dearman R.J. and Kimber I. (2005) Detection of allergen-specific IgE antibody responses. J. Immunotox. 11, 189–199.
25. Hermeling S., Jiskoot W., Crommelin D., Bornaes C. and Schellekens H. (2005) Development of a transgenic mouse model immune tolerant for human interferon beta. Pharm. Res. 22(6) 847–851.
26. Traggiai E., Chicha L., Mazzucchelli L., Bronz L., Piffaretti J.-C., Lanzavecchia A. and Manz M.G. (2004) Development of a human adaptive immune system in cord blood cell-transplanted mice. Science. 304(5667) 104–107.
27. Dearman R.J. and Kimber I. (2001) Determination of protein allergenicity: Studies in mice. Toxicol. Lett. 120 181–186.
28. Koren E., Mytych D., Koscec M., Ferbas J., Gupta S., Moxness M. and Swanson S. (2005) Strategies for the preclinical and clinical characterization of immunogenicity.In 'State of the Art Analytical Methods for the Characterization of Biological Products and Assessment of Comparability', ed. Mire-Sluis A.R. Dev. Biol. Basel, Karger, 122, 195–200.
29. Van Ree R. and Aalberse R.C. (1999) Specific IgE without clinical allergy.. J. Allergy Clin. Immunol. 103 1000–1001.
30. Jin C., Bencúrová M., Borth N., Ferko B., Jensen-Jarolim E., Altmann F. and Hantusch B. (2006) Immunoglobulin G specifically binding plant N-glycans with high affinity could be generated in rabbits but not in mice. Glycobiology 16 (4), 349–357.
31. Walsh G. and Jefferis R. (2006) Post-translational modifications in the context of therapeutic proteins. Nat. Biotech. 24, 1241–1252.
32. Rosenberg A.S. (2003). Immunogenicity of biological therapeutics: A hierarchy of concerns. In '*Immunogenicity of Therapeutic Biological Products*', eds. Brown F. and Mire-Sluis A.R. *Dev. Biol.* Basel, Karger 112, 15–21.
33. Westphal S., Kolarich D., Foetisch K., Lauer I., Altmann F., Conti A., Crespo J.F., Rodríguez J., Enrique E., Vieths S. and Scheurer S. (2003) Molecular characterization and allergenic activity of Lyc e 2 (beta-fructofuranosidase), a glycosylated allergen of tomato. Eur. J. Biochem. 270 (6), 1327–1337.
34. Cheifetz A. and Mayer L. (2005). Monoclonal antibodies, immunogenicity, and associated infusion reactions. Mt. Sinai J. Med. 72 (4), 250–256.
35. Tacey R., Greway A., Smiell J., Power D., Krommonga A., Daha M., Casadevall N. and Kelley M. (2003) The detection of anti-erythropoietin antibodies in human serum and plasma, Part I: Validation of the protocol for a radioimmunoprecipitation assay. J. Immunol. Methods 283, 317–329.
36. Schellekens H. (2002) Immunogenicity of therapeutic proteins: Clinical implications and future prospects. Clin. Ther. 24 (11), 1720–1740.
37. Pendley C., Schantz A. and Wagner C. (2003) Immunogenicity of therapeutic monoclonal antibodies. Curr. Opin. Mol. Ther. 5 (2), 172–179.

Chapter 20

Biosafety, Risk Assessment and Regulation of Plant-Made Pharmaceuticals

Penelope A. C. Sparrow and Richard M. Twyman

Summary

The technology for plant-made pharmaceuticals (PMPs) has progressed significantly over the last few years, with the first commercial products for human use expected to reach the market by 2009 (*see* **Note 1**). As part of the 'next generation' of genetically modified (GM) crops, PMPs will be subject to additional biosafety considerations and are set to challenge the complex and overlapping regulations that currently govern GM plants, plant biologics (*see* **Note 2**) and 'conventional' pharmaceutical production. The areas of responsibility are being mapped out between the different regulatory agencies (Sparrow, P.A.C., Irwin, J., Dale, P., Twyman, R.M., and Ma, J.K.C. (2007) Pharma-Planta: Road testing the developing regulatory guidelines for plant-made pharmaceuticals. *Transgenic Res.*, 2007), with specific guidelines currently being drawn up for the regulation of PMPs. In this chapter, we provide an overview of the biosafety (*see* **Note 3**), risk assessment (*see* **Note 4**) and regulation of this emerging technology. While reference will be made to EU regulations, the underlying principles of biosafety and risk assessment are generic to most countries.

Key words: Plant-made pharmaceuticals (PMPs), Genetically modified (GM), Risk assessment, Biosafety, Regulation, Host plant, Containment, Field release.

20.1. Background: History of Drug Production

20.1.1. Plants and Their Products

Plants have been used in the production of medicines for hundreds of years. In their most basic form, medicines from plants consist of whole plants, crude extracts or infusions. These are still widely used today both in developing countries (ethnobotanical remedies), and increasingly in the herbal medicines market in the industrialized world. Indeed, it is hard to believe that

Loïc Faye and Véronique Gomord (eds.), *Methods in Molecular Biology, Recombinant Proteins From Plants, vol. 483*

DOI: 10.1007/978-1-59745-407-0_20

just 50 years ago in the UK, the majority of medicines were herbals *(1)*. However, over the years our ability to extract and refine plant metabolites and use them as ingredients, alongside chemically synthesized products, has led to the development of many purified pharmaceutical products derived from plants, such as morphine and codeine from poppy (*Papaver somniferum*).

20.1.2. Conventional Drugs

Conventional drugs include chemically synthesized products as well as more complex biological compounds such as antibodies, antigens and enzymes that are produced through the use of recombinant DNA technology in systems such as mammalian cells and bacteria. Now, that plants have joined this select group of production platforms, the development of specific plant-made pharmaceutical (PMP) regulations will build on the regulations that currently govern conventional and herbal drugs, and genetically modified (GM) plants. Unfortunately, given the politically sensitive nature and overall negative public perception of GM crops, PMP regulations are likely to emerge at a level that perhaps supersedes scientific needs!

20.2. Regulatory Background

20.2.1. GM Crops

It is now over 20 years since the first experimental GM plants were grown in the field and more than 10 years since the first commercial GM crops were planted. The International Service for the Acquisition of Agri-Biotech Applications (ISAAA) predicts that by 2015 more than 20 million farmers will plant 200 million hectares of biotech crops in about 40 countries *(2)*. As such, a strong biosafety record and good regulatory system has developed to cover these crops (**Table 20.1**), and this provides a good basis for the regulation of PMPs. The fundamental difference between the so-called 'first generation crops' and PMPs is that first-generation crops have been produced to express input traits (e.g. herbicide resistance) in commodity crops intended for food and feed, while the next generation of GM crops (which includes PMPs) will produce high value output products that are not intended for food or feed and are designed potentially to elicit a response in mammalian systems. In addition, these crops will have been developed to produce maximum yields of the desired transgene [potentially up to a 5000-fold increase in transgene product compared with first-generation crops *(3)*], which therefore distinguishes them from first-generation GM crops and explains the need for increased biosafety considerations. The protein products will then enter the pharmaceutical production environment, and

Table 20.1
Overview of EU environmental and pharmaceutical regulations.

EU Directive	Regulates	What does it cover?	Who regulates it?	Implication for PMPs
Environmental regulations				
98/81/EC	Contained use of GMOs	Required at the institute level, covers laboratory and greenhouse (contained) use of GMOs	Local competent authority In the UK, this would be the Health and Safety Executive (HSE)	Would apply to PMPs at the initial research stage from laboratory through to glasshouse
2001/18/EC	Deliberate release	Experimental field release (part B of 2001/18)	Local competent authority. In the UK, this would be the Department of Environment and Rural Affairs (Defra). Defra takes advice from an Independent 'Advisory Committee on Release to the Environment' (ACRE) that reviews all field release applications. If supported, then all Member States (MS) are invited to comment.	Would apply to all experimental field-grown PMPs, but no products derived from these crops can be used in commercial production
		Commercial field release (part C of 2001/18) *Summary notifications (SNIFs) of all part b and c releases in Europe can be viewed at http:/gmoinfo.jrc.it*	As above. Also overseen by EFSA (European Food Standard Agency) when food or feed crops used; and in all cases where there is disagreement between MSs.	Required by PMPs when at the commercial field release stage.
1829/2003/EC	GM plant application for placing on the market	Covers commodity crops intended for food/feed as well as import, processing and/or cultivation	EFSA	

(continued)

Table 20.1 (continued)

EU Directive	Regulates	What does it cover?	Who regulates it?	Implication for PMPs
Pharmaceutical regulations				
2001/10/EC	Clinical trials	Safety and efficacy testing of drugs	Overseen by the local competent authority at the R&D level and the European Agency for the Evaluation of Medicinal Products (EMEA) at the marketing stage	Also applies to PMPs
2309/93⇒ 726/2004	Mandatory regulation for biotech products	Covers all marketed drugs (i.e. after experimental testing)	Overseen by EMEA	Would currently draw guidance from EMEA draft guidance notes for PMPs 2006.
2001/82/⇒ 2004/28/ EC	Veterinary medicines			
2001/83/⇒ 2004/27/ EC	Human medicines			
2004/24/EC	Herbal medicines	Includes guidance on collection, processing and storage of plants (good agricultural practice)		

GMOs genetically modified organisms, *PMPs* plant-made pharmaceuticals

will be formulated into an end product following the strict current good manufacturing practice (cGMP) associated with the pharmaceutical industry and its regulation (*see* **Note 5**). How this 'clean' cGMP environment responds to the introduction of raw plant material from non-sterile environments such as the field or glasshouse remains to be determined.

20.2.2. Pharmaceutical Regulations

Pharmaceutical products from plants will also need to adhere to the 2309/93/EU regulations for biotech-derived drugs (**Table 20.1**). These regulations are overseen by the relevant national authority during the research and early clinical trial phase, and by the European Medicines Evaluation Agency (EMEA) – which is broadly equivalent to the Food and Drug Administration (FDA) in the US – at the point of commercial application. In 2002, EMEA published draft guidance notes on the quality of biologically active substances produced by stable transgene expression in higher plants *(4)* accompanying a similar document produced by the FDA *(5)*. Although the FDA guidelines are yet to be updated, the EMEA guidelines were revised in 2006 and are still developing *(6)*. Part of the difficulty in finalizing these guidance notes has been standardizing the terminology used between two very different areas; essentially the agricultural biotechnology side (production in plants) and the pharmaceutical side (production in conventional fermenter systems). For example, a number of concepts developed for bioreactors need to be modified and redefined precisely for plants, while others need to be developed from first principles. These include working and master cell banks, batch-to-batch consistency, standard operating procedures for different production systems (field grown, grown in containment and systems based on cultured plant cells) – as well as subsequent downstream processing steps. All these need defining in terms of inputs and outputs. Determining batch-to-batch consistency, quality and 'purity' is also important for cGMP. Because no 'natural' home exists for the entire start-to-finish regulation of these crops and their derived products, there is currently an overlap between authorities and duplication of the information required by the different regulatory bodies. The stage at which each regulatory authority becomes involved, and the limits of their authority, is currently being established.

20.3. Biosafety

Many of the issues surrounding the biosafety of commodity GM crops apply equally to PMPs, that is, it must be possible to ensure that the crops do not mix with commodity crops or other PMP crops, and it must be possible to provide good traceability and identity preservation. From a biosafety and regulatory perspective, the additional factors that need to be addressed for PMPs are to ensure that (a) the desired product can be synthesized, assembled and stored in the host plants and subsequently extracted efficiently and safely under cGMP; and (b) products are consistently expressed in the correct format from batch to batch. These factors effectively cover all stages of the production process from the choice of host plant, through transformation and breeding of the PMP crop, to field production, handling, harvest and transport, all the way through to the end product (**Fig. 20.1**).

20.3.1. Increasing Biosafety, Lowering Risks, Aiding the Regulatory Process

20.3.1.1. Choice of Host Plant and Expression Strategy

The biology of the production host needs to be considered from both the production perspective and according to how it impacts on the environment, food safety and human health. When considering host plants in the context of biosafety and risk assessment, the better scientific option (e.g. plants with the highest yield, best protein stability and better downstream processing ability) may not be the best choice of host from a biosafety and regulatory side. This is summarized in **Fig. 20.2**.

The choice of host crop falls into three main categories: non-food crops, food crops and non-cultivated crops *(7, 8)*. Desirable crops from a biosafety perspective could include male sterile plants (such as male sterile lines of potato), self compatible crops with no wild relatives (such as soybean – in certain countries), or plants with heavy pollen (such as barley and rice, again in countries with no wild relatives). A wealth of knowledge

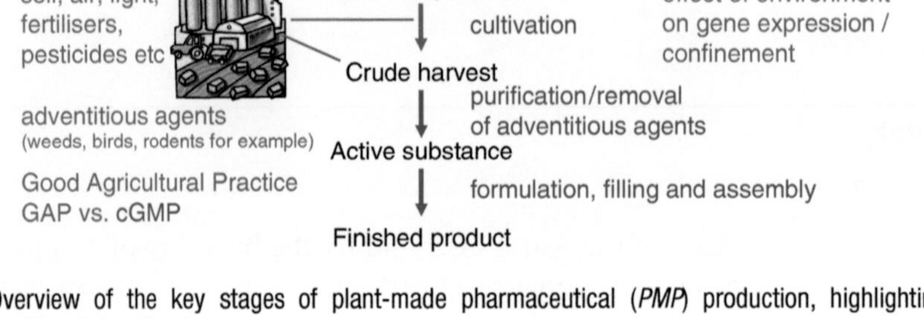

Fig. 20.1. Overview of the key stages of plant-made pharmaceutical (*PMP*) production, highlighting points for consideration. (*See Color Plates*)

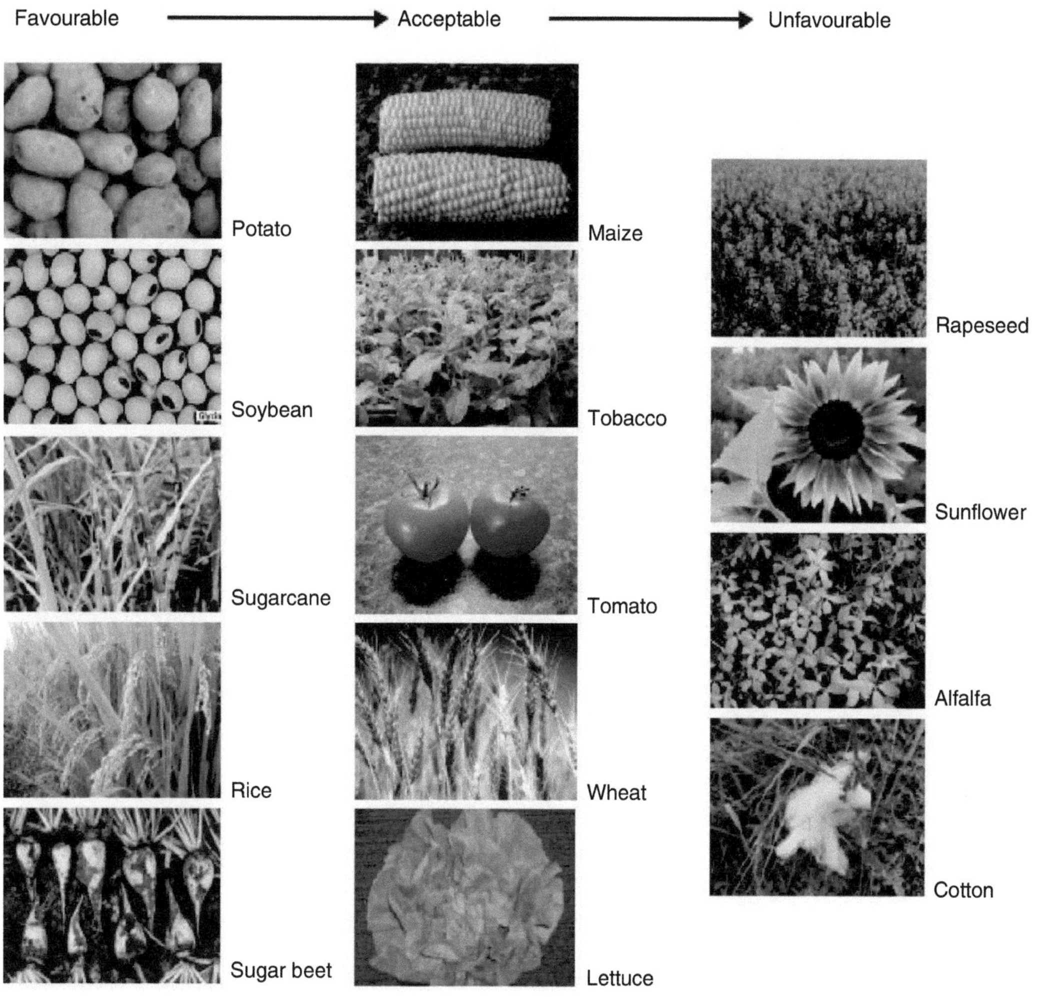

Fig. 20.2. Choice of host crop from a biosafety perspective. (*See Color Plates*)

surrounds the cultivation of food crops and transformation procedures are well defined for a number of major crop species. The use of plants with GRAS (generally recognized as safe) status may also have advantages from a biosafety perspective (*see* **Note 6**). However, the regulators may view such crops as less desirable because they represent major food crops, and therefore the risk of PMPs entering the food chain (even if only a perceived risk) would be higher than it would be when using non-food crops such as tobacco. The Union of Concerned Scientists (UCS) recently advised that only non-food crops should be considered for PMP field production *(9)*.

Non-cultivated crops, like non-food crops, have the advantage that they are not part of the human food chain. However, less is known about the genetics and biology of such species, including whether or not they produce toxins, and their potential for outcrossing. Little if any work will have been done to domesticate such species to make field cultivation more feasible. However, non-cultivated crops such as duckweed (*Lemna minor*) that are grown in containment (*see* **Note 7**) would be advantageous from a biosafety perspective.

The choice of expression site within the plant can also have biosafety implications, as summarized in **Table 20.2**. Targeting expression to the seed reduces the potential exposure of herbivores and other non-target organisms to pharmaceutical products expressed in the leaves as well as leaching of recombinant proteins into the environment from roots. However, one advantage of harvesting leaf tissue, rather than seeds or fruit, is that the plants do not need to flower, thereby reducing the risk of pollen and seed dispersal. Procedures need to be adopted to control seed spillage and seed mixing during transportation and handling, as inadvertent entry into the food chain is the biggest perceived risk for PMPs produced in seed crops. Transparency and clear documentation is vital not only at the production stage but also throughout the development and breeding stages of all PMP crops.

Some plants have already been recommended as not suitable for field production by regulatory bodies such as the US Animal and Plant Health Inspectorate (APHIS) and the Canadian regulatory authorities, e.g. oilseed rape (canola) and alfalfa.

Table 20.2
Host systems for manufacturing plant-made pharmaceuticals (PMPs)

	Yield	Scale-up potential	Protein stability	Biosafety
Leaf	+++	+++	+	++/+
Seed	++	++	+++	++
Fruit/vegetables	+	++	++	++
Non-food	+	++	++	+
Lemna	++	+	++	+++
Plant cells	++/+	+	++	+++
Mushrooms	++	++	?	++

The number of (+) symbols indicates increasing benefit in each category

These plants are considered to be unsuitable because they are bee-pollinated, sexually compatible with local weed species and have multiple year seed dormancy. These traits therefore increase the risk of transgene spread and volunteer plants in subsequent crops. However, such crops may be considered as beneficial hosts if grown in containment.

20.3.1.2. Location

The choice of location for PMP crops will also have implications from a biosafety and regulatory angle. The 'risk' to the environment will be greatly reduced if plants are grown in containment, as it reduces the potential of pollen transfer to neighbouring crops and unintentional exposure to grazing animals, birds and insects. For some PMP products, where a therapeutic protein is considered 'high risk' because it could result in adverse effects if inadvertently eaten, regulatory approval would only be granted if crops were grown under containment conditions. Therapeutic proteins grown under containment conditions are likely to face fewer regulatory hurdles than field-grown PMPs because the risk to the environment is lower.

20.3.1.3. Co-existence

The term co-existence refers to the ability of farmers to choose between conventional, organic or GM crop production, in compliance with the relevant EU regulations on labelling and purity levels. In 2003, the EU adopted recommendations 2003/556/EC on guidelines for co-existence. With commodity crops, a threshold level of 0.9% contamination with GM is generally considered to be acceptable. Whether any adventitious mixing of PMPs will be tolerated is yet to be seen. However, it should be noted that PMP crops are unlikely to form part of a farmer's normal crop rotation. These will be high value, low acreage crops where it is practical to look at novel confinement measures if field production is chosen.

20.3.1.4. Transparency

Transparency, or openness, is an important requirement of many regulatory processes. In most countries where biosafety regulations are in place, the release of genetically modified organisms (GMOs) into the environment requires the applicant to apply for a permit or approval from the competent authorities. Within the EU system, a summary of the notifier's application and the risk assessment report are made available to the public when a product is released into the environment, either as an experimental release (2001/18/EC part B) or when commercialized (2001/18EU part C). These summaries can be accessed on the following web site: http://gmoinfo.jrc.it/. In the US, some agencies (e.g. USDA) provide the public with access to the application and notice, while other agencies (e.g. FDA) provide information only when specifically requested to do so. In countries that do not yet have biosafety regulations in place, but are nevertheless party to the Cartagena Protocol on Biosafety (*see* **Note 8**) the transboundary movement of GMOs for release into the environment requires notification before such movement.

20.3.1.5. Mitigation Measures

Most of the steps required to prevent pharmaceutical crops entering the food chain involve relatively low-tech measures, such as meticulous planning and execution at each stage of the production process. The crop must be grown in isolation from breeding materials to avoid genetic and/or mechanical mixing of material carrying the pharmaceutical trait with conventional breeding materials. In practice, such mixing might be difficult to detect, so effective handling and labelling protocols are essential. Similarly, both small- and large-scale field trials need to be carried out in isolation from conventional agricultural crop trials. Parent seed for commercial production and the commercial crops themselves need be grown in isolation from other plants of the same species or wild relatives to avoid adventitious pollination.

The achievement of a sufficient level of isolation can be difficult, and is challenging for both wind- and insect-pollinated species. Realistically, if mixing with a food crop or wild weed relatives must be avoided then the novel crop should not be grown in areas where the species is normally grown for food or where wild relatives are prevalent.

The mitigation measures appropriate for a particular pharmaceutical crop will depend on many factors including the properties of the molecule, the biology of the crop and the nature of the environment in which it is grown. Some of the potential mitigation methods that have been proposed include the following *(10, 11)*.

1. *Identity preservation*, referring to the use of particular crop varieties such as white tomatoes and purple maize, which are visually distinct from normal commodity varieties and would be unlikely to become mixed with commodity crops accidentally.
2. Where identity preservation is not possible using natural varieties, the use of marker genes to make the crop, or its products (e.g. seeds), visually distinctive from commodity food and feed crops would be a suitable alternative. For example, the use of a fluorescent protein such as DsRed (as shown in **Fig. 5.2**, Chapter 5).
3. The introduction of a bitter or *undesirable taste* so the crop would be unpalatable
4. *Post-harvest expression*
5. *Physical isolation* by distance from sexually compatible crops, weeds and feral species
6. *Barrier crops* to minimize outcrossing

7. *Temporal isolation* to isolate crops genetically, that is, sowing different crops at different times can result in them flowering at different times and reduce the risk of outcrossing
8. Physical *removal of flowers* to prevent the possibility of pollination – this is a frequent isolation measure used in maize, where removal of tassels from the female transgenic parent prevents the release of pollen

Additional protocols should be in place to avoid mixing PMP crops with agricultural crops and their products. These might include dedicated land with security fencing, dedicated agricultural machinery, dedicated storage facilities, secure methods of transporting seeds for establishing the crop, secure methods of transporting the pharmaceutical containing crop product (e.g. seeds) and crop residues, measures to prevent adverse consequences of volunteer plants growing in subsequent years, and partial processing of the pharmaceutical product on the production site.

The field-release of PMP crops grown in the US requires a permit from APHIS. This must include a containment plan for the production, handling and movement of plants in and out of the field. APHIS also reviews all plans for seed production, timing of pollination, harvest, crop destruction, shipment, confinement and the storage and use of equipment. APHIS issues a field test permit either to an individual company or research institution which, in turn, may subcontract with growers. Subcontractors are required to undergo training in permit requirements and implementation.

20.4. Conclusions

The issues currently surrounding the regulation of PMPs are reminiscent of those facing the regulators in the late 1970s and early 1980s when Eli Lilly and Co developed Humulin (recombinant human insulin). At that time, the product broke completely new ground and the FDA and equivalent European agencies had to respond quickly to a new generation of biotechnology products. The surprisingly rapid approval of Humulin reflected in no small way the constant communication between Lilly scientists and the regulators. Similarly, today's PMP regulations are evolving through such contacts, particularly between industrial players and academic consortia such as Pharma-Planta *(12)*. Through such cooperative endeavours, it should be possible to fine-tune the draft guidelines into a workable regulatory system which satisfies the regulators, the producers and the general public.

20.5. Notes

1. Safflower-derived insulin is likely to be one of the first PMPs on the market for human use, reflecting the accelerated approvals process for drugs already approved for use but being developed using a different production platform. The safflower system was developed by the Canadian biotechnology company SemBioSys (www.sembiosys.com).
2. Biologics in this context refers not only to proteins but also to plant metabolites, many of which have pharmaceutical uses. These are usually too complex to synthesize chemically and therefore rely on whole plants or cultured plant cells. Taxol is an example of such a product that has been commercialized as a therapy for cancer.
3. The term 'biosafety' refers to the safe development of products of modern biotechnology (such as recombinant DNA technologies). According to the Cartagena Protocol (*see* **Note 8**), aspects of biosafety should govern the safe use and application of GMOs and their products to safeguard both human and animal health, ensure safe industrial application, and protect natural plant and animal species (flora and fauna) and the environment from any possible negative consequences.
4. GM risk assessments play an important role in the decision-making process surrounding the regulation, notification and permission to handle GMOs. Ultimately, the role of a GM risk assessment is to ensure the safe handling and containment (be that physical and/or genetic) of the GMO, and to assess any potential impacts on the environment and human health. A risk assessment should answer all 'what if' scenarios, based on scientific fact.
5. GMP is a set of regulations and guidelines for the manufacture of active pharmaceutical ingredients, drug/medicinal products, medical devices, diagnostic products and foods. Since every batch of product cannot realistically be tested for suitability, and since sampling such products only ensures that the samples themselves are suitable for use, cGMP takes the alternative approach of regulating the entire manufacturing process and environment, and documenting all aspects of the production process.
6. GRAS status is a designation granted by the FDA that a substance added to food is considered safe, and is therefore exempted from the usual Federal Food, Drug, and Cosmetic Act (FFDCA) food additive tolerance requirements. GRAS status automatically applies to substances used in

food before 1958 and therefore includes all staple food crops. In the context of PMPs, it should be noted that GRAS status takes only oral administration into account, and would not cover topical or injection formats (EMEA draft guidance document on PMPs, http://www.emea.eu.int/pdfs/human/bwp/076402en.pdf).

7. Duckweed is a whole plant system grown in containment and pioneered in its use for PMP production by Biolex Inc. (http://www.biolex.com).

8. The Cartagena Protocol is the internationally recognized protocol for biosafety (http://www.biodiv.org/biosafety/). It asserts that products from new technologies must be based on the precautionary principle, thus allowing countries to balance public health and economic benefits. In the context of PMPs, the most important aspect of the Cartagena Protocol is that it will allow countries to ban imports of GMOs or their products if they feel there is not enough scientific evidence the product is safe.

References

1. Dawson, W. (2005) Herbal medicines and the EU Directive. J. Royal Coll. Phys. Edinburgh 35, 25–27.
2. ISAAA (2006) *Global Status of Commercialized Biotech/GM Crops: 2006.* http://www.isaaa.org/.
3. Spok, A. (2007) Molecular farming on the rise—GMO regulators still walking a tightrope. Trends Biotechnol. 25, 74–82.
4. CPMP (2002) *Points to Consider on Quality Aspects of Medicinal Products Containing Active Substances Produced by Stable Transgene Expression in Higher Plants (CPMP/BWP/764/02).* European Agency for the Evaluation of Medicinal Products (EMEA).
5. FDA (2002) *Guidance for Industry. Drugs, Biologics, and Medical Devices Derived from Bioengineered Plants for use in Humans and Animals.* Food and Drug Administration.
6. CPMP (2006) *Draft Guideline on the Quality of Biological Active Substances Produced by Stable Transgene Expression in Higher Plants.* European Agency for the Evaluation of Medicinal Products (EMEA).
7. Twyman, R.M., Stoger, E., Schillberg, S., Christou, P., and Fischer, R. (2003) Molecular farming in plants: Host systems and expression technology. Trends Biotechnol. 21, 570–578.
8. Twyman, R.M., Schillberg, S., and Fischer, R. (2005) Transgenic plants in the biopharmaceutical market. Expert. Opin. Emerg. Drugs 10, 185–218.
9. Union of Concerned Scientists (2006) *UCS Position Paper: Pharmaceutical and Industrial Crops.* UCS. http://www.ucsusa.org/food_and_environment/genetic_engineering/ucs-position-paper.html
10. Commandeur, U., Twyman, R.M., and Fischer, R. (2003). The biosafety of molecular farming in plants. AgBiotechNet 5, 1–9.
11. Dunwell, J. (2005) *Technologies for Biological Containment of GM and Non-GM Crops.* DEFRA Contract CPEC 47.
12. Sparrow, P.A.C., Irwin, J., Dale, P., Twyman, R.M., and Ma, J.K.C. (2007) Pharma-Planta: Road testing the developing regulatory guidelines for plant-made pharmaceuticals. Transgenic Res. 16, 147–161.
13. Spök, A., Twyman, R.M., Fischer, R., Ma, J. K.C., and Sparrow, P.A.C. (2008) Evolution of a regulatory framework for pharmaceuticals derived from genetically modified plants. Trends in Biotech. 26(9): 506–517.

INDEX

F

G

H

I

L

M

N

O

P

Color Plates

Figure 5.2. Detection of transgenic seeds on a segregating ear of maize using red fluorescent protein (DsRed) as a visible marker.

Fig. 10.1. Generation of transgenic plants. After 4–5 weeks of bombardment, putative transplastomic shoots appear on RMOP selection medium (a). Leaves from polymerase chain reaction (*PCR*)-positive putative transformants are transferred to RMOP selection medium for second round of selection. Within 2–3 weeks, several shoots appear (b). Regenerated shoots are transferred to MSO selection medium for third round of selection where roots appear in about 10 days (c).

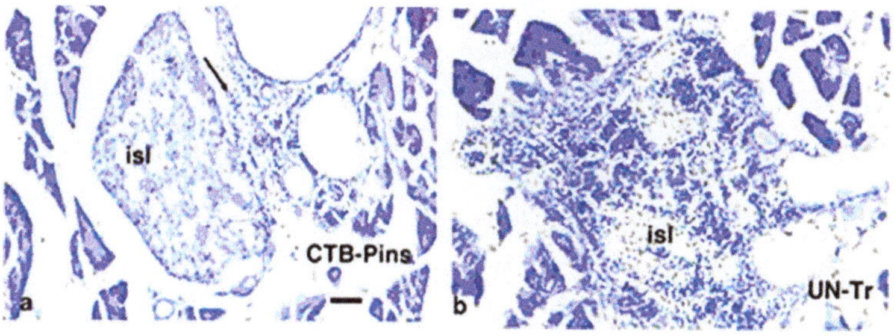

Fig. 10.8. Histochemical stained sections of nonobese diabetic (*NOD*) mouse pancreatic islets following oral delivery of cholera toxin B subunit fused to human proinsulin (*CTB–Pins*). Hematoxylin and eosin (H&E) stained sections of pancreas of mice orally delivered with (a) transgenic plant expressing CTB–Pins showing decreased lymphocytic infiltration in islet (isl) (b) Untransformed (UN-Tr) plant leaf material where severe lymphocytic infiltration in the islet is observed.

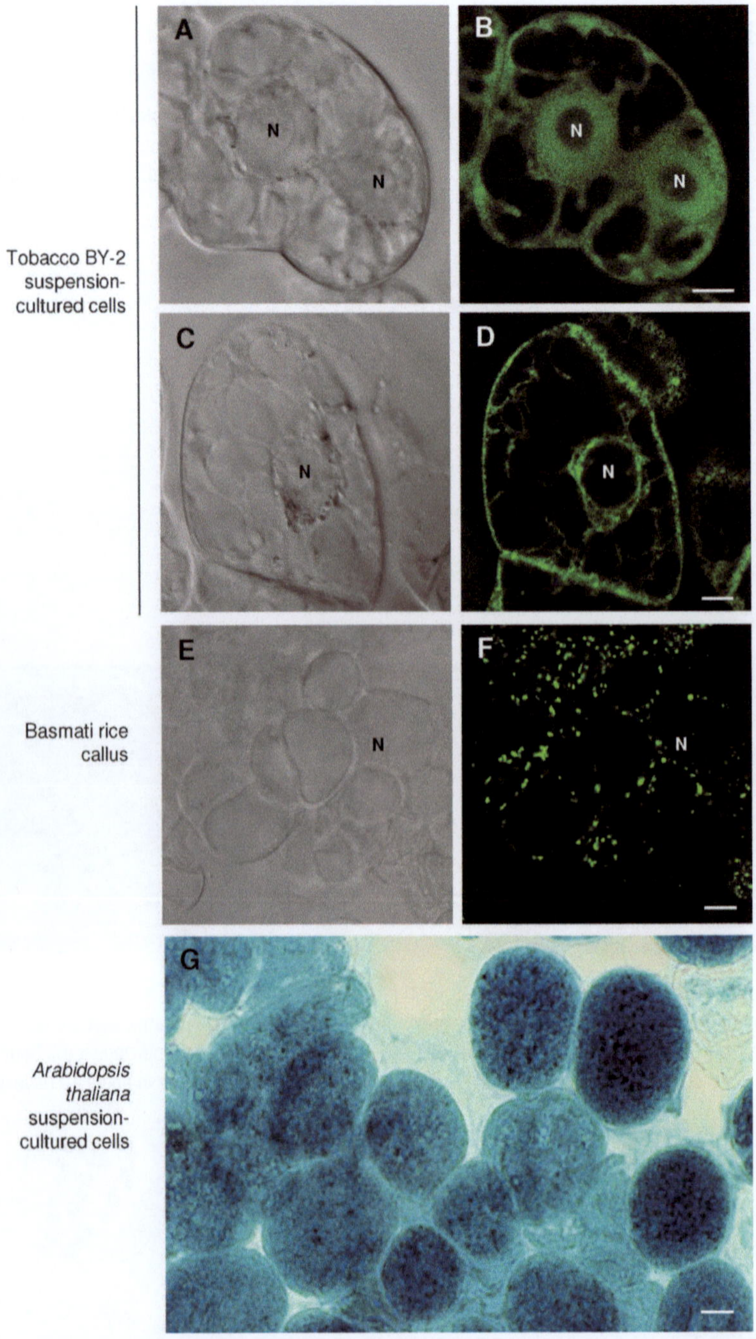

Fig. 9.1. Targeted expression of recombinant proteins in transgenic suspension-cultured cells (**A–F**) Suspension-cultured cells expressing GFP-fusions. Transmitted light (**A,B,E**). green fluorescent protein (*GFP*) expression (**B,D,F**). Bars = 8 μm, N, nucleus. When no signal peptide is added to a recombinant protein sequence, the protein is accumulated in the cytosol (BY-2 cells, **A,B**). When a signal peptide and an endoplasmic reticulum (*ER*) retention sequence are added, recombinant proteins are found mainly or exclusively in the *ER* (BY-2 cells, **C,D**). Proteins targeted to the *ER* bear high-mannose-type N-glycans. Recombinant proteins can also be retained in the Golgi apparatus (Rice cells, **E,F**). (**G**) *Arabidopsis* suspension cultured cells expressing a Gus-fusion. Bar = 20 μm

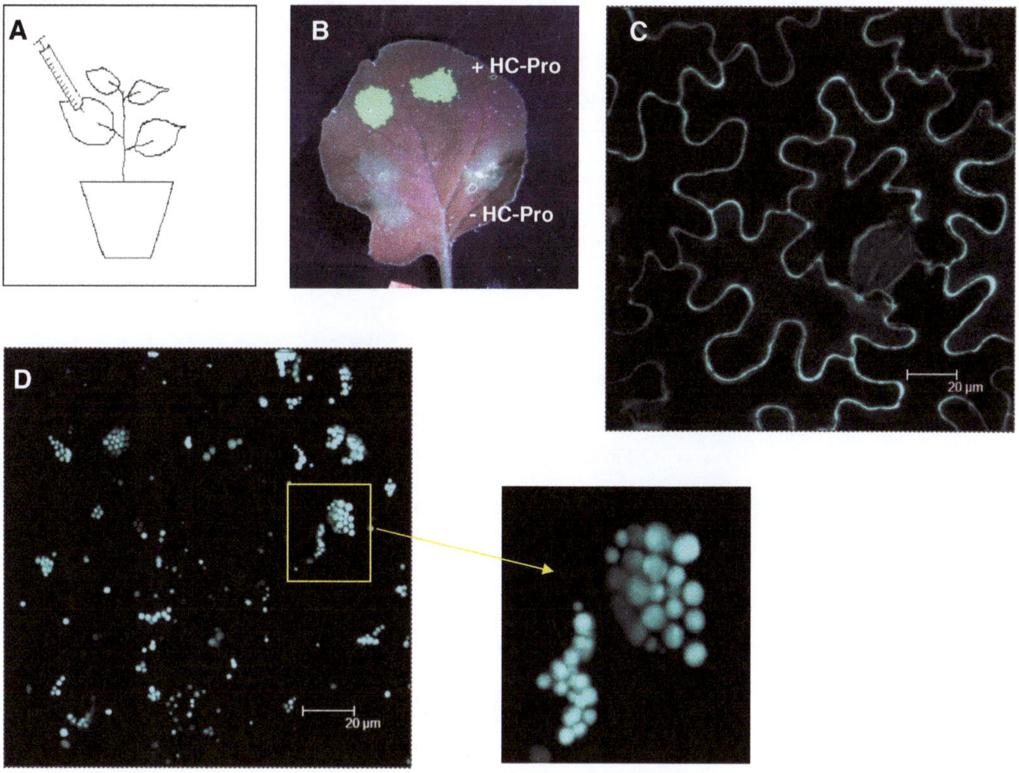

Fig. 11.1. Protein body induction in *Nicotiana benthamiana* leaves. (**A**) Schematic drawing showing young, fully expanded leaves infiltrated with *Agrobacterium* cultures carrying the desired constructs. (**B**) Effect of the silencing suppressor HC-Pro on the levels of Zera-*ECFP* (enhanced cyan fluorescent protein) expression. *Agrobacterium* cultures harbouring Zera-ECFP construct were infiltrated into leaves, either alone (*lower half of the leaf*) or mixed with an *Agrobacterium* culture transformed with the suppressor of silencing HC-Pro (*upper half*). The fluorescence was assessed under UV illumination on the fourth day after agroinfiltration. (**C**) Confocal fluorescent image of epidermal cells agroinfiltrated with SPg-ECFP + HC-Pro. The absence of the Zera partner results in the secretion of the recombinant protein. (**D**) Subcellular fluorescence localization in epidermal cells agroinfiltrated with Zera-ECFP + HC-Pro. Confocal microscope observation reveals that the recombinant protein accumulates in numerous protein bodies (*PBs*) in each cell. A higher magnification of the enclosed area is shown on the right.

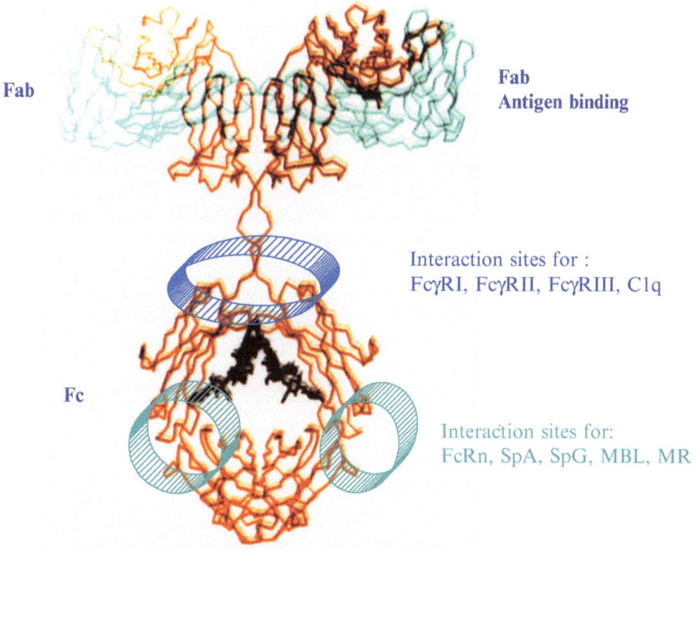

Fig. 13.1. Alpha carbon backbone structure of an IgG1 molecule. Digestion within the hinge region, by papain, releases the Fab (Fragment antigen binding) and Fc (Fragment crystallisable) fragments.

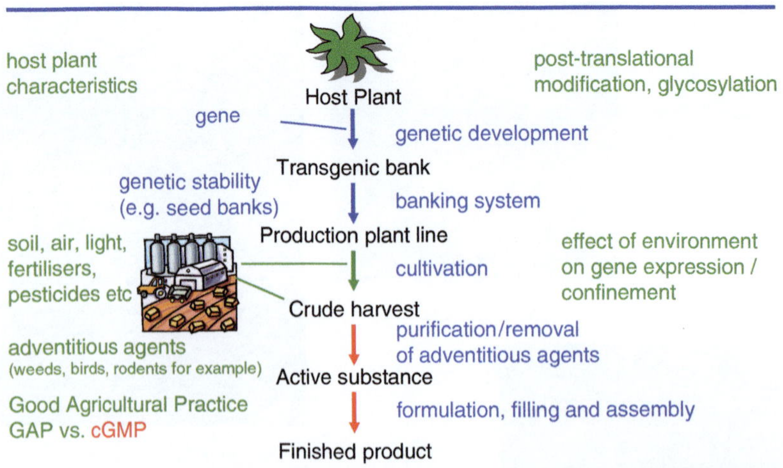

Fig. 20.1. Overview of the key stages of plant-made pharmaceutical (*PMP*) production, highlighting points for consideration.

Fig. 20.2. Choice of host crop from a biosafety perspective.

MIX
Papier aus verantwortungsvollen Quellen
Paper from responsible sources
FSC® C105338

If you have any concerns about our products,
you can contact us on
ProductSafety@springernature.com

In case Publisher is established outside the EU,
the EU authorized representative is:
Springer Nature Customer Service Center GmbH
Europaplatz 3, 69115 Heidelberg, Germany

Printed by Libri Plureos GmbH
in Hamburg, Germany